Universitext

Universitext

Editors (North America): J.H. Ewing, F.W. Gehring, and P.R. Halmos

Aksoy/Khamsi: Nonstandard Methods in Fixed Point Theory
Aupetit: A Primer on Spectral Theory
Berger: Geometry I, II (two volumes)
Bliedtner/Hansen: Potential Theory
Booss/Bleecker: Topology and Analysis
Cecil: Lie Sphere Geometry: With Applications to Submanifolds
Chandrasekharan: Classical Fourier Transforms
Charlap: Bieberbach Groups and Flat Manifolds
Chern: Complex Manifolds Without Potential Theory
Cohn: A Classical Invitation to Algebraic Numbers and Class Fields
Curtis: Abstract Linear Algebra
Curtis: Matrix Groups
van Dalen: Logic and Structure
Devlin: Fundamentals of Contemporary Set Theory
Dimca: Singularities and Topology of Hypersurfaces
Edwards: A Formal Background to Mathematics I a/b
Edwards: A Formal Background to Mathematics II a/b
Emery: Stochastic Calculus
Foulds: Graph Theory Applications
Frauenthal: Mathematical Modeling in Epidemiology
Fukhs/Rokhlin: Beginner's Course in Topology
Gallot/Hulin/Lafontaine: Riemannian Geometry
Gardiner: A First Course in Group Theory
Gårding/Tambour: Algebra for Computer Science
Godbillon: Dynamical Systems on Surfaces
Goldblatt: Orthogonality and Spacetime Geometry
Hlawka/Schoissengeier/Taschner: Geometric and Analytic Number Theory
Howe/Tan: Non-Abelian Harmonic Analysis: Applications of $SL(2,R)$
Humi/Miller: Second Course in Ordinary Differential Equations
Hurwitz/Kritikos: Lectures on Number Theory
Iverson: Cohomology of Sheaves
Jones/Morris/Pearson: Abstract Algebra and Famous Impossibilities
Kelly/Matthews: The Non-Euclidean Hyperbolic Plane
Kempf: Complex Abelian Varieties and Theta Functions
Kostrikin: Introduction to Algebra
Krasnoselskii/Pekrovskii: Systems with Hysteresis
Luecking/Rubel: Complex Analysis: A Functional Analysis Approach
MacLane/Moerdijk: Sheaves in Geometry and Logic
Marcus: Number Fields
McCarthy: Introduction to Arithmetical Functions
Meyer: Essential Mathematics for Applied Fields
Meyer-Nieberg: Banach Lattices
Mines/Richman/Ruitenburg: A Course in Constructive Algebra
Moise: Introductory Problem Course in Analysis and Topology
Montesinos: Classical Tessellations and Three Manifolds

(continued after index)

Alexandru Dimca

Singularities and Topology of Hypersurfaces

With 44 Illustrations

Springer-Verlag
New York Berlin Heidelberg London Paris
Tokyo Hong Kong Barcelona Budapest

Alexandru Dimca
School of Mathematics and Statistics
The University of Sydney
Sydney, NSW 2006
Australia

Mathematics Subject Classifications: 57-02, 55-02

Library of Congress Cataloging-in-Publication Data
Dimca, Alexandru.
    Singularities and topology of hypersurfaces/Alexandru Dimca.
      p. cm.—(Universitext)
    Includes bibliographical references and index.
    ISBN 0-387-97709-0.—ISBN 3-540-97709-0
    1. Algebraic topology.   2. Hypersurfaces.   3. Singularities
(Mathematics)   I. Title.
QA612.D56   1992
514'.2—dc20   91-33986

Printed on acid-free paper.

Production coordinated by Brian Howe and managed by Francine Sikorski; manufacturing supervised by Jacqui Ashri.
Typeset by Asco Trade Typesetting Ltd., Hong Kong.
Printed and bound by R.R. Donnelley & Sons, Harrisonburg, VA.
Printed in the United States of America.

9 8 7 6 5 4 3 2 1

ISBN 0-387-97709-0 Springer-Verlag New York Berlin Heidelberg
ISBN 3-540-97709-0 Springer-Verlag Berlin Heidelberg New York

For Gabriela, John, George, and Maria

Preface

From the very beginning, algebraic topology has developed under the influence of the problems posed by trying to understand the topological properties of complex algebraic varieties (e.g., the pioneering work by Poincaré and Lefschetz). Especially in the work of Lefschetz [Lf2], the idea is made explicit that singularities are important in the study of the topology even in the case of smooth varieties.

What is known nowadays about the topology of smooth and singular varieties is quite impressive. The many existing results may be roughly divided into two classes as follows:

 (i) very general results or theories, like stratified Morse theory and (mixed) Hodge theory, see, for instance, Goresky–MacPherson [GM], Deligne [De1], and Steenbrink [S6]; and
(ii) specific topics of great subtlety and beauty, like the study of the fundamental group of the complement in $\mathbb{P}^2$ of a singular plane curve initiated by Zariski or Griffiths' theory relating the rational differential forms to the Hodge filtration on the middle cohomology group of a smooth projective hypersurface.

The aim of this book is precisely to introduce the reader to some topics in this latter class. Most of the results to be discussed, as well as the related notions, are at least two decades old, and specialists use them intensively and freely in their work. Nevertheless, it is impossible to find an adequate introduction to this subject, which gives a good feeling for its relations with other parts of algebraic geometry and topology.

The first three chapters of this book contain most of the prerequisites, as well as a broad survey of the local topology associated with a hypersurface singularity. The main purpose is to introduce the basic concepts and techniques, to motivate them, and to give natural and interesting examples. Many of the results in this first part are given without proofs, since there are excellent presentations of the local theory, see Arnold, Gusein-Zade, and Varchenko [AGV1], [AGV2], Looijenga [Lg], and Milnor [M5]. On the other hand, we prove here many "known" or "obvious" facts, for which it is difficult to find a reference. Here is a detailed description of this first part of the book.

Chapter 1 starts with a presentation of the basic facts on regular Whitney stratifications, including their relations to μ^*-constant deformations and the first Thom isotopy lemma. The main output of this technique is the possibility of identifying families of singular algebraic varieties in which the topological type is constant, see (1.3.7)–(1.3.9). A short digression on the topology of affine hypersurfaces shows that singularities and stratifications may pop up even if we want to look only at the smooth fibers of a polynomial function (1.4.6). Then we discuss the local conic structure of an analytic set (1.5.1) and the cylindric structure at infinity of an affine algebraic set (1.6.9). We also state here a quite general version of a Zariski theorem of Lefschetz type (1.6.5) and point out the role played by Whitney stratifications. Here, as elsewhere in this book, it is not our purpose to state the most generally known version of this result, but a version that covers all the usual applications and that is also intuitively appealing.

Chapter 2 begins with a brief survey of classical knot theory, which can be regarded in many ways as a precursor of the modern study of the topology associated with a hypersurface singularity. Several key notions, like the group of a knot, the linking number (we even mention an apparently new definition for this numerical invariant in (2.1.18)), the Alexander polynomial, and the Seifert form, are introduced and illustrated by the simplest possible examples. Then we restrict our attention to the knots and links associated to plane curve singularities and discuss the main tools in their study: Puiseux parametrizations, Puiseux pairs, and iterated torus knots.

Next we make a digression and present the homology and cohomology Gysin sequences associated with a smooth hypersurface in a complex manifold (2.2.13)–(2.2.14) and the corresponding Poincaré–Leray residue morphisms. These Gysin sequences are used in (2.2.20) to describe the cohomology of the complement (in a small open ball) of a plane curve singularity in terms of meromorphic differential forms, as in Atiyah, Bott, and Garding [ABG]. This description is a prelude to the much more elaborate considerations in the same spirit as was done in Chapter 6.

The second half of Chapter 2 is devoted to surface singularities and their links. The main theme here is how the topological invariants of such a link can be computed from the resolution graph of the singularity, e.g., the classical theorem by Mumford that gives a presentation of the fundamental group of the link in terms of the resolution graph (2.3.8). We discuss in detail the topology of the links of the most frequently encountered classes of surface singularities, namely, the Hirzebruch–Jung singularities (2.4.1), the simple ADE singularities (2.4.3), the cusp singularities (2.4.5), the triangle singularities (2.4.7), the simple-elliptic singularities (2.4.9), and the weighted homogeneous singularities (2.4.10).

We start Chapter 3 with a discussion of the various fibrations known under the general heading of Milnor fibrations. In particular, we describe the

global affine Milnor fibration associated with a weighted homogeneous polynomial (3.1.11)–(3.1.14). Then we present the main results on the connectivity
of the link of a hypersurface singularity, of the associated Milnor fiber, and of
the boundary of this (compact) Milnor fiber (3.2.1)–(3.2.4) and (3.2.9). In the
case of an isolated singularity, this boundary is diffeomorphic to the corresponding link, and maybe for this reason, this interesting manifold has not
been investigated much until now (see (5.2.28) for a connection between such a
boundary and the complement of a projective hypersurface). We compute,
for a nonreduced hypersurface singularity, the number of connected components of its Milnor fiber (3.2.3).

Using miniversal deformations and morsifications, we next introduce the
vanishing cycles and the Milnor lattice of an IHS (isolated hypersurface singularity). Special attention is devoted to the Thom–Sebastiani construction
(3.3.19)–(3.3.23). This basic technical device is used several times in a crucial
way in the sequel (e.g., in the proof of (4.4.16)).

In the final part of Chapter 3, we give necessary and sufficient conditions
for the link of an IHS to be an integral or a rational homology sphere. In the
case of the Brieskorn–Pham singularities, these conditions can be stated in
terms of an associated graph. We give a proof of an elementary but elusive
claim considered "obvious" in the original paper by Brieskorn [B1], see
(3.4.13). These Brieskorn–Pham singularities are famous for at least two
reasons:

(i) Brieskorn [B1] used these singularities to show that an IHS singularity X
     can be a topological manifold when dim $X \geq 3$. When dim $X = 2$, such
     a phenomenon cannot occur because of Mumford's theorem mentioned
     above (2.3.12).
(ii) Hirzebruch [Hz2] discovered that some of the links associated with
     Brieskorn–Pham singularities are exotic spheres, i.e., manifolds homeo-
     morphic but not diffeomorphic to the usual unit sphere.

We give a brief account of this second topic and note a formal analogy between one of the main formulas in this subject and a recent formula by
Neumann–Wahl [NW] for the Casson invariant of certain links of surface
singularities (3.4.15)–(3.4.18).

The final three chapters form the second and main part of this book. The
underlying idea is to use the local topological information associated with a
singularity, as described in the first part in order to compute some global
topological invariants of complex algebraic varieties (usually hypersurfaces in
a projective or affine space). Most of the results in this part of the book are
given with complete proofs, and examples and exercises continue to abound.

Chapter 4 deals with the computation of the fundamental group of a
hypersurface complement in a projective space. Some simple results even
treat the case of a hypersurface complement in a weighted projective space,
using a result due to Armstrong (4.1.15). Please note that basic information
on weighted projective spaces and on their subvarieties is collected in

Appendix B for the reader's convenience. After some generalities, the problem of computing such fundamental groups is reduced to the case of curve complements in the plane via a Zariski theorem of Lefschetz type (4.1.17). This is a classical and difficult subject started by Zariski in his famous paper [Z1].

Before attacking this problem, we introduce, following an idea of Libgober, the Alexander polynomials for a projective hypersurface (4.1.18)–(4.1.26). As in classical knot theory, these polynomials are invariants easier to compute than the homotopy groups of the corresponding hypersurface complements. There is also a preliminary discussion on presentations of groups. Here we just want to fix the notation and to introduce the reader to some basic non-commutative groups, namely, the binary dihedral groups $\tilde{D}_k$ (4.2.1(iii)), the free product of two groups (4.2.1(iv)), and some braid groups (4.2.6)–(4.2.13). Using a basic result by Serre about groups acting on trees, we give a "topological" construction for the free product $(\mathbb{Z}/p\mathbb{Z}) * (\mathbb{Z}/q\mathbb{Z})$ for $(p, q) = 1$. We recall then the usual version (with an open cover) of the van Kampen theorem (4.2.17), but we also remark that what is sometimes needed in the sequel is a very sophisticated version of it (with a closed covering satisfying a long list of conditions!).

Next we investigate an affine (or local) problem, namely, the computation of the fundamental group of the complement $\mathbb{C}^2 \backslash C$, where C is the affine plane curve given by the equation

$$C: x^p - y^q = 0.$$

Here we follow a paper by Oka [O5], but make some of the details more explicit. In this way, we get a presentation for the group $\pi_1(\mathbb{C}^2 \backslash C)$ in which the relations are the prototypes for the so-called "monodromy relations" to be introduced later. Then we discuss the Zariski–van Kampen theorem that gives a presentation for the fundamental group of a plane curve complement in terms of generators and monodromy relations (4.3.15). Several instances where this fundamental group is abelian are described, namely, when the curve is smooth, when the curve has an inflection point of order equal to the degree (4.3.17), when the curve has only nodes as singularities (4.1.13) and (4.3.18), or when the curve is irreducible and has just one singular point with high multiplicity (4.3.8), (4.3.11), and (4.3.21). In the last section of Chapter 4, we discuss some noncommutative fundamental groups of plane curve complements. The main two examples treated here come from the original paper by Zariski [Z1], namely, the three cuspidal quartic curve (4.4.8) and the sextic curve with six cusps situated on a conic (4.4.16). In fact, in (4.4.16) we treat a more general example due to Oka, but with a different idea of proof based on Némethi [N2] and on Serre's result mentioned above. At the end we give probably the first example of an irreducible curve having just one singularity and still with a noncommutative associated fundamental group (4.4.21).

In Chapter 5 we consider mainly the integral homology and cohomology of projective complete intersections. We start this discussion with some facts on the topology of a complex projective space and on the line bundles on such a space. Then we discuss the cohomology properties of a projective complete intersection, making no assumption on its singular locus (5.2.6), (5.2.11), (5.2.12), and (5.2.17). We present a criterion for a hypersurface to have the same rational homology groups as a projective space (5.2.22) and give some examples of this situation. To each (singular) projective complete intersection, one can associate a smooth manifold, namely, the boundary of a good "tubular" neighborhood. In the hypersurface case, we relate this manifold to the boundary of the corresponding Milnor fiber (5.2.28) and to the hypersurface complement (5.2.31).

Next we consider in detail the cohomology structure of a smooth projective complete intersection (5.3.1), (5.3.6), and (5.3.7). Using the Hirzebruch Index Theorem (5.3.11), we can easily compute the signature of an even-dimensional complete intersection. The cup product on the middle cohomology group can then be determined, once we know its parity, via a method due to Libgober–Wood (5.3.16). We end our discussion of the smooth case by comparing the topological classification to the diffeomorphism classification for the two-dimensional smooth complete intersections (5.3.31)–(5.3.34). For this discussion we follow closely Ebeling [E4].

The final section of Chapter 5 deals with projective complete intersections with isolated singularities and is based on my paper [D2]. The main result (5.4.3) relates the integral homology of such a complete intersection to the Milnor lattices of its singularities. For the reader's convenience, we have collected some useful facts on integer bilinear forms, discriminant groups, Dynkin diagrams, and related notions in Appendix A. As an illustration of this technique, we compute the homology of all the normal cubic surfaces in $\mathbb{P}^3$ in (5.4.8), as well as for some hypersurfaces introduced by Barthel and this author [BD] that have the same integral homology like a projective space (5.4.12) and (5.4.13). Last we discuss some related topics, namely, the homology of some projective cones (5.4.18)–(5.4.21) and the cup product on the middle cohomology group (5.4.24)–(5.4.27). We end this chapter with an example (5.4.28) showing the difficulty of understanding the topology of affine hypersurfaces by compactifying them. We use this also as an opportunity to present to the reader a useful classical trick from algebraic topology (related to the Smith theory), see (5.4.31) and (5.4.34).

The last chapter essentially aims at computing the Betti numbers of a singular projective hypersurface using rational differential forms defined on its complement. Most of the results in this chapter are also valid for hypersurfaces in weighted projective spaces. We start with a discussion of the rational differential forms defined on a hypersurface complement and find out that such a form can be written quite explicitly (6.1.16) and (6.1.17). Then we recall two basic results by Grothendieck that say that the de Rham cohomology of an affine (or Stein) hypersurface complement can be computed using rational (or polar meromorphic) differential forms (6.1.21) and (6.1.22). These special differential forms induce in a natural way a polar filtration on the cohomology of such a complement. This cohomology has also a Hodge filtration as part of its mixed Hodge structure (basic information on mixed Hodge structures is collected in Appendix C). A fundamental fact due to Deligne and this author [DD] is that these two filtrations are related by an inclusion in the projective case (6.1.31). In the local case, results by Karpishpan and this author show that the situation is similar, but slightly more complicated (6.1.39) and (6.1.40). We present next some spectral sequences that converge to the cohomology of a hypersurface complement, in the same spirit as in my papers [D5], [D8], and [D9]. When the hypersurface under consideration is smooth, this approach goes back to Griffiths' fundamental paper [G]. In this smooth case, we give examples (6.2.18) that show some of the advantages of having a method to produce explicit bases for the cohomology groups of hypersurface complements. We concentrate then on the first difficult case, namely, to understand the topology associated to a weighted homogeneous polynomial f such that the affine hypersurface $\{f = 0\}$ has a one-dimensional singular locus. With every irreducible component of this singular locus, one can associate a transversal singularity, which is in fact an isolated hypersurface singularity together with a finite cyclic group of automorphisms. There is a conjectural formula for the Euler characteristics of the weighted hypersurface and of the Milnor fiber associated to the polynomial f (which surely holds in most of the interesting cases) involving only the weights and the degree of the polynomial f, as well as some local invariants coming from the transversal singularities, see (3.19) in [D5] and relevant parts of [D8].

In this book we treat the more difficult problem of computing the corresponding Betti numbers and Alexander polynomials. It is well known that these topological invariants depend not only on such local invariants of the transversal singularities, but also on the position of these singularities, see Zariski [Z1]. To understand this strange behavior, we derive a useful exact sequence (6.3.15) in which everything can be, at least in principle, computed explicitly in terms of differential forms. The efficacy of our method is then tested on several examples (which had been considered by several authors using different methods, based essentially on the resolution of singularities), namely, nodal hypersurfaces (6.4.5), sextic curves with six cusps (6.4.9) and some of their higher dimensional generalizations (6.4.14), and two configurations of nine lines introduced by Artal-Bartolo (6.4.15). As another application we give a sufficient condition for the triviality of the Alexander polynomial of a hypersurface with an unique singularity (6.4.17).

Finally, in this chapter, we point out that for a hypersurface with only isolated singularities the usual Betti numbers and the intersection Betti numbers (i.e., the dimensions of the middle perversity intersection cohomology groups) determine each other by very simple formulas (6.4.22) and (6.4.23).

This book is designed for a large audience of graduate students and professional mathematicians. Most of it is addressed to workers in algebraic geometry and algebraic and differential topology. However, there are also parts for those with an interest in pure algebra (e.g., the group-theoretic discussion in Chapter 4 or Appendix A) or in complex manifolds (parts of Chapters 5 and 6).

This book can serve a graduate student either as additional material to a general course in algebraic or differential topology (giving natural and interesting examples that are missing from many textbooks on these subjects!), or as new motivation for a student dealing with some abstract parts of algebraic geometry (e.g., mixed Hodge structures). The open questions raised in the text may stimulate the student's imagination in his or her work toward a doctorate. There are plenty of exercises to give additional information and to help the reader to test his or her understanding of the material.

As in any book devoted to a subject that has evolved over a century at least, it was not possible to avoid many references to results needed, but not proved, in this book. Generally, we have avoided the long and technically complicated proofs and offered the reader the shorter and more illuminating ones (when they are available). The reference system in the book uses two numbers, e.g., (3.7) to refer to a result within the same chapter; and three numbers, e.g., (4.3.7) to refer to a result in a different chapter (whose number appears then in the first position).

Discussions with many mathematicians have provided valuable information, new ideas, and the necessary encouragement during the preparation of this manuscript. It is a great pleasure to mention here E. Brieskorn, P. Deligne, A Durfee, W. Ebeling, H. Esnault, A. Granville, A. Guivental, H. Hamm, F. Hirzebruch, I. Karpishpan, D.T. Lê, A. Libgober, P. Orlik, T. Petrie, D. Siersma, J. Steenbrink, A. Varchenko, C.T.C. Wall, and S.S.-T. Yau. The author is grateful to the Max-Planck-Institut für Mathematik in Bonn and to the Institute for Advanced Study in Princeton for financial support and the exceptional scientific atmosphere that made this project possible.

Sydney, Australia   Alexandru Dimca

Contents

CHAPTER 1

Whitney Stratifications

§1. Some Motivations and Basic Definitions

A (smooth) manifold (i.e., a C^∞-manifold without boundary, of constant dimension) enjoys the next well-known *homogeneity property*, see, for instance, [M4], p. 22.

(1.1) Proposition. *Let M be a connected manifold and let x, y $\in$ M be two points. Then there is a diffeomorphism h: M $\to$ M that is smoothly isotopic to the identity and carries x to y.*

In fact, it is quite easy to construct the diffeomorphism h above (at least for nearby points x and y) by integrating a suitable vector field on M. Exactly the same argument gives the following *relative version* of the result.

(1.2) Proposition. *Let M be a manifold, let A $\subset$ M be a connected submanifold, and let x, y $\in$ A be two points. Then there is a diffeomorphism h: M $\to$ M smoothly isotopic to the identity and such that h(A) = A, h(x) = y.*

We may say informally that the pair (M, A) *looks exactly the same at the points x and y*. In fact, the pair (M, A) looks at any point $a \in A$ exactly like the *model pair* $(\mathbb{R}^m, R^d \times 0)$ at the origin of $\mathbb{R}^m$, where $m = \dim M$, $d = \dim A$. This remark is just the definition of a submanifold, see, for instance, [GG], p. 9.

When we consider more general subsets $A \subset M$, there is no reason for the homogeneity property (1.2) to hold. A basic example to keep in mind is the following.

(1.3) Example. Let $M = \mathbb{C}^n$ be the complex affine space and let A be the union of the coordinate hyperplanes, i.e., A is the affine hypersurface given by the equation $f = 0$, where $f(x) = x_1, \ldots, x_n$.

The points in A can be divided into classes according to various properties.

Let us consider two such possibilities, one essentially algebraic, the other essentially topological.

(1.3(i)) **Multiplicity Partition.** If a hypersurface X is defined in some open subset $U \subset \mathbb{C}^n$ by an analytic equation $g = 0$, we recall that the *multiplicity of X at a point x* is denoted by $\text{mult}_x X$ and is equal to the order of the first nonvanishing term in the Taylor expansion of g at the point x. In particular, note that $\text{mult}_x X = 1$ if and only if x is a nonsingular point of X.

In our case, it is easy to see that the set

$$A_k = \{a \in A; \text{mult}_a A = k\}$$

coincides with the set

$$\{a = (a_1, \ldots, a_n) \in \mathbb{C}^n; \text{exactly } k \text{ coordinates } a_i \text{ vanish}\}$$

for $k = 1, \ldots, n$.

We recall now two basic equivalence relations defined for *reduced* analytic set germs $(A, 0)$ at the origin of $\mathbb{C}^n$.

(1.4) **Definition.** (i) Two germs $(A, 0)$ and $(A', 0)$ as above are called *analytically equivalent* if there is a germ of an analytic isomorphism $\varphi: (\mathbb{C}^n, 0) \to (\mathbb{C}^n, 0)$ such that $\varphi(A, 0) = (A', 0)$. Compare with [D4], p. 23.

(ii) Two germs $(A, 0)$ and $(A', 0)$ as above are called *topologically equivalent* if there is a germ of a homeomorphism $\varphi: (\mathbb{C}^n, 0) \to (\mathbb{C}^n, 0)$ such that $\varphi(A, 0) = (A', 0)$.

It is clear that analytically equivalent implies topologically equivalent, and it is easy to construct counterexamples for the converse implication (consider the plane curve singularities $x^4 + 2\lambda x^2 y^2 + y^4$). It is also easy to see that the multiplicity is an *analytic invariant*, i.e., if $(A, 0)$ and $(A', 0)$ are two analytically equivalent hypersurface singularities, then $\text{mult}_0(A) = \text{mult}_0(A')$. And it is a famous *conjecture* by Zariski [Z5] that the multiplicity is a topological invariant for hypersurfaces. But the strongest result proved until now states that the multiplicity is in this case a C^1-invariant, see [Em] for details. This dicussion implies that the pair $(\mathbb{C}^n, A)$ in our example does not look the same at the points $x \in A_k$ and $y \in A_l$ for $k \neq l$ (i.e., there is no diffeomorphism h as in (1.2)).

(1.3(ii)) **Embedded Local Fundamental Groups.** For any analytic set germ $(A, 0)$ at the origin of $\mathbb{C}^n$ we can consider a basic topological invariant, namely, the (embedded) *local fundamental group* $\pi_1^{\text{loc}}(A, 0)$ which is, by definition, the fundamental group of the complement of A in a small enough ball, i.e.,

$$\pi_1^{\text{loc}}(A, 0) = \pi_1(B_\varepsilon \backslash A),$$

where

$$B_\varepsilon = \{x \in \mathbb{C}^n; |x| < \varepsilon\}.$$

It follows from the conic structure of the analytic sets (see (4.1) below) that this definition is independent of $\varepsilon > 0$ for ε small enough. Moreover, using D. Prill's concept of good neighborhoods [P], it follows easily that $\pi_1^{loc}(A, 0)$ is a topological invariant (i.e., if $(A, 0)$ and $(A', 0)$ are topologically equivalent analytic germs, then $\pi_1^{loc}(A, 0) = \pi_1^{loc}(A', 0)$).

In our example, it is easy to show that

$$\pi_1^{loc}(A, a) = \pi_1((\mathbb{C}^*)^k) = \mathbb{Z}^k \qquad \text{for any} \quad a \in A_k.$$

It follows that for $x \in A_k$ and $y \in A_l$ with $k \neq l$ even a homeomorphism h as in (1.2) cannot exist. On the other hand, it is an easy *exercise* to show that for x, $y \in A_k$ one can find an explicit linear isomorphism $h: \mathbb{C}^n \to \mathbb{C}^n$ with $h(A) = A$ and $h(x) = y$.

In conclusion, both the algebraic criterion (i) and the topological criterion (ii) have led to the same *partition* $A = \bigcup_{k=1,n} A_k$ of our singular set A into smooth (semialgebraic) submanifolds A_k such that A is homogeneous along each of them. It is a remarkable fact that such partitions exist for large classes of sets A, and that they can be defined by some basic algebro–topological invariants as in Example (1.3).

It is now time to give some technical definitions. Let M be a smooth manifold and let $A \subset M$ be a *locally closed* (abbreviated in the sequal to l.c.) subset.

(1.5) Definition. A *stratification* of A is a locally finite partition $\mathscr{A}$ of A into smooth submanifolds of M, called *strata*.

This means that we have a disjoint decomposition $A = \bigcup_{i \in I} A_i$ such that any point in A has a neighborhood intersecting finitely many strata A_i.

Now let X, Y be disjoint submanifolds in $\mathbb{R}^m$ such that $0 \in X \cap \overline{Y}$.

(1.6) Definition. Y is said to be *Whitney regular over X at the point 0* if for any sequences of points $\{x_n\}$ in X and $\{y_n\}$ in Y converging to 0 and satisfying the two conditions:

(i) the sequence of tangent spaces $T_{y_n} Y$ (regarded as linear subspaces in $\mathbb{R}^m = T_{y_n}\mathbb{R}^m$) converges to a subspace T in the corresponding Grassmannian;
(ii) the sequence of lines $\overline{x_n y_n}$ converges to a line l in the Grassmannian of lines through the origin in $\mathbb{R}^m$;

one has $l \subset T$.

(Sometimes this is called the Whitney *b*-regularity condition, see [Ma], [W1], and [Tr] for several regularity conditions and the relations among them.)

Most of the time we consider complex analytic submanifolds X and Y in an ambient space $\mathbb{C}^m$. Definition (1.6) also makes sense when we regard such X and Y with their complex structures (i.e., $T_{y_n} Y$ will be the complex tangent space and $\overline{x_n y_n}$ will be a complex line). It is easy to check that Y is Whitney regular over X at 0 in this complex version if and only if Y is Whitney regular over X at 0 in the real version, i.e., when X and Y are regarded as smooth real submanifolds in $\mathbb{C}^m = \mathbb{R}^{2m}$. Y is said to be *Whitney regular over X* (or the pair (Y, X) is said to be *Whitney regular*) when Y is Whitney regular over X at any point in $X \cap \bar{Y}$.

It is easy to check that this condition (1.6) is invariant under diffeomorphisms and hence we can extend Definition (1.5) (using coordinate charts) to any disjoint submanifolds of M.

(1.7) Definition. A *Whitney stratification* $\mathscr{A}$ of A is a stratification $\mathscr{A}$ such that for any pair of strata $X, Y \in \mathscr{A}$, Y is Whitney regular over X. In this situation, we may also say that the stratification $\mathscr{A}$ is Whitney regular.

(1.8) Examples. (i) Let M be a smooth manifold and let $A \subset M$ be a closed submanifold. Then $M = (M \backslash A) \cup A$ is a Whitney stratification for M.

(ii) The stratification $A = \bigcup_{k=1,n} A_k$ described in Example (1.3) is Whitney regular.

(iii) Consider the surface $W: x^2 - zy^2 = 0$ in $\mathbb{C}^3$ (the famous *Whitney umbrella* or pinch point from classical algebraic geometry).

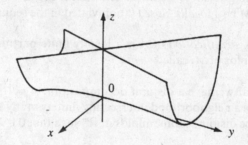

Consider the partition by multiplicity of W. It is easy to see that

$$W_2 = \{a \in W; \text{mult}_a\, W = 2\} = W_{\text{sing}}$$

(the *singular part* of W) is exactly the z-axis defined by $x = y = 0$ and that

$$W_1 = \{a \in W; \text{mult}_a\, W = 1\} = W \backslash W_{\text{sing}} = W_r$$

is exactly the *regular part* of W.

To show that W_1 is *not* Whitney regular over W_2 at the point 0, consider the sequences of points $x_n = (0, 0, 1/n) \in W_2$ and $y_n = (0, 1/n^2, 0) \in W_1$. Then $T = \lim T_{y_n} W_1$ is given by the (complex) equation $z = 0$, while l is clearly the z-axis. Hence indeed $l \not\subset T$.

This shows that a Whitney stratification of an algebraic set $A \subset \mathbb{C}^n$ is more subtle than the multiplicity partition (sometimes called the *Samuel stratification*) of A.

(1.9) **Remark.** When both X and Y are *semialgebraic* submanifolds of $\mathbb{R}^m$ (see [GWPL], p. 17, for a definition and basic properties of semialgebraic sets) we may replace in Definition (1.6) the limits along sequences of points with limits along real analytic paths (this was remarked on, for instance, by Bruce in [Bc]).

This reduction is based on the Curve Selection Lemma, see [M5], p. 25, and can be quite useful in practice, as shown by the next result.

Note also that we use the term "semialgebraic set" in the complex setting too, where some people use the term "constructible set," see, for instance, Mumford [Mu2], p. 37.

(1.10) **Lemma.** *Any semialgebraic stratum Y is Whitney regular over a zero-dimensional stratum X.*

Proof. We can clearly assume that $X = \{0\}$ is a point, namely, the origin in the ambient space $\mathbb{R}^m$. Let $c: [0, \varepsilon) \to \mathbb{R}^m$ be a real analytic path such that $c(0) = 0$ and $c(t) \in Y$ for $t \in (0, \varepsilon)$. Then each component $c_i(t)$ of $c(t)$ has a power series expansion

$$c_i(t) = c_i t^{a_i} + \cdots$$

with $c_i \in \mathbb{R}^*$, a_i some positive integer, and the dots standing for higher order terms. Let $k = \min\{a_i; i = 1, \ldots, m\} > 0$ and assume for simplicity that one has $a_1 = \cdots = a_j = k < a_{j+1} \leq \cdots \leq a_m$. Then we have

$$l = \lim_{t \to 0} \frac{c(t)}{|c(t)|} = \frac{\bar{c}}{|\bar{c}|} \quad \text{with} \quad \bar{c} = (c_1, \ldots, c_j, 0, \ldots, 0),$$

under the identification of a line in $\mathbb{R}^m$ with a unit vector spanning it.

It is easy to see that we also have

$$\frac{\bar{c}}{|\bar{c}|} = \lim_{t \to 0} \frac{\dot{c}|t|}{|\dot{c}(t)|},$$

and since $\dot{c}(t) \in T_{c(t)}Y$ for all $t \in (0, \varepsilon)$ this implies that $\bar{c}/|\bar{c}| \in T = \lim_{t \to 0} T_{c(t)}Y$ and hence $l \subset T$. Note that the last limit always exist, unlike in the sequence case! Moreover, the example of the logarithmic spiral (Y = spiral, $X = \{0\}$, the origin) shows that (1.10) is false in general.

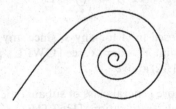

(In polar coordinates $(\rho, \theta) \in \mathbb{R}_+ \times \mathbb{R}$, we have $Y: \rho = e^{-\theta}$.) Another proof of (1.10) can be obtained from the following fundamental result of Whitney [Wh1], [Wh2] and Lojasiewicz [Lo]. For X, Y semialgebraic subsets in $\mathbb{R}^m$ we define $W(X, Y)$ to be the union of the singular part X_{sing} with the set of points x in $X \setminus X_{\text{sing}}$ such that $Y \setminus Y_{\text{sing}}$ is not Whitney regular over $X \setminus X_{\text{sing}}$ at the point x.   □

(1.11).**Theorem** (Whitney, Lojasiewicz). *$W(X, Y)$ is a semialgebraic subset in X with* $\dim W(X, Y) < \dim X$.

(1.12) **Corollary.** *Any semialgebraic subset A in $\mathbb{R}^m$ has a Whitney stratification with finitely many semialgebraic strata.*

Proof. Let $d = \dim A$. We construct a filtration $A = B_d \supset B_{d-1} \supset \cdots \supset B_0 \supset B_{-1} = \varnothing$ of A by semialgebraic sets B_i closed in A, and such that $A_i = B_i \setminus B_{i-1}$ is a semialgebraic submanifold in $\mathbb{R}^m$ of dimension i (or is empty). Take $B_{d-1} = A_{\text{sing}}$ and assume that $B_d, \ldots, B_i$ have been constructed for some $i \leq d - 1$. If $\dim B_i < i$ we set $B_{i-1} = B_i$ and are done. If $\dim B_i = i$, then we put

$$B_{i-1} = \overline{\left\{ \bigcup_{j=i+1}^{d} W(B_i, A_j) \right\}},$$

where the closure is taken in A.

Then B_{i-1} is semialgebraic and of dimension $\leq i - 1$ by (1.11). The stratification $\mathscr{A} = \{A_i\}_{i=0,1,\ldots,d}$ obtained in this way is clearly Whitney regular.   □

(1.13) **Remark.** (i) With the above notations, assume that $A^0 \subset A$ is an open semialgebraic subset and that $\mathscr{A}^0 = \{A_j^0\}_{j \in J}$ is a Whitney regular stratification of A^0 with semialgebraic strata.

Then there is a canonical way to *extend* the stratification $\mathscr{A}^0$ to a stratification $\mathscr{A}$ of the whole set A. Let $A^1 = A \setminus A^0$ and apply the above procedure to A^1, the only difference being that we take

$$B_{i-1} = \overline{\left\{ \bigcup_{j=i+1}^{d} W(B_i, A_j) \cup \bigcup_{j \in J} W(B_i, A_j^0) \right\}}.$$

(ii) Let $\mathscr{A}$ be a Whitney regular stratification of the l.c. set A in M. We can get a *filtration* of A by setting

$$F^s A = \bigcup_{\substack{X \in \mathscr{A} \\ \dim X \leq s}} X.$$

The subsets $F^s A$ are *closed* in A for any s, since any pair (X, Y) satisfying Definition (1.6) verifies $\dim Y > \dim X$, see [GWPL], p. 12. This general fact motivates the proof of (1.12) above.

Note also that the more general class of subanalytic subsets of an analytic manifold admits Whitney stratifications [Ha], [Hi].

At the end of this section we would like to mention one standard way to obtain Whitney stratifications, see, for instance, [GWPL], p. 21.

(1.14) **Proposition.** *Let G be an algebraic (real or complex) group acting algebraically on an algebraic manifold M. Let $A \subset M$ be a locally closed subset which is the union of a finite number of G-orbits A_i for $i = 1, \ldots, n$. Then $A = \bigcup A_i$ is a Whitney regular stratification.*

Proof. Each orbit A_i is a semialgebraic subset in M and as such it has at least one nonsingular point $x \in A_i$ [GWPL], p. 19. But the orbit A_i clearly has a homogeneity property as in (1.2) under the action of the group G. This implies that A_i is a smooth submanifold in M.

For any two distinct orbits A_i and A_j with $A_i \subset \overline{A}_j$, there is a point $x \in A_i$, such that A_j is Whitney regular over A_i at the point x according to Whitney's result (1.11). Using again the homogeneity under the G-action, it follows that A_j is Whitney regular over A_i. $\qquad\qquad\square$

We now give some concrete examples of this construction which show that a lot of familiar mathematical objects are related to Whitney stratifications.

(1.15) **Examples.** (i) Let $G = \mathrm{Gl}(n, \mathbb{C}) \times \mathrm{Gl}(p, \mathbb{C})$ be a product of general linear groups and let $M = \mathrm{Hom}(\mathbb{C}^n, \mathbb{C}^p)$ be the vector space of all linear maps. The G-action is defined by the rule

$$(g, h) \cdot u = h \circ u \circ g^{-1}.$$

Let $e_1, \ldots, e_n$ (resp. $f_1, \ldots, f_p$) be the standard basis for $\mathbb{C}^n$ (resp. $\mathbb{C}^p$) and for each positive integer $0 \le k \le \min(n, p)$ consider the linear map $u_k \in M$ defined by $u_k(e_i) = f_i$ (for $i = 1, \ldots, k$) and $u_k(e_j) = 0$ (for $j > k$). Then the orbit $G \cdot u_k$ coincides exactly with the submanifold in M defined by

$$\{u \in M; \mathrm{rank}(u) = k\}.$$

When $n = p$ we can take A to be the hypersurface

$$\{u \in M; \det(u) = 0\}$$

and then $A = \bigcup_{k=0, n-1} (G \cdot u_k)$ is a Whitney stratification.

(ii) Let $G = \mathrm{Gl}(n, \mathbb{C})$ and $M = \mathrm{Sym}^2(\mathbb{C}^n)$, the space of all quadratic forms $q: \mathbb{C}^n \to \mathbb{C}$ or, if we prefer, the space of all $n \times n$ symmetric matrices over $\mathbb{C}$. The G-action is the usual one, namely,

$$(g \cdot q)(x) = q(g^{-1}(x)).$$

A list for the *normal forms* of quadratic forms is given by $q_k(x) = x_1^2 + \cdots + x_k^2$ for $k = 0, \ldots, n$ and hence $M = \bigcup_k (G \cdot q_k)$ is a Whitney stratification.

More details about these two examples (e.g., computation of the dimension of the orbits, description of the closure of the orbits) can be found in [D4], Chap. 5, §1.

§2. Topological Triviality and μ^*-Constant Deformations

To state formally the homogeneity property of the stratified sets analogous to property (1.2), we need some preliminaries. Given two stratified sets $(A, \mathscr{A})$ and $(B, \mathscr{B})$ we can define the *product* stratified set $(A \times B, \mathscr{A} \times \mathscr{B})$, the strata in $\mathscr{A} \times \mathscr{B}$ being precisely all the products $A_i \times B_j$ for $A_i \in \mathscr{A}$ and $B_j \in \mathscr{B}$. When A is a smooth submanifold, then (if not stated otherwise) A will be regarded as a stratified set with just one stratum, i.e., $\mathscr{A}_A = \{A\}$.

(2.1) **Definition.** Let $(A, \mathscr{A})$ be a stratified subset in the manifold M, let x be a point in A, and let X be the stratum in $\mathscr{A}$ containing x. We say that $\mathscr{A}$ is *topologically locally trivial at x* if there is a stratified set $(S, \mathscr{S})$, a point $s_0 \in S$ which is a stratum in $\mathscr{S}$, and a *local* homeomorphism $h: A \to X \times S$ such that:

(i) $h(x') = (x', x_0)$ for any point $x' \in X$;
(ii) h maps the strata in $\mathscr{A}$ into the strata in $\mathscr{A}_X \times \mathscr{S}$.

The stratified set $(A, \mathscr{A})$ is said to be *topologically locally trivial* if it is so at every point $x \in A$.

The homeomorphism h above is *local* in the sense that it is defined on a (small) neighborhood of the point x in A, i.e., it is a *germ of a homeomorhism* at the point x. In the same vein, it is better to regard (S, s_0) as a germ of a stratified set (in an obvious sense) and in the language of germs (2.1) says that one has a homeomorphism $(A, x) \sim (X, x) \times (S, s_0)$.

The importance of such a result lies in the fact that the germ (X, x) is trivial, i.e., $(X, x) \simeq (\mathbb{R}^d, 0)$ where $d = \dim X$, and hence all the information about the local structure of A at the point x is contained in the (lower dimensional) germ (S, s_0).

(2.2) **Theorem.** *Any Whitney stratified set $(A, \mathscr{A})$ is topologically locally trivial.*

For the proof of this basic result we refer to [GWPL], p. 61. Here we intend to explain and exemplify it.

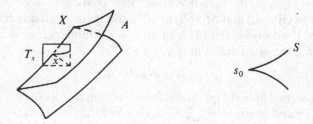

First, note that to obtain the stratified germ (S, s_0) which appears in (2.1) we may just take a small transversal T_x (i.e., a small submanifold transversal to X

at the point x and such that dim T_x + dim X = dim M) and define $(S, s_0) = (T_x \cap A, x)$. Since T_x is transversal to X, it follows easily from the Whitney Regularity Condition (1.6) that T_x is also transversal to all the other strata in $\mathscr{A}$ in a neighborhood of x in A. Hence replacing A with this neighborhood (having the obvious induced stratification) we see that $S = T_x \cap A$ has a Whitney stratification with strata $S_i = T_x \cap A_i$.

The key point here is the elementary observation that the Whitney regularity is preserved under transverse intersections, see [GWPL], p. 13, for a more general result. It is also clear that $s_0 = x$ is a stratum in $\mathscr{S}$ since $\{x\} = X \cap T_x$. The homeomorphism h is obtained by integrating certain vector fields, defined along each stratum of A and having certain compatibility conditions when passing from a stratum A_i to an adjacent stratum $A_j \subset \bar{A_i}$ (without this precaution the resulting h would not be continuous). These are the so-called *stratified controlled vector fields*.

Note also that in moving the transversal T_x parallel to itself we get the fibers of a projection $A \xrightarrow{\rho} X$, similar to the normal bundle (tubular neighborhood) projection of a submanifold. We remark that the obvious diagram (of space and map germs at x)

(2.3)
$$
\begin{array}{ccc}
(A, x) & \xrightarrow{\;\;h\;\;} & (X, x) \times (S, x) \\
& {\scriptstyle\rho}\searrow \qquad \swarrow{\scriptstyle \mathrm{pr}_1} & \\
& (X, x) &
\end{array}
$$

is commutative.

(2.4) **Corollary.** *All the germs of slices* $(T_x \cap A, x)$ *are topologically equivalent (by homeomorphisms preserving the induced stratifications) for x varying in the same connected component of a stratum X.*

(2.5) **Example.** Recall the notations from (1.15(i)). Let X be the stratum $G \cdot u_{n-2}$ in A and let $x = u_{n-2}$. Then, as explained in [D4], p. 45, a transversal T_x is given by all the linear maps (identified with $n \times n$ matrices)

$$
u(a, b, c, d) = \left[\begin{array}{c|cc}
I_{n-2} & & \\
\hline
& a & b \\
& c & d
\end{array} \right]
$$

for $(a, b, c, d) \in \mathbb{C}^4$.

The slice $(T_x \cap A, x)$ is clearly isomorphic to the vertex $(Q, 0)$ of the conic $Q: ad - bc = 0$ in $\mathbb{C}^4$. Note that when the stratified set $(A, \mathscr{A})$ comes from a group action as in (1.14), the homogeneity result (2.2) can be proved elementarily, see, for instance, [D4], p. 37.

To give another point of view on the significance of the basic theorem (2.2), we discuss now the problem of Whitney regularity over a one-dimensional stratum. Unlike Lemma (1.10) which treated the zero-dimensional case, this is a subtle and difficult problem. We assume, for simplicity, that in addition to the one-dimensional stratum X, there is just one other stratum Y and that $X \cup Y$ is a complex analytic hypersurface in $\mathbb{C}^{n+1}$. Since our problem is clearly a local one, we consider only germs at the origin of $\mathbb{C}^{n+1}$ and choose coordinates $x_1, \ldots, x_n$, t at the origin such that

$$X: x_1 = \cdots = x_n = 0,$$

$$\overline{Y} = Y \cup X : F(x, t) = 0,$$

for some analytic function germ F.

Motivated by our discussion of Theorem (2.2), it is natural to look at the transversal hyperplanes

$$T_\lambda : t - \lambda = 0$$

and at the intersections

$$\overline{Y}_\lambda = \overline{Y} \cap T_\lambda : f_\lambda(x) = 0,$$

regarded as hypersurface singularities at the origin of $\mathbb{C}^n$, with $f_\lambda(x) = F(x, \lambda)$. In this way we have obtained a one-parameter family of IHS (isolated hypersurface singularities), since, of course, $Y_\lambda = Y \cap T_\lambda$ is smooth in the neighborhood of the origin if we assume that Y is Whitney regular over X at 0. This family $(\overline{Y}_\lambda, 0)$ can be regarded as a deformation of the IHS $(\overline{Y}_0, 0)$.

We recall that for an IHS $Z: f = 0$ at the origin of $\mathbb{C}^n$ a basic numerical invariant is its *Milnor number* which is defined by

$$(2.6) \qquad \mu(Z, 0) = \mu(f) = \dim_{\mathbb{C}} \frac{\mathcal{O}_n}{J_f},$$

where $\mathcal{O}_n = \mathbb{C}\{x_1, \ldots, x_n\}$ is the $\mathbb{C}$-algebra of all the convergent power series in $x_1, \ldots, x_n$ and $J_f = (\partial f / \partial x_1, \ldots, \partial f / \partial x_n)$ is the *Jacobian ideal* of f, generated by all the partial derivatives of f. We have the following basic results relating the Milnor numbers and the topological equivalence (1.4) of IHS.

(2.7) **Proposition** [Lê2]. *The Milnor number is a topological invariant for IHS, i.e., given two IHS $(Z_1, 0)$ and $(Z_2, 0)$ in $\mathbb{C}^n$ which are topologically equivalent, then $\mu(Z_1, 0) = \mu(Z_2, 0)$.*

We recall that a family $Z_\lambda : f_\lambda = 0$, $\lambda \in I$ of IHS is called μ-*constant* (resp. topologically constant) when $\mu(Z_\lambda, 0)$ is constant for any $\lambda \in I$ (resp. $(Z_\alpha, 0)$ is topologically equivalent to $(Z_\beta, 0)$ for any $\alpha, \beta \in I$). Two IHS $(Z_0, 0)$ and $(Z_1, 0)$ are said to be μ-*equivalent* (or to belong to the same μ-*constant class*) if there is a μ-constant family $(Z_t, 0)$, $t \in [0, 1]$ connecting the two IHS.

(2.8) **Theorem** (Lê–Ramanujan [LêR]. *A μ-constant deformation of IHS in $\mathbb{C}^n$ is topologically constant, for $n \neq 3$.*

For the excluded case $n = 3$, we refer to the recent paper [Sz] and to the papers listed in the references there.

After this brief survey, we come back to our problem and our notations.

(2.9) Proposition. *If Y is Whitney regular over X at the origin, then the family $(\overline{Y}_\lambda, 0)$ of IHS defined above is μ-constant, for $|\lambda|$ small enough.*

Proof. If Y is Whitney regular over X at the origin 0, it follows by (1.11) that the same it true for points x near 0. Hence a small neighborhood A of 0 in $\mathbb{C}^{n+1}$ is Whitney stratified by $(A \setminus \overline{Y}) \cup (A \cap Y) \cup (A \cap X)$. $\qquad\square$

Now apply (2.4) and (2.7).

(2.10) Example (Whitney Umbrella Revisited). Let $F(x, y, t) = x^2 - ty^2 + y^3$. Then $\overline{Y} : F = 0$ is nothing other than the Whitney umbrella (1.8(iii)) written in a different coordinate system ($z = t - y$). Then $\mu(\overline{Y}_\lambda, 0) = 1$ for $\lambda \neq 0$ and $\mu(\overline{Y}_0, 0) = 2$. Hence Y is not Whitney regular over X at the origin, a fact proved directly in (1.8(iii)).

It would clearly be useful to have a converse result to (2.9) giving numerical tests for Whitney regularity. To state the result we introduce the μ^*-invariants of an IHS $Z : f = 0$ at the origin of $\mathbb{C}^n$, following B. Teissier [T1]. Let $L \subset \mathbb{C}^n$ be a *generic* linear subspace of dimension l. Then $(Z \cap L, 0)$ is an IHS in $(L, 0) \simeq (\mathbb{C}^l, 0)$ whose topological type does not depend on L, for L generic.

It follows by (2.7) that the number

$$\mu^l(Z, 0) = \mu(Z \cap L, 0)$$

is well defined. Note that $\mu^1(Z, 0) = \text{mult}_0(Z) - 1$ and that $\mu^n(Z, 0) = \mu(Z, 0)$, the usual Milnor number defined in (2.6).

(2.11) Definition. The sequence of numbers

$$\mu^*(Z, 0) = (\mu^1(Z, 0), \dots, \mu^n(Z, 0))$$

is called the μ^*-invariant of the IHS $(Z, 0)$.

Using an alternative definition (more algebraic) of the sequence $\mu^*(Z, 0)$, Teissier, [T1], p. 315, was able to prove the next result.

(2.12) Proposition. *The μ^*-invariant of an IHS is an analytic invariant, i.e., given two IHS $(Z_1, 0)$ and $(Z_2, 0)$ which are analytically equivalent, it follows that $\mu^*(Z_1, 0) = \mu^*(Z_2, 0)$.*

One of the main reasons for considering the μ^*-invariants is the following beautiful result.

(2.13) **Theorem** ([T1], [BS2]). *The two statements below are equivalent:*

(i) *Y is Whitney regular over X at 0;*
(ii) *the family of IHS $(\overline{Y}_\lambda, 0)$ is μ^*-constant for $|\lambda|$ small enough, i.e., the μ^*-invariants $\mu^*(\overline{Y}_\lambda, 0)$ are the same for all λ small enough.*

The next example, due to Briançon–Speder [BS1], was very important in the development of the subject, since it has provided negative answers to several "natural" conjectures in [T1].

(2.14) **Example** (μ-Constant $\not\Rightarrow \mu^*$-Constant). Let $F(x, y, z, t) = x^5 + txy^6 + y^7 z + z^{15}$. Note that each of the polynomials f_λ are in this case *weighted homogeneous* of degree $d = 15$ with respect to the weights $w_1 = \mathrm{wt}(x) = 3$, $w_2 = \mathrm{wt}(y) = 2$, $w_3 = \mathrm{wt}(z) = 1$. And there is a well-known formula

$$\mu(Z, 0) = \frac{(d - w_1)\cdots(d - w_n)}{w_1 \cdots w_n}$$

(see, for instance, [D4], p. 112) computing the Milnor number of a weighted homogeneous IHS $(Z, 0)$ in terms of the degree d and of the weights w_i. Using this formula in our case we get $\mu(\overline{Y}_\lambda, 0) = 364$, i.e., the family $(\overline{Y}_\lambda, 0)$ is μ-constant. But this family is not μ^*-constant.

Indeed, the equation of a generic plane H in $\mathbb{C}^3$ can be taken to be $z = ax + by$ with $a, b \in \mathbb{C}^*$, and the equation for $H \cap \overline{Y}_\lambda$ becomes

$$\bar{g}_\lambda(x, y) = x^5 + \lambda xy^6 + y^7(ax + by) + (ax + by)^{15}.$$

For $\lambda \neq 0$, $\bar{g}_\lambda$ is a *semiweighted homogeneous* singularity with leading term $\bar{f}_\lambda = x^5 + \lambda xy^6$ of degree $d = 15$ with respect to the weights $\mathrm{wt}(x) = 3$, $\mathrm{wt}(y) = 2$. Using the fact that $\mu(\bar{g}_\lambda) = \mu(\bar{f}_\lambda)$ (see, for instance, [D4], p. 116) we get

$$\mu^2(\overline{Y}_\lambda, 0) = \mu(H \cap \overline{Y}_\lambda, 0) = 26 \qquad \text{for} \quad \lambda \neq 0.$$

For $\lambda = 0$, $\bar{g}_0$ is still semiweighted homogeneous, but with a *new* leading term $\bar{f}_0 = x^5 + by^8$. Hence

$$\mu^2(\overline{Y}_0, 0) = \mu(H \cap \overline{Y}_0, 0) = 28$$

and therefore the family $(\overline{Y}_\lambda, 0)$ is not μ^*-constant.

(2.15) **Remark.** There are however some important classes of IHS for which *any* μ-constant deformation is μ^*-constant. The include:

(i) plane curve singularities and their suspensions;
(ii) homogeneous singularities and their suspensions; and
(iii) all the IHS with $\mu < 14$.

Proof. By suspension of an IHS $(Z, 0)$ given by $f = 0$ in $\mathbb{C}^n$ we mean any IHS $(\overline{Z}, 0)$ given by the equation

$$(2.16) \qquad \bar{f}(x, y) = f(x) + y_1^2 + \cdots + y_k^2 = 0 \qquad \text{in } \mathbb{C}^{n+k}.$$

It is easy to see that any deformation $\bar{f}_\lambda$ of the function germ $\bar{f}$ can be written in suitable coordinates as coming from a deformation f_λ of f, namely, by the formula

$$\bar{f}_\lambda(x, y) = f_\lambda(x) + y_1^2 + \cdots + y_k^2.$$

Moreover, since Milnor numbers are preserved under suspension, the family $\bar{f}_\lambda$ is μ-constant if and only if this is true for the family f_λ.

It is easy to see that the μ^*-invariant of $(\bar{Z}, 0)$ can be computed from the μ^*-invariant of $(Z, 0)$ by the following formulas:

$$\mu^s(\bar{Z}, 0) = \mu^{s-k}(Z, 0) \text{ for } s \geq k + 1 \quad \text{and} \quad \mu^s(\bar{Z}, 0) = 1 \text{ for any } s = 1, \ldots, k.$$

We now discuss the above cases (i), (ii), and (iii) separately.

(i) It is known that any μ-constant family of plane curve singularities is topologically trivial (2.8) and that the Zariski conjecture about the topological invariance of the multiplicity is true in the curve case, see [T2], Appendix, for more details. On the other hand, for a plane curve singularity the μ^*-invariant consists of essentially the multiplicity and the Milnor number, so that we have clearly a μ^*-constant family as soon as we have a μ-constant one. To prove the statement for a suspension as in (2.16) with $n = 2$, we note that

$$\mu^{k+2}(\bar{Z}, 0) = \mu(Z, 0), \qquad \mu^{k+1}(\bar{Z}, 0) = \text{mult}(Z, 0) - 1$$

and that $\mu^s(\bar{Z}, 0) = 1$ for $s = 1, 2, \ldots, k$. By the above discussion, it is clear that any such μ-constant family is μ^*-constant.

(ii) This follows from the fact that the Milnor number of a *homogeneous* IHS $(Z, 0): f = 0$ (by which we mean that f is a homogeneous polynomial, say of degree d in $\mathbb{C}[x_1, \ldots, x_n]$) depends only on n and d. More precisely,

$$\mu(Z, 0) = (d - 1)^n.$$

Since any generic section $(Z \cap L, 0)$ is, in this case, again homogeneous of degree d, it follows that

$$\mu^s(Z, 0) = (d - 1)^s \qquad \text{for any} \quad s = 1, \ldots, n.$$

It is known [GK] that any μ-constant deformation f_λ of our homogeneous singularity f is given by the formula $f_\lambda = f + g_\lambda$, where g_λ is a polynomial containing only monomials of degree $\geq d$. It follows that all the associated IHS $Z_\lambda: f_\lambda = 0$ are semiweighted homogeneous (as well as their linear sections $(Z_\lambda \cap L, 0)$) and the above computation of μ^s still holds true, by [D4], p. 116. Hence the family $(Z_\lambda, 0)$ is μ^*-constant. Recalling (2.14), we see that the class of *weighted* homogeneous IHS behaves quite differently than the usual homogeneous IHS. The reader will find in this book some other places where the treatment of the larger class of weighted homogeneous IHS is much more difficult than the treatment of the smaller class of homogeneous IHS.

(iii) Using the classification of IHS (see [AGV1], p. 246, or, better, [W2] where the *contact* classification is given), we know that any IHS with $\mu < 14$ is in one of the classes of *simple singularities* A_k, D_k, E_6, E_7, or E_8, or in one of the *unimodal families* $\tilde{E}_6$, $\tilde{E}_7$, $\tilde{E}_8$, $T_{p,q,r}$, or belongs to the 14 exceptional families. Note that with respect to the *contact classification* (which is different from the right equivalence classification given in [AGV1] as soon as we go beyond the simple singularities) the only families in the list above which contain a *modulus* (i.e., continuous parameter) are the simple-elliptic singularities $\tilde{E}_6, \tilde{E}_7, \tilde{E}_8$. It follows that any μ-constant family of IHS f_λ with $\mu < 14$ falls into one of two classes:

(α) All the germs f_λ belong to a single contact equivalence class. Then the family is μ^*-constant by (2.12).

(β) All the germs in the family belong to one of the families $\tilde{E}_6, \tilde{E}_7$, or $\tilde{E}_8$. Then the result follows from (i) and (ii) above.   □

Before ending this section we mention a situation in classical algebraic geometry where the last two μ^*-invariants play an essential role. Let $V \subset \mathbb{P}^{n+1}$ be a hypersurface of degree $d > 1$ which has only isolated singularities. Let $a_1, \ldots, a_k$ be the set of these singular points. The dual hypersurface $\hat{V}$ is a hypersurface in the dual projective space $\hat{\mathbb{P}}^{n+1}$, see, for instance, [D4], p. 204. It is natural to try to compute its degree $\hat{d} = \deg(\hat{V})$ in terms of the degree d, and some local invariants associated with the singularities (V, a_i) for $i = 1, \ldots, k$.

The answer is the following (for a proof we refer to Kleiman [Kl]).

(2.17) **Proposition.**

$$\hat{d} = (d-1)^n d - \sum_{i=1,k} (\mu^{n+1}(V, a_i) + \mu^n(V, a_i)).$$

When V is a plane curve, we get the following nice formula

(2.18) $$\hat{d} = (d-1)d - \sum_{i=1,k} (\mu(V, a_i) + \text{mult}(V, a_i) - 1).$$

(2.19) **Exercise.** (i) Assume that V is a plane curve of degree d having δ nodes and κ cusps as its singularities. Prove the following classical Plücker formula (see also [GH], p. 280)

$$\hat{d} = d(d-1) - 2\delta - 3\kappa.$$

(ii) Let V be a quartic curve ($d = 4$) having three cusps. Show that the dual curve $\hat{V}$ is a nodal cubic. (An equation for such a curve V is given in (4.4.2) later in this book.)

For another relation between dual varieties and Milnor numbers the reader may consult [D4], p. 214.

§3. The First Thom Isotopy Lemma

One of the basic applications of Whitney stratifications is to prove that the topological type does not change in certain deformations of algebraic varieties.

The simplest situations (no singularities) are covered by the following well-known result [Eh], [L2].

(3.1) **Proposition** (Ehresmann's Fibration Theorem). *Let* $f: E \to B$ *be a proper submersion between the manifolds* E *and* B. *Then* f *fibers* E *locally trivially over* B, *i.e., for every point* $b \in B$ *there is a neighborhood* U *of* b *and a diffeomorphism* $\varphi: f^{-1}(b) \times U \to f^{-1}(U)$ *such that* $f \circ \varphi = \mathrm{pr}_2 = $ *the second projection*.

Moreover, if $A \subset E$ is a closed submanifold such that $f|A: A \to B$ is still a submersion, then f fibers the pair (E, A) locally trivially over B, i.e., the diffeomorphism φ above can be chosen to carry $(f^{-1}(b) \cap A) \times U$ onto $f^{-1}(U) \cap A$. In particular, in the above situation, all the fibers $E_b = f^{-1}(b)$ (resp. all the pairs (E_b, A_b) with $A_b = E_b \cap A$) are diffeomorphic to one another as soon as we assume B to be connected.

(3.2) **Example** (Smooth Hypersurfaces in $\mathbb{P}^n$). Let $P(n, d)$ denote the vector space of all the homogeneous polynomials in $\mathbb{C}[x_0, \ldots, x_n]$ of degree d and note that $D = \dim P(n, d) = \binom{n + d}{d}$. Consider the incidence relation hypersurface

$$Z = \{(x, f) \in \mathbb{P}^n \times P(n, d); f(x) = 0\}$$

and let $p: Z \to P(n, d)$ be the second projection. It is easy to show that Z is a smooth variety (using the first projection Z can be regarded as a vector bundle over $\mathbb{P}^n$) and that $f \in P(n, d)$ is a regular value for p if and only if the associated hypersurface $V_f = p^{-1}(f)$ is smooth. Let $f_0, \ldots, f_{D-1}$ be a basis of the vector space $P(n, d)$ and consider the *Veronese embedding*

(3.3) $$v_d: \mathbb{P}^n \to \mathbb{P}^{D-1}, \qquad v_d(x) = (f_0(x): \cdots : f_{D-1}(x))$$

and its image $V_{n,d} = v_d(\mathbb{P}^n)$, called the Veronese variety, see [L2].

Then the fibers of p may be clearly identified with the *hyperplane sections* of the Veronese variety $V_{n,d}$ and the critical values of p correspond precisely to (the affine cone over) the *dual* of the Veronese variety, denoted $\hat{V}_{n,d}$, see [L2] for details.

Let $U = P(n, d) \backslash C(\hat{V}_{n,d})$ be the set of regular values of p and note that U is a *connected* manifold [Mu2], p. 68. Now apply Ehresmann's Fibration Theorem (3.1) with $E = \mathbb{P}^n \times U$, $B = U$, $A = Z \cap E$, and $f = $ the second projection.

We thus get the next result

(3.4) **Corollary.** *The diffeomorphism type of the pair* $(\mathbb{P}^n, V_f)$ *is the same for all smooth hypersurfaces* V_f *of a given degree d in* $\mathbb{P}^n$.

What can be said about the *singular* hypersurfaces? First, we have to state the analog of (3.1) which holds in the more general context of Whitney stratified sets. As a preliminary remark, note that a pair (E, A), as in the second part of (3.1), is one of the simplest examples of Whitney stratified sets (recall 1.8(i)).

(3.5) Theorem (Thom's First Isotopy Lemma). *Let A be a locally closed subset in the smooth manifold M and let $\mathscr{A}$ be a Whitney stratification of A. Let $f: M \to B$ be a smooth mapping of manifolds such that $f|A_i: A_i \to B$ is a submersion and $f|(\bar{A}_i \cap A)$ is proper for any stratum $A_i \in \mathscr{A}$. Then f fibers A over B topologically locally trivially in the stratified sense, i.e., the trivialization homeomorphism $\varphi: (f^{-1}(b) \cap A) \times U \to f^{-1}(U) \cap A$ preserves the strata of the obvious stratifications induced by $\mathscr{A}$ on both the source and the target of φ.*

For the proof of this result (which is again based on the integration of some controlled stratified vector fields) we refer to [GWPL], p. 58. The importance of Thom's First Isotopy Lemma comes from the fact that maps f with the properties stated in (3.5) exist in abundance in algebraic geometry.

(3.6) Proposition. *Let $f: X \to Y$ be a (regular) map of algebraic varieties. Then there exist Whitney stratifications $\mathscr{X}$ of X and $\mathscr{Y}$ of Y, both consisting of a finite number of semialgebraic strata, such that for any stratum $X_i \in \mathscr{X}$ there is a stratum $Y_j \in \mathscr{Y}$ such that $f(X_i) \subset Y_j$ and $f|X_i: X_i \to Y_j$ is a submersion.*

Note that if, in addition, f is proper, we may apply Theorem (3.5) with $B = Y_j$ and $A = f^{-1}(B)$ (with the obvious induced stratification).

Proof of (3.6). When $Y = \{y\}$, then (3.6) is nothing other than (1.12) in the complex case. The case dim $Y = 0$, i.e., Y is a finite number of points can be clearly reduced to the previous one.

Assume now that dim $Y = 1$ and that Y is *irreducible*. There are two distinct cases to consider:

(i) dim $f(X) = 0$.
   Then it is enough to take the Whitney stratification $\mathscr{Y}$ of Y to consist of the points in $f(X)$, the singular points Y_{sing} and the remaining open set Y_0 (a complement of a finite number of points). Using the case dim $Y = 0$, we are clearly done.
(ii) dim $f(X) = 1$.
   Let $\mathscr{X}'$ be a Whitney stratification for X and let $X'_1, \ldots, X'_k$ be all the strata in $\mathscr{X}'$ with $\overline{f(X'_j)} = Y$. It follows easily (see, for instance, [Mu2], p. 43) that there is a Zariski open set $Y_0 \subset Y \setminus Y_{\text{sing}}$ such that $Y_0 \subset f(X'_j)$, and Y_0 consists only of regular values of $f|X_j$ for $j = 1, \ldots, k$.

Moreover, we may take Y_0 such that

$$Y_0 \cap f(X'_j) = \varnothing \qquad \text{for} \quad j \notin [1, k].$$

Define $X_i = X_i' \cap f^{-1}(Y_0)$ for $i = 1, \ldots, k$, and note that $\{X_1, \ldots, X_k\}$ form a Whitney stratification of the Zariski open subset $U = f^{-1}(Y_0)$ of X. Using Remark (1.13(i)), this can be extended to a Whitney stratification $\mathscr{X}$ of the whole space X. Let $\mathscr{Y}$ consist of Y_0 plus the points in $Y \setminus Y_0$. It is clear that the stratifications $\mathscr{X}$ and $\mathscr{Y}$ satisfy the claims in (3.6).   □

We leave the reader to formalize the proof in the general case (no new ideas are needed).

(3.7) **Corollary.** *There exists a (finite semialgebraic) Whitney stratification $\mathscr{A}$ of the vector space $P(n, d)$ such that two pairs $(\mathbb{P}^n, V_f)$ and $(\mathbb{P}^n, V_g)$ are topologically equivalent when f and g belong to the same connected component of a stratum in $\mathscr{A}$.*

Recalling that a semialgebraic set has finitely many components, see [M5], Appendix A, it follows that altogether there are only finitely many distinct topological types among the pairs $(\mathbb{P}^n, V_f)$ for $f \in P(n, d)$. This finiteness should be contrasted with the presence of *moduli* (i.e., continuous parameters) for the *algebraic* (or *analytic*) *classification* of the hypersurfaces in $\mathbb{P}^n$.

Recall now the map $p: Z \to P(n, d)$ from Example (3.2) and the description of the set X of critical values of p as the affine cone $C(\hat{V}_{n,d})$ over the Veronese variety. It is well know that the varieties $\hat{V}_{n,d}$ and X are irreducible [L2], and hence $X_{\text{reg}} = X \setminus X_{\text{sing}}$ is a connected smooth submanifold in $P(n, d)$. It is also known that $f \in X_{\text{reg}}$ if and only if the associated hypersurface V_f has exactly one singularity x_f and this singularity is of type A_1, i.e., in local coordinates $(u_1, \ldots, u_n)$ it is given by

$$(V_f, x_f): u_1^2 + \cdots + u_n^2 = 0.$$

Such a singularity is also called a *node*. Hence $Z' = p^{-1}(X_{\text{reg}})$ is essentially the union of all these hypersurfaces and has an obvious Whitney stratification

$$Z' = Z_0' \cup Z_1' \quad \text{with} \quad Z_0' = \bigcup_{f \in X_{\text{reg}}} (V_f \setminus \{x_f\}) \quad \text{and} \quad Z_1' = \bigcup_{f \in X_{\text{reg}}} \{x_f\}.$$

(To check the Whitney condition, note that it is not necessary to apply (2.13), since what we have here is an *analytically* trivial family of IHS.)

Applying (3.5) to the restriction $p|Z': Z' \to X_{\text{reg}}$ we get the following:

(3.8) **Corollary.** *The topological type of the pair $(\mathbb{P}^n, V_f)$, where V_f is a hypersurface of degree d having exactly one A_1 singular point, does not depend on the defining equation f.*

We can try to extend this result as follows. Fix a certain list Y of isomorphism classes (or, more generally, of μ-constant classes) of IHS

$$Y: (Y_1, 0), \ldots, (Y_k, 0).$$

Let $B(Y)$ be the set of polynomials f in $P(n, d)$ such that the hypersurface V_f

has exactly k singular points $x_f^1, \ldots, x_f^k$ such that $(V_f, x_f^i) \simeq (Y_i, 0)$ for all $i = 1, \ldots, k$, where $\simeq$ means analytic isomorphism (or, μ-equivalence) of IHS.

In the second case (where we deal with μ-classes), assume that our singularities have the property that any μ-constant deformation is μ^*-constant, see (2.15). Then it is reasonable to expect that $B(Y)$ will be a smooth semi-algebraic submanifold in $P(n, d)$, and $A(Y) = p^{-1}(B(Y))$ is Whitney stratified by taking

$$A(Y)_0 = \bigcup_{f \in B(Y)} V_f \backslash \{x_f^1, \ldots, x_f^k\},$$

$$A(Y)_i = \bigcup_{f \in B(Y)} \{x_f^i\} \quad \text{for} \quad i = 1, \ldots, k.$$

The *main problem* is to decide whether the base $B(Y)$ is connected (and in the affirmative case we get a result similar to (3.8)). The answer to this problem is very subtle and depends very much on the relations between n, d, and the list Y.

We mention the following special cases:

(i) $n = 2$, d arbitrary, all $(Y_i, 0)$ are nodes.
    The connectivity of $B(Y)$ in this case is the famous Severi problem and this was completely solved only recently by J. Harris [Hs].
(ii) $n = 2$, $d = 6$, $Y = 6 \text{ cusps} = 6A_2$.
    It was remarked by Zariski that in this case $B(Y)$ has two connected components: The sextic curves corresponding to one of these components have all of their six cusps situated on a conic, while for the other component this property does not hold. And Zariski [Z1] showed that, in this case, the topological types $(\mathbb{P}^2, C_6)$ corresponding to these two types of sextic curves C_6 are distinct. We shall come back to this classical example of Zariski in Chapter 4.

The general philosophy is that $B(Y)$ is connected if d is big enough compared to the number and the complexity of the IHS in the list Y. A concrete result along this line is the following.

(3.9) **Proposition.** *Let s_i be the $\mathcal{K}$-determinacy order of the singularity $(Y_i, 0)$ and assume that*

$$d \geq s_1 + \cdots + s_k + k - 1.$$

Then $B(Y)$ is a connected semialgebraic smooth submanifold in $P(n, d)$ and the topological type of the pair $(\mathbb{P}^n, V_f)$ is constant for $f \in B(Y)$.

Proof. Consider the map

$$q: B(Y) \to (\mathbb{P}^n)^k, \qquad q(f) = (x_f^1, \ldots, x_f^k).$$

It is clearly enough to show that q is a fiber bundle with typical fiber F, where F is a smooth connected semialgebraic subset in $P(n, d)$. Fix a point $a = (a_1, \ldots, a_k) \in (\mathbb{P}^n)^k$ and let us investigate the corresponding fiber $q^{-1}(a)$.

Choosing the coordinates on $\mathbb{P}^n$ in a suitable way, we can assume that $a_i = (1 : \bar{a}_i)$ for some point $\bar{a}_i \in \mathbb{C}^n$ and all $i = 1, \dots, k$.

For the definition of the order of $\mathscr{K}$-determinacy and related concepts we refer to [D4]. Let $J = J^{s_1}(n, 1) \times \cdots \times J^{s_k}(n, 1)$ be a product of jet spaces of orders $s_1, \dots, s_k$. Let $S \subset J$ be the product of the $\mathscr{K}$-orbits corresponding to the singularities $(Y_i, 0)$, $i = 1, \dots, k$. Then S is a connected smooth submanifold in J, since the contact groups are connected.

Let
$$P_a = \{f \in P(n, d); f(a_i) = 0 \text{ for } i = 1, \dots, k\}.$$

It is an easy exercise for the reader to show that the inequality $d \geq s_1 + \cdots + s_k + k - 1$ implies the following:

(i) $\dim P_a = \dim P(n, d) - k$;
(ii) the linear map

$$j: P_a \to J, \qquad j(f) = (j_{a_1}^{s_1}(f), \dots, j_{a_k}^{s_k}(f)),$$

which associates to a polynomial f its s_i-jet at the point a_i (identified with the point $\bar{a}_i \in \mathbb{C}^n$), is surjective.

Hence $j^{-1}(S)$ is a connected smooth submanifold in P_a and if $g \in j^{-1}(S)$, then V_j has clearly isolated singularities at the points a_i isomorphic to $(Y_i, 0)$ for all $i = 1, \dots, k$. It is easy to show that there exists a Zariski open subset $F_a \subset j^{-1}(S)$ such that $f \in F_a$ if and only if the hypersurface V_f has no other singularities. Note that this F_a is still connected and smooth and one clearly has $q^{-1}(a) = F_a$.

It is also clear from this description that the union $(F_a)_{a \in (\mathbb{P}^n)^k}$ gives rise to a fiber bundle with total space $B(Y)$, thus ending the proof. $\qquad\qquad\square$

(3.10) **Remark.** Severi's problem for curves mentioned above shows that the inequality in (3.9) is not a necessary condition (in that case, $s_i = 2$ for all i). On the other hand, note that the Severi problem has a *negative answer* in higher dimensions. There are indeed examples of nodal quintics in $\mathbb{P}^4$ with 100 nodes and with various topological types for $(\mathbb{P}^4, V)$, see van Geemen–Werner [GW].

§4. On the Topology of Affine Hypersurfaces

This section contains an application of Thom's Isotopy Lemma. We have singled it out for its generality and it is one of the few places in our book where we deal with *affine* hypersurfaces (which are much more complicated to treat than the projective ones, especially if we restrict ourselves, in both cases, to the smooth hypersurfaces, e.g., refer to Example (5.4.29) in this book for a concrete situation).

The point of view we take now is slightly different from the previous section, namely, we fix a polynomial function $f: \mathbb{C}^n \to \mathbb{C}$ and look at the topology of the various fibers $X_t = f^{-1}(t)$.

(4.1) Proposition. *There exists a finite set of points $B_f \subset \mathbb{C}$ such that the restriction $f: \mathbb{C}^n \setminus f^{-1}(B_f) \to \mathbb{C} \setminus B_f$ is a smooth locally trivial fibration.*

Proof (Compare with [V], [Bt1], [HL]). The main problem is that the map f above is not proper for $n > 1$. To get a proper map we can proceed as follows. Assume that $\deg(f) = d$ and let

$$\tilde{f}(x_0, x_1, \ldots, x_n) = x_0^d f\left(\frac{x_1}{x_0}, \ldots, \frac{x_n}{x_0}\right)$$

be the *homogenization* of the polynomial f. Consider the hypersurface in $\mathbb{P}^n \times \mathbb{C}$ defined by

$$X = \{(x, t) \in \mathbb{P}^n \times \mathbb{C}; \tilde{f}(x) - t x_0^d = 0\}.$$

Let $X_\infty = X \cap (H_\infty \times \mathbb{C}) = \{x \in H_\infty; f_d(x) = 0\} \times \mathbb{C}$ be the part of X "sitting at infinity," where $H_\infty: x_0 = 0$ is the hyperplane at infinity in $\mathbb{P}^n$ and $f = f_0 + f_1 + \cdots + f_d$ is the decomposition of f into homogeneous components ($f_d \neq 0$).

Note that the singular part of X is precisely the set $A \times \mathbb{C}$, where

$$A = A_f = \left\{x \in H_\infty; \frac{\partial f_d}{\partial x_1}(x) = \cdots = \frac{\partial f_d}{\partial x_n}(x) = f_{d-1}(x) = 0\right\}.$$

On the other hand, A can be obviously regarded as being the singular part of the closure $\overline{X}_t$ in $\mathbb{P}^n$ of any of the smooth fibers X_t of f.

Case 1. $\dim A \leq 0$ (i.e., $A = \varnothing$ or a finite set).
Define

$$A'_f = \left\{x \in H_\infty; \frac{\partial f_d}{\partial x_1}(x) = \cdots = \frac{\partial f_d}{\partial x_n}(x) = 0\right\}$$

and note that $A'_f \times \mathbb{C}$ is the singular part of X_∞. Let $\mathscr{A}' = (A'_i)_{i=2,\ldots,p}$ be a Whitney regular stratification of A'_f such that $A_f \subset A'_f$ is a union of strata in $\mathscr{A}'$ and $\{V(f_d) \setminus A'_f, A'_2, \ldots, A'_p\}$ is a Whitney stratification for $V(f_d)$ in $H_\infty = \mathbb{P}^{n-1}$. Consider the stratification B of X having as strata

$$B_0 = X \setminus X_\infty, \quad B_1 = X_\infty \setminus (A'_f \times \mathbb{C}) \quad \text{and} \quad B_i = A'_i \times \mathbb{C} \text{ for any stratum } A'_i \text{ in } \mathscr{A}'.$$

If $A'_i \cap A = \varnothing$, it is clear that $A'_i \times \mathbb{C}$ is contained in the smooth part of X and hence B_0 (which is an open stratum) is Whitney regular over $A'_i \times \mathbb{C}$. Since $B_1 = (V(f_d) \setminus A'_f) \times \mathbb{C}$, it follows that B_1 is also Whitney regular over $A'_i \times \mathbb{C}$. Similarly, for a stratum $A'_i \times \mathbb{C}$ over a stratum $A'_j \times \mathbb{C}$. It follows that the only failure of the Whitney regularity condition may appear for the pair of strata of type $(B_0, a \times \mathbb{C})$, where a is a point in A.

But this situation is exactly the situation from (2.13), so that we have to check the μ^*-constant property. Since the μ^*-invariant is clearly semicontinuous (in the algebraic sense), it follows that for each $a \in A$ there is a finite set $B_a \subset \mathbb{C}$ such that X is μ^*-constant along $a \times \mathbb{C}$ except for the points $a \times B_a$.

If we set $B_f = \bigcup_{a \in A} B_a \cup \{$critical values of $f\}$, then B_f is a finite set satisfying the claim in (4.1). Indeed, we have just to apply Thom's First Isotopy Lemma to the situation

$$\mathrm{pr}_2 \colon X \setminus \mathrm{pr}_2^{-1}(B_f) \to \mathbb{C} \setminus B_f.$$

Case 2. dim $A > 0$.

To treat this case, we may use the more sophisticated analogues of (2.13) presented in [LêT]. We leave the details for the interested reader.   □

(4.2) **Example.** Broughton has remarked, [Bt1], that the polynomial $f(x, y) = x^2 y - x$ has no critical value, however, $B_f = \{0\}$. To explain why $t = 0$ is a special value, we have to apply the stratification procedure described above to this polynomial f. In this case, we clearly have

$$A = \{(x; y) \in \mathbb{P}^1; 2xy = x^2 = 0\} = \{(0:1)\} = a_1$$

and hence we are in Case 1.

The family of IHS $(\overline{X}_t, a_1 \times t)$ has the following equation (in the coordinates $y = 1, z = x_0$):

$$g_t(x, z) = x^2 - xz^2 - z^2 t = 0.$$

For $t \neq 0$ this is an A_1-singularity ($\mu = 1$), while for the special value $t = 0$ we get an A_3-singularity ($\mu = 3$) and hence there is a jump in the Milnor number at $t = 0$.

(4.3) **Remark.** When $A_f = \varnothing$, note that we have no problems with the Whitney regularity and hence we can take $B_f = $ critical values of f. This equality also follows from the fact that a polynomial f with $A_f = \varnothing$ is a quasi-tame polynomial, see [N1], [N3], and [D7] for details. For such polynomials, the *generic* fiber X_t for $t \in \mathbb{C} \setminus B_f$ has the homotopy type of a bouquet of spheres of dimension $n - 1$, the number of these spheres being given by a *global* Milnor number

$$(4.4) \qquad \mu_{\mathrm{global}}(f) = \dim_\mathbb{C} \frac{\mathbb{C}[x_1, \ldots, x_n]}{(\partial f / \partial x_1, \ldots, \partial f / \partial x_n)}.$$

Since this formula is not so handy (due to the fact that the polynomial ring $\mathbb{C}[x_1, \ldots, x_n]$ is not a local ring as is the convergent power series ring $\mathbb{C}\{x_1, \ldots, x_n\}$ which appeared in (2.6)), we offer the following formula

$$(4.5) \qquad \mu_{\mathrm{global}}(f) = (d - 1)^n - \sum_{x \in A_f'} \mu(\overline{X}_t^\infty, x).$$

Here $\overline{X}_t^\infty = \overline{X}_t \cap H_\infty = \{x \in H_\infty, f_d(x) = 0\}$ is the part at infinity of the fibers of f, and A_f' is a finite set under our assumption. This formula will be an obvious corollary of our results in Chapter 5, §4.

A related result is the following generalization of a nice result by Hà–Lê [HL].

(4.6) Proposition. *Let* $f: \mathbb{C}^n \to \mathbb{C}$ *be a polynomial function such that*:

(i) A_f *is a finite set, say* $\{a_1, \ldots, a_p\}$;
(ii) *any of the* IHS *families* $(\overline{X}_t, a_i \times t)$ *consists only of singularities for which any μ-constant deformation is μ^*-constant.*

   Then B_f is the complement of the maximal infinite set $U_f \subset \mathbb{C}$ such that $\chi(X_t)$ is constant for $t \in U_f$.

Proof. Since the fibers of any fibration have the same Euler characteristic, it is clear that $\mathbb{C} \backslash B_f \subset U_f$. To prove the equality it is enough to remark that we have the following formula for computing $\chi(X_t)$ (with V_{smooth}^d being a smooth hypersurface in $\mathbb{P}^n$ of degree d):

$$\chi(X_t) = \chi(\overline{X}_t) - \chi(\overline{X}_t^\infty) = \chi(V_{\text{smooth}}^d) + (-1)^n \sum_{x \in A_f} \mu(\overline{X}_t, x) - \chi(\overline{X}_t^\infty).$$

As above, we refer to Chapter 5 for details. From this formula we see that U_f is precisely the Zariski open set in $\mathbb{C}$ over which all the families $(\overline{X}_t, x)$ are μ-constant (use the semicontinuity of the Milnor number). It follows from (ii) that all these families are then μ^*-constant and, hence by (2.13), the Whitney regularity conditions are fulfilled. Note that for $n = 2$ both conditions (i) and (ii) above are fulfilled (recall (2.15(i))) and this is precisely the case treated in [HL]. □

(4.7) Exercise. Consider the polynomial function

$$f: \mathbb{C}^3 \to \mathbb{C}, \qquad f(x, y, z) = x + x^{d-1}y + y^{d-a}z^a,$$

where the positive integers d and a satisfy the inequalities $d - 1 > a \geq 1$.
   Show that:

(i) For $a = 1$, the set A_f consists of the point $a_1 = (0 : 0 : 1)$. The corresponding family $(\overline{X}_t, a_1 \times t)$ is μ^*-constant for $t \in \mathbb{C}$ and hence we can take $B_f = \varnothing$ in this case.
(ii) For $a > 1$, the set A_f consists of two points, namely, a_1 and $a_2 = (0 : 1 : 0)$. The family $(\overline{X}_t, a_2 \times t)$ is μ^*-constant for $t \in \mathbb{C} \backslash \{0\}$ and hence we can take $B_f = \{0\}$ in this case.
(iii) Compute the Euler characteristic of the generic fiber in these two cases.

   The reader may like to compare this exercise to Exercise (5.4.33) later in this book.

§5. Links and Conic Structures

This section is a short discussion about the links and the conic structure of the complex algebraic and analytic sets. For additional information we refer the reader to the very accessible paper by Durfee [Df6]. Our discussion in this section is a local one, so we can work in a neighborhood U of the origin in $\mathbb{C}^n$.

Let A be a closed analytic set in U such that $0 \in A$. Let $\mathscr{A} = \{A_i\}_{i=1,p}$ be the Whitney stratification of A such that $\{0\} = A_1$ is a stratum (this is easy to achieve by (1.10)). Let $\rho\colon \mathbb{C}^n \to \mathbb{R}$ be the square of the usual distance function, namely,

$$\rho(x) = |x_1|^2 + \cdots + |x_n|^2.$$

It is easy to see (using condition (1.6)) that there exists an $\varepsilon_0 > 0$ such that all the spheres $S_\varepsilon = \rho^{-1}(\varepsilon)$ for $\varepsilon_0 \geq \varepsilon > 0$ are *transversal* to all the strata A_i in $\mathscr{A}$ with $0 \in \bar{A}_i$.

It follows that the map

$$\rho\colon B_{\varepsilon_0}\backslash\{0\} \cap A \to (0, \varepsilon_0),$$

where $B_{\varepsilon_0} = \{x \in \mathbb{C}^n; \rho(x) < \varepsilon_0\}$, satisfies all the conditions in the Thom First Isotopy Lemma (3.5). In particular, the topological type of the space $S_\varepsilon \cap A = \rho^{-1}(\varepsilon)$ is the same for all $\varepsilon \in (0, \varepsilon_0)$. We denote it by $L(A, 0)$ and call it the *link of the set A at the point* 0.

Since local triviality over the contractible space $(0, \varepsilon_0)$ means triviality, we have, in fact, a homeomorphism

$$\varphi\colon L(A, 0) \times (0, \varepsilon_0) \to (B_{\varepsilon_0}\backslash\{0\}) \cap A$$

such that $\rho \circ \varphi(x, t) = t$. This homeomorphism can be extended in an obvious way to the *cone* $C(L(A, 0))$ obtained from the product $L(A, 0) \times [0, \varepsilon_0]$ by collapsing $L(A, 0) \times \{0\}$ to a point *. Note that we can regard B_{ε_0} as a Whitney stratified set by taking $A_0 = B_{\varepsilon_0}\backslash A$ and $A_1, \ldots, A_p$ as above the strata covering A. And again we can apply (3.5) to the map $\rho\colon B_{\varepsilon_0}\backslash\{0\} \to (0, \varepsilon_0)$.

In this way we get the following basic result.

(5.1) **Theorem** (The Conic Structure of Analytic Sets). *If* 0 *is not an isolated point of* A, *then we have a homeomorphism of pairs*

$$(\bar{B}_{\varepsilon_0}, \bar{B}_{\varepsilon_0} \cap A) \simeq C(S_{\varepsilon_0}, L(A, 0))$$

with $L(A, 0) = S_{\varepsilon_0} \cap A$ *and the identification* $\bar{B}_{\varepsilon_0} = C(S_{\varepsilon^0})$ *being the natural one.*

The case when 0 is an isolated singular point of A is due to Milnor [M5], p. 18, while the general case is due to Burghelea–Verona [BV]. See also [Df6], where it is shown that the link $L(A, 0)$ (and hence also the cone $C(L(A, 0))$) depends only on the isomorphism class of the germ $(A, 0)$ and not on the embedding $(A, 0) \subset (\mathbb{C}^n, 0)$.

(5.2) **Corollary.** *Any complex (algebraic or analytic) variety* V *is locally contractible.*

This result has many important consequences for the *algebraic topology* of V, e.g.:

(i) all the cohomology theories (singular, Alexander, Čech, see [Sp]) give the same result when applied to V;

(ii) any connected variety V has a universal covering space $\tilde{V}$, see [Sp], p. 80.

(5.3) **Exercise.** Let $(A, 0)$ be the germ at the origin in $\mathbb{C}^3$ of the Whitney umbrella considered in (1.8). Show that the corresponding link $L(A, 0)$ is homeomorphic to the join $S^1 * S^1$ of two circles (the definition for the join of two topological spaces is recalled in Chapter 3, §3). *Hint.* Consider the circles $K_1: x = y = 0$ ($|z| = $ const.) and $K_2: x = z = 0$ ($|y| = $ const.) contained in the link $L(A, 0)$. Show that the difference $L(A, 0) \backslash (K_1 \cup K_2)$ can be identified to the product $S^1 \times S^1 \times (0, 1)$.

In the next two chapters we make a detailed study of the link $L(A, 0)$ and of the embedding $L(A, 0) \subset S_\varepsilon$ in the case when $(A, 0)$ is an IHS.

§6. On Zariski Theorems of Lefschetz Type

We start with a definition.

(6.1) **Definition.** A germ of a complex analytic set $(X, 0)$ at the origin $0 \in \mathbb{C}^n$ is called a *complete intersection singularity* if all the irreducible components of $(X, 0)$ have the same dimension, say m, and $(X, 0)$ can be set-theoretically defined as the zero set of $n - m$ analytic functions.

This means that in a small enough open ball B_ε centered at the origin of $\mathbb{C}^n$ one has

$$X \cap B_\varepsilon = \{x \in B_\varepsilon; f_1(x) = \cdots = f_{n-m}(x) = 0\}$$

for some $f_1, \ldots, f_{n-m} \in \mathcal{O}_n$. Note that we do not require that $f_1, \ldots, f_{n-m}$ generate the ideal

$$I_X = \{g \in \mathcal{O}_n; g|X = 0\} \subset \mathcal{O}_n$$

associated to the germ $(X, 0)$.

(6.2) **Example.** (i) It is well known that the ring $\mathcal{O}_n$ is a factorial ring, [KK], p. 80. Using this it is easy to show that any hypersurface germ $(X, 0)$ (i.e., $(X, 0)$ is a pure one-codimensional set germ) can be defined by one equation. Indeed, we may assume that $(X, 0)$ is irreducible and let $g \in I_X$ be a nonzero element. Then g can be written as a product of prime elements $g_1, \ldots, g_p$. At least one of these factors, say g_1, has to vanish on X (I_X is a prime ideal) and it is clear that g_1 generates the ideal I_X. Hence any hypersurface singularity is a complete intersection singularity.

(ii) If $(X, 0)$ is the germ of a smooth complex submanifold in $\mathbb{C}^n$, then clearly $(X, 0)$ is a complete intersection singularity.

(iii) Not all analytic set germs are complete intersections. We shall show in

(3.2.13) that the germ

$$(X, 0) = (H_1, 0) \cup (H_2, 0) \subset (\mathbb{C}^4, 0),$$

which is the union of the planes

$$H_1: x_1 = x_2 = 0 \qquad \text{and} \qquad H_2: x_3 = x_4 = 0$$

in $\mathbb{C}^4$, is not a complete intersection.

(6.3) **Definition.** Let P be a smooth variety of dimension n and let $X \subset P$ be a closed subvariety. We say that X is a *locally complete intersection* if for any point $x \in X$, the germ (X, x) is a complete intersection singularity in $(P, x) \simeq (\mathbb{C}^n, 0)$, as defined in (5.1).

We also recall a definition from algebraic topology.

(6.4) **Definition** ([Sp], p. 404). A continuous map $f: X \to Y$ of path-connected topological spaces is said to be a *q-equivalence* ($q \geq 1$) if the induced homomorphisms

$$\pi_i(f): \pi_i(X, x) \to \pi_i(Y, f(x))$$

are isomorphisms for $i < q$ and epimorphisms for $i \leq q$, the point $x \in X$ being arbitrary.

All these new notions appear in the following statement.

(6.5) **Theorem** (Zariski Theorem of Lefschetz Type). *Let $X \subset \mathbb{P}^n$ be a closed subvariety which is locally a complete intersection of dimension m.*

Let Z be another closed subvariety in $\mathbb{P}^n$. Let $\mathscr{A}$ be a Whitney stratification of X such that $X \cap Z$ is a union $\bigcup_{i=1,p} A_i$ of strata $A_i \in \mathscr{A}$. Then the inclusion

$$j_H: (X \backslash Z) \cap H \to X \backslash Z$$

is an $(m-1)$-equivalence for any hyperplane H in $\mathbb{P}^n$ which is transversal to all the strata A_i in $X \cap Z$.

A proof of this result (using essentially Morse theory on manifolds with boundary) can be found in Hamm [H4]. For several other related results and references we refer to the book by Goreski–MacPherson [GM] and to the paper by Hamm–Lê [HLê2].

We would like to point out the fact that the result (6.5) and its analogues are usually stated in a *vague way*: "for a generic hyperplane H in $\mathbb{P}^n$, the inclusion j_H is an $(m-1)$-equivalence." But for many applications, see [D7] and [N2] for some examples, it is *crucial* to know what "generic" means in a given concrete situation. We now list some special cases of particular interest of (6.5):

(i) The classical "*Lefschetz hyperplane section theorem*" is obtained by taking $Z = \varnothing$. Note that this holds for *any* hyperplane H in $\mathbb{P}^n$.

(ii) The "*affine Lefschetz hyperplane section theorem*" is obtained by taking $Z: x_0 = 0$, i.e., Z is the hyperplane at infinity which is added to $\mathbb{C}^n$ to obtain $\mathbb{P}^n$.

(iii) Zariski's theorem about the fundamental groups of the complements of projective hypersurfaces [Z4] (or its more general version [HLê1]) is obtained by taking $X = \mathbb{P}^n$ and $Z = $ a hypersurface in $\mathbb{P}^n$.

For the sake of completness, we also recall the following notion and related basic result.

(6.6) Definition. A closed algebraic subvariety X in $\mathbb{C}^n$ is called an *affine variety*. A closed analytic subvariety X in $\mathbb{C}^n$ is called a *Stein variety*.

It is clear that any affine variety is Stein.

(6.7) Examples. (i) Let $H = \{x \in \mathbb{C}^n; f(x) = 0\}$ be a hypersurface in $\mathbb{C}^n$, where f is a polynomial (or, more generally, an entire function). Then $\mathbb{C}^n \backslash H$ is an affine variety (or, respectively, a Stein space). Indeed, $\mathbb{C}^n \backslash H$ is clearly isomorphic to the hypersurface $\{(x_1, \ldots, x_n, y) \in \mathbb{C}^{n+1}; f(x) \cdot y - 1 = 0\}$.

(ii) Let $V_f = \{x \in \mathbb{P}^n; f(x) = 0\}$ be a projective hypersurface of degree d. Then $\mathbb{P}^n \backslash V_f$ is affine. Indeed, recall the Veronese embedding (3.3):

$$v_d \colon \mathbb{P}^n \to \mathbb{P}^{D-1}$$

and let H_f be the hyperplane in $\mathbb{P}^{D-1}$ corresponding to the polynomial f.

Then $\mathbb{P}^n \backslash V_f$ is identified via v_d with the closed algebraic set $V_{m,d} \backslash H_f$ in $\mathbb{P}^{D-1} \backslash H_f \simeq \mathbb{C}^{D-1}$.

(6.8) Theorem (Hamm [H3]). *A Stein variety X of complex dimension m has the homotopy type of a* CW-*complex of (real) dimension m.*

We note that the special case when X is an affine variety is due to Karchyauskas [Ks]. When X is a Stein space the above CW-complex may be infinite (e.g., take $X = \{x \in \mathbb{C}; \sin x \neq 0\}$). For affine algebraic varieties the situation is different as shown by the following result, see [BV].

(6.9) Theorem (Cylindric Structure at Infinity of Algebraic Sets). *Let $X \subset \mathbb{C}^n$ be a closed algebraic variety. Then there exists $R_0 > 0$ such that for all $R \geq R_0$ we have a homeomorphism of pairs*

$$(\mathbb{C}^n \backslash B_R, (\mathbb{C}^n \backslash B_R) \cap X) \cong [R, \infty) \times (S_R, S_R \cap X),$$

where $B_R = \{x \in \mathbb{C}^n; |x| < R\}, S_R = \partial \bar{B}_R$.

Proof. Let $\mathscr{X} = \{X_i\}$ be a Whitney stratification of X, consisting of *finitely* many semialgebraic strata X_i. It is easy to see that the restriction of the distance function $\rho(x) = |x|$ to each of the strata X_i has only *finitely* many critical values (if necessary, have a look at [M5], p. 16). Take R_0 to be a positive real number strictly greater than all these critical values.

Note that we can apply Thom's First Isotopy Lemma to the map

$$\rho: \mathbb{C}^n \setminus \bar{B}_{R_0} \to (R_0, \infty),$$

where $Y = \mathbb{C}^n \setminus \bar{B}_{R_0}$ is stratified as $Y = (Y \setminus X) \cup \bigcup_i (X_i \cap Y)$. The interval (R_0, ∞) is contractible, and hence the locally trivial fibration ρ is in fact trivial. In particular, it follows that X has the same homotopy type with $X \cap \bar{B}_R$, which is a compact polyhedron [Hi2]. $\square$

Since algebraic varieties have by definition finite coverings with affine varieties, we get the following corollary.

(6.10) Corollary. *Any algebraic variety X has the homotopy type of a finite CW-complex (or compact polyhedron). In particular, the fundamental group $\pi_1(X, x)$ (any $x \in X$) and the homology and cohomology groups $H_i(X; A)$, $H^i(X; A)$ (any $i \geq 0$) with coefficients in a principal ideal domain A are all finitely generated (as a group and as A-modules, respectively). Usually $A = \mathbb{Z}, \mathbb{Q}, \mathbb{C}$.*

This follows from a well-known property of the fundamental group, see [Sp], p. 141, and from the definition of the homology and cohomology groups for a polyhedron, see [Sp], Chaps. 4 and 5.

The next example shows that the higher homotopy groups of an (affine) variety may be *infinitely* generated.

(6.11) Example. Consider the following smooth affine surface in $\mathbb{C}^3$:

$$X: (xy)^2 + z^2 - 1 = 0.$$

It follows from the general results of Oka [O1], that X has the same homotopy type as the *join* $X_1 * X_2$ where $X_1 = \{(x, y) \in \mathbb{C}^2; (xy)^2 - 1 = 0\}$ has two connected components, each of them homotopy equivalent to S^1 and $X_2 = \{z \in \mathbb{C}; z^2 - 1 = 0\}$ consists of two points. See also (3.3.19).

Hence X is homotopy equivalent to the space

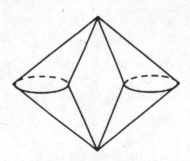

which can be deformed in an obvious way to $Y = S^1 \vee S^2 \vee S^2$. We can represent the universal covering space $\tilde{Y}$ of Y as the real line $\mathbb{R}$ with a copy of $S^2 \vee S^2$ attached to each integer point $m \in \mathbb{Z} \subset \mathbb{R}$. It follows that $\pi_2(Y) = \pi_2(\tilde{Y})$ is generated by this countable collection of 2-spheres.

With the notations from (6.9), it is natural to call the intersection $X_R = X \cap S_R$ for $R \geq R_0$ the *link at infinity* of the affine variety X.

It is natural to try to compare the topology of the spaces X and X_R. We have the following useful result due to Hamm [H4].

(6.12) Proposition. *Assume that X is locally a complete intersection of dimension m. Then the inclusion $X_R \to X$ is an $(m-1)$-equivalence.*

Proof. Use Theorem 2 in [H4] and the result (6.9) on the cylindric structure at infinity.   □

(6.13) Example. Consider the hypersurface X defined in $\mathbb{C}^4$ by the following equation

$$x + x^{d-1}y + y^{d-a}z^a + t^d = 0,$$

where d and a are positive integers, $d > a$. It follows from [D7] that this hypersurface X is simply-connected. Using (6.12), it follows that

$$\pi_1(X_R) = 0.$$

CHAPTER 2

Links of Curve and Surface Singularities

§1. A Quick Trip into Classical Knot Theory

We start with a brief account of knot theory for the following two reasons. First, the links of (plane) curve singularities—which are usually regarded as the simplest class of singularities to investigate—form a special class of knots, the so-called *algebraic links*. Second, many of the fundamental concepts related to the local topology of a higher dimensional IHS (e.g., Seifert matrix, intersection form, Milnor fibration, Alexander polynomial) have been considered first in relation to knot theory.

We hope the reader will get a better understanding of these concepts by encountering them first in their most simple and intuitive setting. For more details and the proofs of the results stated in this section we refer to standard books on knot theory, e.g., the excellent introduction by D. Rolfsen [R].

However, in our definition, we restrict our attention to the class of *tame* links, which is big enough to contain all the algebraic links.

(1.1) **Definition** (Compare with [R], p. 48). A *link* L is a closed smooth submanifold in the three-dimensional sphere S^3 such that each connected component L_i of L is diffeomorphic to the circle S^1. A connected link L is usually called a *knot*.

By the well-known fact that any one-dimensional closed connected smooth manifold is diffeomorphic to S^1 (see, for instance, [M4], p. 55), it is enough to ask in the above definition that $\dim L_i = 1$ for all i.

Two links L, L' are *equivalent* if there is a homeomorphism $h: S^3 \to S^3$ with $h(L) = L'$. Usually we regard the sphere S^3 in one of the following two explicit ways:

$$S^3 = \{(x, y) \in \mathbb{C}^2; |x|^2 + |y|^2 = 1\}$$

or

$$S^3 = \mathbb{R}^3 \cup \{\infty\},$$

the one point compactification of $\mathbb{R}^3$ (the identification $S^3 \setminus \{\infty\} \to \mathbb{R}^3$ being given by the stereographic projection.

(1.2) **Examples.** (i) *The Trivial Knot K_0.* We can define a trivial knot $K_0 = K_x$ in S^3 by the complex equation $x = 0$.

It is a standard fact that the sphere S^3 may be written as a union of the following two full 2-tori:

$$T_1 = \{(x, y) \in \mathbb{C}^2; |x|^2 + |y|^2 = 1, |x| \le |y|\},$$

$$T_2 = \{(x, y) \in \mathbb{C}^2; |x|^2 + |y|^2 = 1, |x| \ge |y|\},$$

with a two-dimensional torus $T_1 \cap T_2$ as the common boundary. There is clearly a retraction

$$r: S^2 \backslash K_x \to K_y, \qquad r(x, y) = \left(\frac{x}{|x|}, 0\right),$$

where

$$K_y = \{(x, y) \in S^3; y = 0\} \simeq S^1.$$

And the above decomposition of S^3 implies that in fact K_y is a *deformation retract* of $S^3 \backslash K_x$ and hence the two spaces have the same homotopy type (see [Sp], p. 30, for the concept of deformation retract).

Graphically we may picture the circles K_x and K_y as follows:

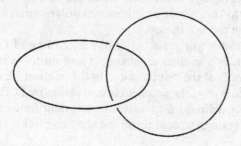

i.e., as a pair of *linked circles* (the linking number here being 1, see (1.7) below for a definition).

(ii) *The Trivial Link $L_{0,m}$ with $m \ge 1$ Components.* Under the natural inclusion $\mathbb{R}^3 \subset S^3$ we can represent the link $L_{0,m}$ by the picture

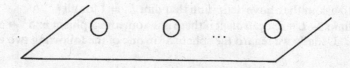

i.e., $L_{0,m}$ consists of m circles situated in a horizontal plane R^2. It follows from (2.9) below that this link has no algebraic *complex* equation. An interesting exercise in [R], p. 65, asks us to show that the complement $S^3 \backslash L_{0,m}$ has the homotopy type of the join $mS^1 \vee (m - 1)S^2$.

A useful hint (communicated to me by A. Parusinski) for the proof of this statement is the remark that the complement $\bar{B}\backslash S$ has the homotopy type of $S^1 \vee S^2$, where

$$\bar{B} = \{(x, y, z) \in \mathbb{R}^3; x^2 + y^2 + z^2 \le 1\},$$

$$S = \{(x, y, z) \in \mathbb{R}^3; z = 0, x^2 + y^2 = \tfrac{1}{9}\}.$$

More precisely, the space

$$\partial \bar{B} \cup \{(x, y, z) \in \mathbb{R}^3; x = 0, (y - \tfrac{1}{2}) + z^2 = \tfrac{1}{4}\} = S^2 \vee S^1$$

is a deformation retract of $\bar{B}\backslash S$, see the following picture.

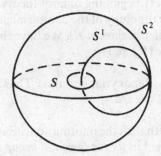

(iii) *The (p, q)-Torus Knot $K_{p,q}$.* Let us regard the circle S^1 as the set $\{z \in \mathbb{C}; |z| = 1\}$ and consider the embedding

$$j: S^1 \times S^1 \to S^3, \qquad j(x, y) = (x/\sqrt{2}, y/\sqrt{2}).$$

Let p, q be two positive integers with $(p, q) = 1$ and consider the embedding

$$e_{p,q}: S^1 \to S^1 \times S^1, \qquad z \mapsto (z^p, z^q)$$

(where is the condition $(p, q) = 1$ used?). Then the (p, q)-torus knot $K_{p,q}$ is the image of the composition $j \circ e_{p,q}$.

Consider also the algebraic knot

$$K'_{p,q} = \{(x, y) \in S^3; x^q + y^p = 0\}.$$

It is easy to show that the knots $K_{p,q}$ and $K'_{p,q}$ are equivalent. The case $p = 2$, $q = 3$ gives rise to the famous *trefoil knot*, see also [Mu2], pp. 13–14.

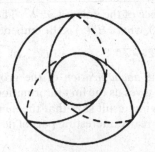

(iv) *The Fibers of the Hopf Bundle.* Consider the Hopf bundle $p: S^3 \to \mathbb{P}^1$, $(x, y) \mapsto (x : y)$ and fix a set of m points in $\mathbb{P}^1$: $z_i = (x_i : y_i)$ for $i = 1, \ldots, m$. Let $L_i = p^{-1}(z_i)$ and $L^m = L_1 \cup \cdots \cup L_m$. Then L^m is an algebraic link, since it can be defined by the following equation:

$$L^m = \{(x, y) \in S^3; (y_1 x - x_1 y), \ldots, (y_m x - x_m y) = 0\},$$

L^m consists of m circles, each pair of them having linking number 1. By an easy application of Ehresmann's Fibration Theorem (1.3.1), the equivalence class of L^m does not depend on the choice of the m points in $\mathbb{P}^1$. In particular, L^2 is equivalent to the union $K_x \cup K_y$ from (i) above.

It was clear from the very beginning of knot theory that to study a knot K it is useful to consider the topology of the complement $S^3 \backslash K$ (which is clearly an invariant of the equivalence class of K). We have the following basic result on these complements, see [R], p. 116.

(1.3) **Asphericity Theorem** (Papakyriakopoulos). *The knot complement* $S^3 \backslash K$ *is a* $K(\pi, 1)$ *Eilenberg–MacLane space.*

This means informally that all the information about the homotopy type of the space $S^3 \backslash K$ is contained in its *fundamental group* $\pi = \pi_1(S^3 \backslash K)$, which is also called the *group of the knot* K. The Asphericity Theorem is no longer true for links: recall our example (1.2(ii)) above.

It is easy to compute the homology (with $\mathbb{Z}$-coefficients) of link complements.

(1.4) **Lemma.** *Let L be a link with m components. Then*

$$H_i(S^3 \backslash L) = \begin{cases} \mathbb{Z}, & i = 0, \\ \mathbb{Z}^m, & i = 1, \\ \mathbb{Z}^{m-1}, & i = 2, \\ 0, & i > 2. \end{cases}$$

Proof. By Alexander Duality (see [Sp], p. 296) we have

$$H_i(S^3 \backslash L) = H^{3-i}(S^3, L).$$

Use then the exact sequence of the pair $j: L \subset S^3$. The homomorphism $H^k(j)$ is nontrivial only for $k = 0$, where $H^0(j)$ is the embedding

$$\mathbb{Z} \to \mathbb{Z}^m, \qquad a \mapsto (a, \ldots, a).$$

Since $H_1(S^3 \backslash L)$ is just the *abelianization* of the group $\pi_1(S^3 \backslash L)$, see [Gn], p. 48, or [Sp], p. 394, (1.4) gives us the first information about the group π. On the other hand, the following result says that the *noncommutative* part of the group of the knot K is the really interesting part of it. $\square$

(1.5) **Theorem.** *A knot K has a trivial group (i.e., $\pi_1(S^3 \setminus K) \simeq \mathbb{Z}$) if and only if K is equivalent to the trivial knot K_0.*

In fact, if K is equivalent to the trivial knot K_0, it follows from (1.2(i)) that $S^3 \setminus K$ has the homotopy type of S^1 and hence $\pi_1(S^3 \setminus K) = \pi_1(S^1) = \mathbb{Z}$. The converse in (1.5) is much more subtle, see [R], p. 103.

(1.6) **Example.** For the (p, q)-torus knot $K_{p,q}$ defined in (1.2(iii)) we have

$$\pi_{p,q} = \pi_1(S^3 \setminus K_{p,q}) = \{a, b; a^p = b^q\},$$

i.e., $\pi_{p,q}$ is the group generated by two elements a, b with just one relation $a^p = b^q$.

We can derive this result using a covering of $S^3 \setminus K_{p,q}$ by two open sets (obtained by a slight enlargement of the full tori T_1 and T_2 from (1.2(i)) minus the knot $K_{p,q}$) and applying the van Kampen theorem, see, for details, [R], pp. 51–53. Note that the groups $\pi_{p,q}$ (even $\pi_{2,3}$) have a complicated group structure, see, for instance, [Td].

In general, there is a standard way to write generators and relations for the group of any knot which is given by *a picture* (the so-called *Wirtinger presentation*). Note that there are nonequivalent knots having the same group, see [R], p. 62. For links L we can obtain some basic numerical invariants by considering the linking numbers $\mathrm{lk}(L_i, L_j)$ of the various components L_i of L.

(1.7) **Definition** (see [R], p. 133). Let J and K be two disjoint and *oriented* knots in S^3. Choose a 2-chain C in S^3 such that $\partial C = [J]$ (this is possible since $H_1(S^3) = 0$). Then the intersection $C \cdot [K]$ is a zero cycle in $H_0(S^3) \simeq \mathbb{Z}$, thus corresponding to an integer which we call the *linking number* of the knots J and K and denote by $\mathrm{lk}(J, K)$.

Note that the linking number is symmetric (i.e., $\mathrm{lk}(J, K) = \mathrm{lk}(K, J)$) and changes sign if we reverse the orientation of one of the knots K or J. Since we consider in these notes only unoriented knots, it follows that only the number $|\mathrm{lk}(J, K)|$ is intrinsically defined.

Another natural definition (apparently new) of this number is given as follows.

(1.8) **Exercise.** Let $X = S^3 \setminus (J \cup K)$ and consider the cup product

$$c: H^1(X) \times H^1(X) \to H^2(X).$$

By (1.4) it follows that c is a skew-symmetric bilinear form on $\mathbb{Z}^2$. Any such form is described in a suitable basis by a matrix

$$C = \begin{pmatrix} 0 & -a \\ a & 0 \end{pmatrix} \qquad \text{for some} \quad a \in \mathbb{N},$$

see [La], p. 380. Show that $a = |\text{lk}(J, K)|$. A special case of this will be proved below, see (2.20).

The definition of the linking number given in (1.7) makes natural the following basic concept of knot theory.

(1.9) Definition. For a link L, any orientable compact surface $M \subset S^3$ with boundary $\partial M = L$ is called a *Seifert surface* for L.

It can easily be shown that any link has a Seifert surface, see [R], p. 120. Moreover, any such surface M is obtained from a compact orientable surface $\tilde{M}$ without boundary by taking out m open discs, where m is the number of components of L. The *genus $g(M)$ of the surface M* is by definition the genus of the closed surface $\tilde{M}$ and we can define the *genus $g(L)$ of the link L* to be

$$g(L) = \min\{g(M), M \text{ is a Seifert surface for } L\}.$$

We have the following nice result:

(1.10) Proposition. *For a knot K we have $g(K) = 0$ if and only if K is (equivalent to) the trivial knot K_0.*

Proof. It is clear that the trivial knot embedding $K_0 = S^1 \subset S^3$ can be factored $S^1 \subset S^2 \subset S^3$ and hence a Seifert surface for K_0 is the upper (or lower) hemisphere in S^2. Therefore $g(K_0) = g(S^2) = 0$.   □

On the other hand, it is known that $g(K) = 0$ implies that K bounds a 2-disc M (since all the knots in S^2 are trivial, see [R], p. 9). Moreover, since we deal only with smooth objects, M is a flat 2-ball in S^3 and K is the trivial knot by the Basic Unknotting Theorem, see [R], p. 36.

One of the main reasons for considering the Seifert surfaces of links is the following. We have seen in (1.4) that the homology $H_.(S^3 \backslash L)$ of the complement of the link L is a very poor invariant for L, giving only the number of components of L.

On the other hand, homology (or cohomology) is much easier to compute than the fundamental group $\pi_1(S^3 \backslash L)$, which is a deep invariant of L. It is a natural idea to look for some topological objects associated to the link L and which have more interesting homology groups (i.e., with richer information about L).

Assume from now on that L is a *knot*, denoted as usual by K. For each positive integer k we consider the *cyclic covering of order k*: $\tilde{X}_k \to X$. This is the covering of $X = S^3 \backslash K$ corresponding to the normal subgroup in $\pi_1(X)$ equal to the kernel of the following composition of homomorphisms:

$$\pi_1(X) \xrightarrow{\text{Hurewicz}} H_1(X) \simeq \mathbb{Z} \to \mathbb{Z}/k\mathbb{Z}.$$

Similarly, we let $\tilde{X} \to X$ denote the *infinite cyclic covering* corresponding to the kernel of

$$\pi_1(X) \xrightarrow{\text{Hurewicz}} H_1(X) \simeq \mathbb{Z}.$$

To convince the reader that the spaces $\tilde{X}$ and $\tilde{X}_k$ have nontrivial homology, we give the following example.

(1.11) **Example.** For the trefoil knot $K = K_{2,3}$ we have $H_1(\tilde{X}) = \mathbb{Z}^2$ and $H_1(\tilde{X}_k)$ is given by the following table, where the first line contains the values of k modulo 6 (see [R], pp. 146–150),

k	0	1	2	3	4	5
$H_1(\tilde{X}_k)$	$\mathbb{Z}^3$	$\mathbb{Z}$	$\mathbb{Z} \oplus \mathbb{Z}/3\mathbb{Z}$	$\mathbb{Z} \oplus (\mathbb{Z}/2\mathbb{Z})^2$	$\mathbb{Z} \oplus \mathbb{Z}/3\mathbb{Z}$	$\mathbb{Z}$

Classically, the information contained in $H_1(\tilde{X}_k)$ was expressed in the so-called kth-*torsion numbers*, see [R], p. 145.

The Seifert surfaces for K are important for these computations, since there is a canonical way to build the covering spaces $\tilde{X}_k$ and $\tilde{X}$ starting with a Seifert surface M for K (and an open bicollar of $\mathring{M} = M \backslash K$ in X), see [R], pp. 128–131.

Note that the group of covering transformations $\text{Deck}(\tilde{X}/X)$ is infinite cyclic. Let T denote one of its two generators. Let $A = \mathbb{Z}[t, t^{-1}]$ be the ring of Laurent polynomials in t with integer coefficients and note that $H_1(\tilde{X})$ has a natural A-module structure, if we set

$$t \cdot c = H_1(T)(c) \qquad \text{for any} \quad c \in H_1(\tilde{X}).$$

It is easy to see that A is a *Noetherian domain* and that $H_1(\tilde{X})$ is a finitely generated A-module. *Hint.* For the first claim use the surjective homomorphism $\mathbb{Z}[x, y] \to A$, $x \mapsto t$, $y \mapsto t^{-1}$, and for the second the fact that $H_.(X)$ is finitely generated over $\mathbb{Z}$, X having the homotopy type of a finite CW-complex.

(1.12) **Definition.** The A-module $H_1(\tilde{X})$ is called the *Alexander invariant* of the knot K.

It is easy to see that the equivalent knots K and K' have isomorphic associated A-modules $H_1(\tilde{X})$ and $H_1(\tilde{X}')$. This is why the word "invariant" is used in (1.12).

(1.13) **Example.** Let $K = K_{p,q}$ be the torus knot of type (p, q). Then we can show that $H_1(\tilde{X}) \simeq A/(\Delta_{p,q}(t))$ where

$$\Delta_{p,q}(t) = \frac{(t^{pq} - 1)(t - 1)}{(t^p - 1)(t^q - 1)}$$

(use the condition $(p, q) = 1$ to show that $\Delta_{p,q}$ is indeed a polynomial in t).

A reference for (1.13) will be given below, see (2.8).

(1.14) **Remark.** In the definition of the Alexander invariant above we have considered only the first homology group $H_1(\tilde{X})$, since this is the only one carrying an interesting A-module structure. Indeed, it can be shown for any knot K that

$$H_0(\tilde{X}) = A/(t-1) = \mathbb{Z} \quad \text{and} \quad H_i(\tilde{X}) = 0 \quad \text{for all} \quad i > 1,$$

see [R], p. 171.

On the other hand, there is a similar $A_k = A/(t^k - 1)$-module structure on $H_1(\tilde{X}_k)$, but apparently these A_k-modules have not been investigated so far.

The following result shows that there is great freedom for the Alexander invariant, see [R], p. 171.

(1.15) **Theorem** (Seifert). *Let $\Delta \in A$ be a polynomial such that*

$$\Delta(1) = \pm 1 \quad and \quad \Delta(t) = \Delta(t^{-1}) \cdot t^d \quad where \quad d = \deg \Delta.$$

Then there is a knot $K \subset S^3$ such that its Alexander invariant is isomorphic to $A/(\Delta)$.

It is possible to compute the Alexander invariant of a knot starting with the knot group $\pi_1(X) = G$. Let $C = [G, G] = \ker(G \to H_1(X))$ be the commutator subgroup. Then the covering map $p: \tilde{X} \to X$ induces a morphism

$$p_\#: \pi_1(\tilde{X}) \to \pi_1(X) \quad \text{with} \quad \text{im } p_\# = C.$$

Hence there is an associated $\mathbb{Z}$-module isomorphism

$$q: H_1(\tilde{X}) \tilde{\to} C/[C, C] = C'.$$

We define an A-module structure on C' such that q becomes an A-module isomorphism. Let $x \in G$ be an element such that $[x]$ is a generator for $G/C = \mathbb{Z}$. Then define

$$t \cdot [c] = [xcx^{-1}] \quad \text{for all} \quad [c] \in C'.$$

This induces a well-defined A-module structure on C' and (with suitable choices of the generators of the infinite cyclic groups involved) it is clear that q is an A-isomorphism, see also [R], p. 174.

Another basic invariant of a knot K associated with a Seifert surface M for K is the Seifert form (or *Seifert matrix*) which is defined as follows. Consider a bicollar of $\mathring{M} = M \backslash K$, i.e., an embedding $\mathring{M} \times [-1, 1] \subset S^3$, coming from a tubular neighborhood of $\mathring{M}$ in S^3. For any cycle $x \in H_1(\mathring{M})$ we let x^+ (resp. x^-) denote the cycle in $H_1(\mathring{M} \times \{1\})$ (resp. $H_1(\mathring{M} \times \{-1\})$) given by $x \times \{1\}$ (resp. $x \times \{-1\}$).

(1.16) Definition. The $\mathbb{Z}$-bilinear form

$$S_M: H_1(\mathring{M}) \times H_1(\mathring{M}) \to \mathbb{Z}, \qquad S_M(x, y) = \mathrm{lk}(x, y^-),$$

is called the *Seifert form* of K associated with the Seifert surface M. If S_M^R denotes the Seifert form as defined in [R], p. 201, it is easy to see that the forms S_M and S_M^R are the transpose of each other, i.e.,

$$S_M(x, y) = S_M^R(y, x) \qquad \text{for all} \quad x, y.$$

We have chosen this definition of the Seifert form S_M because it is compatible with the definition of the Seifert form associated with an isolated hypersurface singularity, see Chapter 3, §3, for more details.

Note that $\mathring{M}$ is homotopy equivalent to a *bouquet of circles* S^1 and hence $H_1(\mathring{M})$ is a free $\mathbb{Z}$-module of finite rank $2g$, where $g = g(M)$ is the genus of M. If B is a basis of this $\mathbb{Z}$-module, then the $2g \times 2g$ matrix $V = V_{M,B}$ describing the Seifert form S_M with respect to the basis B is called a *Seifert matrix* for the knot K. In terms of Seifert matrices to pass from S_M^R to S_M is just to take the transpose V^{τ} of a matrix V. Hence all the formulas below involving the Seifert matrix V are obtained from the corresponding formulas in [R] by replacing V with V^{τ}. It is important to keep in mind that in general V is neither symmetric nor skew-symmetric.

But on the homology group $H_1(\mathring{M})$ we have the (skew-symmetric bilinear) *intersection form*

$$I_M: H_1(\mathring{M}) \times H_1(\mathring{M}) \to \mathbb{Z}.$$

Since the Seifert surface M of the knot K looks like (abstractly, not embedded in S^3) a Riemann surface $\tilde{M}$ with a small disc deleted

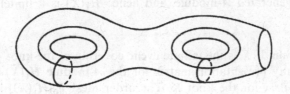

it is easy to see that I_M is a *unimodular* form. In other words, in a suitable basis for $H_1(\mathring{M})$ (which is pictured above), I_M has an associated matrix of the form

$$\left(\begin{array}{c|c} 0 & I_g \\ \hline -I_g & 0 \end{array} \right)$$

with I_g being the $g \times g$ identity matrix. And we have the following basic relation between the Seifert form S_M and the intersection form I_M

(1.17) $I_M(x, y) = S_M(y, x) - S_M(x, y) \qquad \text{for any} \quad x, y \in H_1(\mathring{M}).$

In particular, any Seifert matrix V satisfies

(1.18) $\det(V^{\tau} - V) = \pm 1.$

Next we recall some basic facts about presentation matrices and Fitting ideals. Let A be a commutative ring with a unit element and let M be a *finitely presented* module over A. This means that there is an exact sequence (a so-called "presentation")

$$A^t \xrightarrow{p} A^s \to M \to 0$$

(s = number of generators for M and t = number of relations among these generators). Note that when A is Noetherian, any finitely generated module M has a finite presentation.

(1.19) Definition. With the above notations, the kth *Fitting ideal of M* denoted by $F_k(M)$ is the ideal in A generated by all $(s - k) \times (s - k)$ minors in the matrix P, representing the linear map p with respect to some bases in A^t and A^s.

Such a matrix P is called a *presentation matrix* for the A-module M and it is easy to show that the ideal $F_k(M)$ does not depend on the choices involved (of the presentation, of the bases, ...), for details, see [R], p. 203, and [Lg], p. 59. The 0th Fitting ideal $F_0(M)$ is sometimes called the *order ideal*. When $A = \mathbb{Z}$, and M is a finite abelian group, it is easy to see that $F_0(M)$ is generated by $|M|$, the order of the group M, which justifies the above terminology. Note also that Cramer's rule implies that $F_0(M) \subset \operatorname{Ann}(M) = \{a \in A; a \cdot m = 0$ for all $m \in M\}$. When $s = t$, $F_0(M)$ is just the principal ideal $(\det P)$. Coming back to knot theory, recall that $A = \mathbb{Z}[t, t^{-1}]$ is a Noetherian ring and $H_1(\tilde{X})$ is a finitely generated A-module, and hence $H_1(\tilde{X})$ is a finitely presented A-module.

(1.20) Definition. If $\tilde{X}$ is the infinite cyclic covering of the knot complement $S^3 \backslash K$, then any presentation matrix for the A-module $H_1(\tilde{X})$ is called an *Alexander matrix* for the knot K. The order ideal $F_0(H_1(\tilde{X}))$ is called the *Alexander ideal* of K and, if this ideal is principal (and this is in fact the case), any generator of it is called an *Alexander polynomial* for the knot K.

The main result connecting several of the notions introduced above is the following, see [R], p. 207.

(1.21) Theorem. *If V is a Seifert matrix for a knot K, then*

$$V - tV^{\tau}$$

is an Alexander matrix for K.

(1.22) **Corollary.** *The Alexander ideal of a knot K is principal and an Alexander polynomial for K is given by*

$$\Delta(t) = \det(V - tV^\tau).$$

Consider now the finite cyclic coverings $\tilde{X}_k \to X$ where $X = S^3 \backslash K$, and note that they can be extended to *ramified* cyclic coverings $Y_k \to S^3$, the ramification (or branching) locus being exactly the knot K. It is easy to see that

$$\text{rk } H_1(\tilde{X}_k) = \text{rk } H_1(Y_k) + 1,$$

$$\text{Tors } H_1(\tilde{X}_k) = \text{Tors } H_1(Y_k) \qquad \text{for all } k.$$

Therefore we can study the compact spaces Y_k (which are 3-manifolds) instead of the noncompact 3-manifolds $\tilde{X}_k$.

(1.23) **Proposition.** *If the knot K has an invertible Seifert matrix V (i.e., det V = ± 1), then*

$$P_k = (V(V^{-1})^\tau)^k - I$$

is a presentation matrix for $H_1(Y_k)$ as a $\mathbb{Z}$-module.

Here I is the identity matrix, see, for details, [R], p. 213. Hence practically all the *commutative invariants* of the knot K may be computed from a Seifert matrix V for K.

Finally we discuss a particularly interesting class of knots, containing the class of algebraic knots.

(1.24) **Definition.** A knot $K \subset S^2$ is *fibered* if there is a fibration map $\varphi: S^3 \backslash K \to S^1$ such that K has a tubular neighborhood $T \simeq S^1 \times D^2$ for which the diagram

$$T \backslash K \xrightarrow{\quad\sim\quad} S^1 \times D^2 \backslash \{0\}$$

$$\varphi \searrow \qquad \swarrow \varphi_0$$

$$S^1$$

is commutative, where $D^2 = \{y \in \mathbb{C}, |y| < 1\}$ is the unit 2-disc and $\varphi_0(x, y) = y/|y|$.

Under this assumption, the fiber $F = \varphi^{-1}(1)$, $1 \in S^1$, is the interior of a Seifert surface $M = \bar{F}$ for K. A similar definition to (1.24) can be given for *fibered links*.

The fibered knots are characterized by the following basic result, [R], p. 324.

(1.25) **Theorem** (Stallings). *A knot K is fibered if and only if the commutator group $[\pi_1(S^3 \backslash K), \pi_1(S^3 \backslash K)]$ is finitely generated and free.*

Proof (for the "only if" part). If K is a fibered knot, then the exact homotopy sequence of the fibration φ gives (see [Sp], p. 377)

$$0 = \pi_2(S^1) \to \pi_1(F) \to \pi_1(S^3 \backslash K) \to \pi_1(S^1) = \mathbb{Z} \to 0.$$

This shows that the above commutator group can be identified to $\pi_1(F)$, and this group is a finitely generated free group since F has the homotopy type of a bouquet of circles S^1 (as any open 2-manifold).

As does any fibration over S^1, the fibration φ has a *monodromy homeomorphism* $h: F \to F$. Let T be a matrix for $H_1(h): H_1(F) \to H_1(F)$, the associated *monodromy operator*, see [M5], p. 67. $\square$

It is not difficult to prove the following two results, see [R], p. 326 and p. 334.

(1.26) **Proposition.** *The matrix*

$$tI - T$$

is an Alexander matrix for the knot K. In particular, the characteristic polynomial of the monodromy operator T

$$\Delta(t) = \det(tI - T)$$

is an Alexander polynomial for the knot K.

(1.27) **Proposition.** *There is a natural identification between the covering space $\tilde{X}_k$ corresponding to K and the space obtained from $F \times [0, 1]$ by identifying the point $(y, 1)$ to $(h^k(y), 0)$ for all $y \in F$.*

(1.28) **Corollary.** *For a fibered knot K with monodromy operator T the matrix $P_k = T^k - I$ is a presentation matrix for the abelian group $H_1(Y_k)$.*

(1.29) **Example.** Consider the trefoil knot $K_{2,3}$. It follows from (2.2) below that this is a fibered knot with monodromy operator

$$T = \begin{pmatrix} 0 & 1 \\ -1 & 1 \end{pmatrix},$$

see (3.4.19) or [R], p. 333, where a different convention for writing the matrix associated to a linear map is used.

It follows that an Alexander polynomial for the trefoil knot is $\Delta(t) = t^2 - t + 1$ and we have the following table (here $k \in \mathbb{Z}/6\mathbb{Z}$):

k	0	1	2	3	4	5
$P_k = T^k - 1$	$\begin{pmatrix} 0 & 0 \\ 0 & 0 \end{pmatrix}$	$\begin{pmatrix} -1 & 1 \\ -1 & 0 \end{pmatrix}$	$\begin{pmatrix} -2 & 1 \\ -1 & -1 \end{pmatrix}$	$\begin{pmatrix} -2 & 0 \\ 0 & -2 \end{pmatrix}$	$\begin{pmatrix} -1 & -1 \\ 1 & -2 \end{pmatrix}$	$\begin{pmatrix} 0 & -1 \\ 1 & -1 \end{pmatrix}$

Note that the results in this table imply (via (1.28)) the results in Example (1.11). Therefore for a fibered knot K all the commutative invariants can be computed from its monodromy operator (matrix) T.

We end this section with a basic remark on links.

(1.30) Remark. The equivalence class of a link L is not determined in general by the equivalence classes of its components L_i $(i = 1, \ldots, m)$ and by the set of linking numbers $\mathrm{lk}(L_i, L_j)$ for all $1 \leq i < j \leq m$.

Consider, for instance, the trivial knot with three components

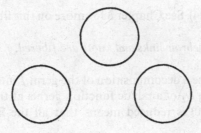

and the *Borromean rings*

In both cases all the components L_i are trivially embedded in S^3 and all the linking numbers are zero. However, these two links are known to be non-equivalent, see [R], p. 66.

§2. Links of Plane Curve Singularities

In this section we discuss basic invariants of algebraic knots, like Puiseux pairs and Alexander polynomials. We use Gysin sequences and Poincaré–Leray residues to compute the cohomology of a plane curve singularity complement in terms of meromorphic differential forms.

Let $C: f = 0$ be a *reduced* curve singularity at the origin of $\mathbb{C}^2$. Since the holomorphic function germ f has in this case an isolated singularity, we can assume when convenient that f is a polynomial (by finite determinacy results, see [D4], p. 81). This motivates the following:

(2.1) Definition. A link $L \subset S^3$ is called *algebraic* if L is equivalent to the link $L(C, 0) = C \cap S_\varepsilon^3 \subset S_\varepsilon^3$ associated to a plane curve singularity $(C, 0)$, for ε chosen small enough as in (1.5.1).

By the general results of Milnor [M5] it follows that the map

$$\varphi: S_\varepsilon^3 \backslash L(C, 0) \to S^1, \qquad \varphi(x) = f(x)/|f(x)|,$$

is a fibration as in (1.24). See Chapter 3 for more on this fibration.

(2.2) Corollary. *All algebraic links and knots are fibered.*

Let $f = f_1, \ldots, f_m$ be a decomposition of the germ f into irreducible factors f_i in the factorial ring $\mathcal{O}_2$ of analytic function germs at the origin of $\mathbb{C}^2$. The assumption that $(C, 0)$ is reduced means that all the factors f_i above are distinct.

First we consider the irreducible case ($m = 1$). An important tool in studying the irreducible curve singularity $(C, 0)$ is a *Puiseux parametrization*

$$p: (\mathbb{C}, 0) \to (C, 0)$$

which is, abstractly speaking, a normalization or a resolution of singularities for the germ $(C, 0)$. In down-to-earth terms, p is just a map germ inducing an analytic isomorphism $\mathbb{C}\backslash\{0\} \to C\backslash\{0\}$ of punctured germs.

To be more explicit, $p(t) = (x(t), y(t))$ where x, y are convergent power series in t verifying

$$f(x(t), y(t)) = 0 \qquad \text{for all } t.$$

It can be shown that we can take

$$(2.3) \qquad \begin{cases} x(t) = t^n, \\ y(t) = \sum_{k \geq n} a_k t^k \qquad \text{with} \quad n = \text{mult}(C, 0). \end{cases}$$

Formally (and classically) we may write this in the following condensed form:

$$(2.3') \qquad\qquad y = \sum_{k \geq n} a_k x^{k/n}.$$

Note that $p: (\mathbb{C}, 0) \to (C, 0)$ is a local homeomorphism (i.e., C is a topological 2-manifold at 0) and hence the only interesting topological facts come from the embeddings $(C, 0) \subset (B_\varepsilon, 0)$ or $L(C, 0) \subset S_\varepsilon^3$. Note also that 0 is a smooth (i.e., nonsingular) point on C if and only if in the formula $(2.3')$ we have $a_k \neq 0$ only for those k which are multiplies of n.

Assume that this is not the case and let k' be the smallest number k such that $a_{k'} \neq 0$ but k'/n is not an integer. Then we can write $k'/n = m_1/n_1$ for *unique* positive integers m_1, n_1 with $(m_1, n_1) = 1$. The pair (m_1, n_1) is called the first *Puiseux pair* of the singularity $(C, 0)$. The exponents which follow k'/n may all be of the form q/n_1 and then we have just one Puiseux pair of $(C, 0)$. If this is not the case, let k'' be the smallest number with $a_{k''} \neq 0$ and such that k''/n is not of the form q/n_1 for some $q \in \mathbb{N}$. Then we can *uniquely* write $k''/n = m_2/n_1 n_2$ for positive integers m_2, n_2 with $(m_2, n_2) = 1$. The pair (m_2, n_2) is then the *second Puiseux pair* of the singularity $(C, 0)$.

This process stops after a finite number of steps (see [BK] for details) and in this way we get the *sequence of Puiseux pairs*

$$(2.4) \qquad P(C, 0) = \{(m_1, n_1), (m_2, n_2), \ldots, (m_s, n_s)\}$$

of the plane curve singularity $(C, 0)$.

(2.5) **Example.** If $y = x^{3/2} + x^{5/2} + x^{11/4}$, then

$$P(C, 0) = \{(3, 2), (11, 2)\}.$$

Next, to any sequence of pairs of positive integers

$$P = \{(a_1, b_1), \ldots, (a_s, b_s)\} \qquad \text{with} \quad (a_i, b_i) = 1 \quad \text{for all} \quad i = 1, \ldots, s,$$

we can associate in a standard way (please refer to the careful discussion in [BK], pp. 433–438, for details) an *iterated torus knot* of type P, denoted by $K(P)$.

Roughly speaking, start with a torus knot K of type (a_1, b_1) as in (1.2(iii)). Next note that a tubular neighborhood T of the knot K in S^3 is a full 2-torus, and hence we may identify its boundary ∂T with $S^1 \times S^1$ (this identification has to be treated carefully, since there are two distinct "natural" possibilities). Consider on this torus ∂T a torus knot of type (a_2, b_2) and then repeat the procedure until all the pairs in P are used.

The main result about this construction is the following, see, for instance, [Lê1].

(2.6) **Proposition** (K. Brauner). *The knot $L(C, 0)$ associated to the irreducible curve singularity $(C, 0)$ is equivalent to the iterated torus knot $K(P(C, 0))$ associated to the sequence of Puiseux pairs of the singularity $(C, 0)$.*

Therefore the embedded topological type of the singularity $(C, 0)$ is completely characterized by the sequence $P(C, 0)$ of Puiseux pairs.

Let us come back to the definition of the Alexander polynomials (1.20) and consider the following extension of it. The ring $A = \mathbb{Z}[t, t^{-1}]$ is not principal, but any family of elements in A has a greatest common divisor. In other words, for any ideal $I \subset A$ there is a *minimal* principal ideal $\bar{I}$ containing I.

(2.7) **Definition** ([Lê1]). For a knot K, let $\tilde{X} \to S^3 \backslash K$ be the corresponding infinite cyclic covering and consider $H_1(\tilde{X})$ as an A-module in the usual way.

A generator of the principal ideal

$$\overline{F_{k-1}(H_1(\tilde{X}))}$$

is called a kth *Alexander polynomial* for the knot K and is usually denoted by $\Delta_k, k \geq 1$.

Note that Δ_1 is just an Alexander polynomial Δ as defined in (1.20). By definition it follows that

$$\Delta_{k+1} \text{ divides } \Delta_k \text{ for all } k \geq 1.$$

The choice for Δ_k can be made unique by asking that $\Delta_k \in \mathbb{Z}[t]$ and $\Delta_k(0) \neq 0$. With these preliminaries, we can state the following basic results on algebraic knots proved by Lê, see [Lê1].

(2.8) **Proposition.** *The (first) Alexander polynomial* Δ_1 *of the knot* $L(C, 0) = C \cap S_\varepsilon^3$ *associated to the plane curve singularity* $(C, 0)$ *with Puiseux pairs*

$$P(C, 0) = \{(m_1, n_1), \ldots, (m_s, n_s)\}$$

is given by the product

$$\Delta_1(t) = P_{\lambda_1, n_1}(t^{v_2}) \ldots P_{\lambda_s, n_s}(t^{v_{s+1}}),$$

where

$$P_{\lambda, n}(t) = \frac{(t^{\lambda n} - 1)(t - 1)}{(t^\lambda - 1)(t^n - 1)},$$

$v_i = n_i \ldots n_s$ *for* $i = 1, \ldots, s$ *and* $v_{s+1} = 1$

$$\lambda_1 = m_1, \qquad \lambda_i = m_i - m_{i-1} n_i + \lambda_{i-1} n_i n_{i-1} \qquad for \quad i = 2, \ldots, s.$$

(2.9) **Corollary.** *For algebraic knots* K *and* K' *the following statements are equivalent:*

(i) *K and K' are equivalent;*
(ii) *K and K' have the same Alexander polynomial.*

(2.10) **Proposition.** *The quotient* Δ_1/Δ_2 *of the first two Alexander polynomials of the algebraic knot* $L(C, 0)$ *is the minimal polynomial of the monodromy operator*

$$T: H_1(F) \to H_1(F)$$

of the singularity $(C, 0)$. *Moreover, this minimal polynomial has distinct roots and hence T has finite order (in particular, T is diagonalizable over $\mathbb{C}$).*

Note that (2.9) is false for arbitrary knots and (2.10) is false for algebraic links with several components, see [R], p. 62, and, respectively, [AC1] and [Df2], [Wd1]. However, a version of (2.9) is true for algebraic links, see [Ym]. Moreover, to be completely accurate, we have to point out that the definition of the Alexander polynomials in [Lê1] is different from ours, using the "Free

Differential Calculus" of Fox. But it is well known, see [R], p. 211, that the two definitions agree.

Consider now the case of reducible plane curve singularities. Assume that the singularity $(C, 0)$ has m branches

$$C_i: f_i = 0 \qquad \text{for} \quad i = 1, \ldots, m.$$

Then the link $L(C, 0) = C \cap S_\varepsilon^3$ consists exactly of m knots $K_i = C_i \cap S_\varepsilon^3$ which, due to the presence of the complex structure, also have a *natural orientation*. In particular, the linking numbers $\mathrm{lk}(K_i, K_j)$ can be defined as in (1.7) using these orientations.

We have the following result (see [BK], p. 442) ruling out the unpleasant situation from (1.30).

(2.11) Theorem. *The topological type of the singularity $(C, 0)$ (or, equivalently, the class of the link $L(C, 0)$) is completely determined by the sequences of Puiseux pairs $P(C_i, 0)$, $i = 1, \ldots, m$, of all the branches C_i and by the linking numbers $\mathrm{lk}(K_i, K_j)$ for all the pairs $1 \le i < j \le m$.*

We remark that the linking number of two algebraic knots may be computed purely algebraically as follows (and, in particular, is always positive).

(2.12) Proposition (See, for instance, [BK], p. 231 and p. 440). *Let $C_i: f_i = 0$ $(i = 1, 2)$ be two distinct irreducible plane curve singularities. The following three positive integers coincide:*

(i) $\mathrm{lk}(L(C_1, 0), L(C_2, 0))$;

(ii) $\dim_{\mathbb{C}} \dfrac{\mathcal{O}_2}{(f_1, f_2)} = (C_1, C_2)_0$; *and*

(iii) $\mathrm{ord}(f_1 \circ p_2)$ *where $p_2: (\mathbb{C}, 0) \to (C_2, 0)$ is a Puiseux parametrization for the singularity $(C_2, 0)$.*

In the remaining part of this section, we try to give a clear understanding of the homology and cohomology groups of the complement $X = S^3 \backslash L$ where L is an algebraic link.

For the moment we do not assume L to be algebraic and look at the *integral* homology group $H_1(X)$. For each connected component L_i of L choose a point $a_i \in L_i$, let T_i be a small transversal to L_i at the point a_i, and let l_i be a small circle in T_i around the point a_i:

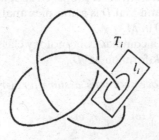

Then it is easy to show that $H_1(X)$ is freely generated as a $\mathbb{Z}$-module by the cycles $[l_i]$. This construction is a special case of the more general one which we describe next.

Let M be a smooth oriented manifold with dim $M = n$ and let $D \subset M$ be a closed oriented submanifold with dim $D = n - 2$ (when D has several components we require that all of them have this dimension). Let $T \subset M$ be a closed tubular neighborhood of D in M. Note that T is then also oriented and can be identified with the total space of the unit disc bundle associated to the normal bundle of D in M. For details, see [GG], p. 69, and [Sp], p. 91. Consider the exact homology sequence of the pair $(M, M \setminus D)$:

$$\cdots \to H_k(M \setminus D) \to H_k(M) \to H_k(M, M \setminus D) \xrightarrow{\partial} H_{k-1}(M \setminus D) \to \cdots.$$

By excision, [Sp], p. 189, we have an isomorphism

$$H_k(M, M \setminus D) \simeq H_k(T, T \setminus D)$$

and then, using the Thom isomorphism, [Sp], p. 259, we get an isomorphism

$$\theta \colon H_{k-2}(D) \xrightarrow{\sim} H_k(T, T \setminus D).$$

Explicitly, if $p \colon T \to D$ denotes the projection of this disc bundle T, we have

$$\theta[c] = [p^{-1}(c)] \qquad \text{for any cycle } [c]$$

in $H_{k-2}(D)$.

Combining these facts we get the following *homology Gysin sequence:*

(2.13) $\cdots \to H_k(M \setminus D) \xrightarrow{i_*} H_k(M) \to H_{k-2}(D) \xrightarrow{\bar{\theta}} H_{k-1}(M \setminus D) \xrightarrow{i_*},$

where $i \colon M \setminus D \to M$ is the inclusion and $\bar{\theta}[c] = [\bar{p}^{-1}(c)]$, with $\bar{p} \colon \partial T \to D$ the restriction of p to the boundary ∂T. This explicit description of $\bar{\theta}$ can be found, for instance, in Griffiths [G]. In our case, take $M = S^3$, $D = L$, and get

$$H_0(L) \xrightarrow{\bar{\theta}} H_1(X).$$

But $H_0(L)$ is the $\mathbb{Z}$-module freely generated by the 0-cycles $[a_i]$ and $\bar{\theta}[a_i]$ is clearly the cycle $[l_i]$ (with a suitable orientation which is not relevant here).

There is also a *cohomological Gysin sequence* (same proof as above) which reads

(2.14) $\cdots \to H^k(M) \xrightarrow{i^*} H^k(M \setminus D) \xrightarrow{R} H^{k-1}(D) \to H^{k+1}(M) \xrightarrow{i^*} \cdots.$

The homomorphism R which appears here is called the *Poincaré–Leray residue map* because of the following basic property. Assume now in addition that M is a complex manifold and that D is a complex analytic hypersurface given by a global equation $f = 0$ in M.

Let $u \in H^k(M \setminus D; \mathbb{C})$ be a complex cohomology class which has a representative ω, a closed analytic k-form on $M \setminus D$.

(2.15) **Lemma.** *With the above notations, assume furthermore that*

$$\omega = \frac{df}{f} \wedge \omega_1 + \omega_2$$

for some analytic forms ω_1 and ω_2 defined on the manifold M. Then

$$R(u) = [2\pi i(\omega_1|D)] \in H^{k-1}(D; \mathbb{C}),$$

where $i = \sqrt{-1}$ and the map R here comes from the cohomological Gysin sequence (2.14) with $\mathbb{C}$ coefficients.

The factor $(2\pi i)$ is necessary here in order that the sequence (2.14) is the dual of the sequence (2.13), i.e., to have the following basic formula

(2.16) $$\int_{\bar{\theta}(c)} \omega = \int_c R(\omega) \quad \text{for any} \quad [\omega] \in H^k(M\backslash D; \mathbb{C})$$

and any cycle $[c] \in H_{k-1}(D)$.

The proofs of (2.15) and (2.16) can be found in [G], p. 490, modulo the factor $2\pi i$. Besides the nice formula (2.16), there is one more basic reason to use this factor $2\pi i$ as we do. Namely, let Y be a manifold such that its integral homology $H_k(Y)$ and cohomology $H^k(Y)$ are torsion free and finitely generated for some $k \geq 0$. Consider the obvious diagram of pairings, where the vertical maps are the natural inclusions

$$
\begin{array}{ccc}
H_k(Y) \times H^k(Y) & \xrightarrow{\langle \ \rangle} & \mathbb{Z} \\
\downarrow & & \downarrow \\
H_k(Y; \mathbb{C}) \times H^k(Y; \mathbb{C}) & \xrightarrow{\langle \ \rangle} & \mathbb{C}
\end{array}
$$

We can then identify

(2.17) $$H^k(Y) = \{u \in H^k(Y; \mathbb{C}); \langle c, u \rangle \in \mathbb{Z} \text{ for all } c \in H_k(Y)\}.$$

Then, if we like the residue map (complex version)

$$R: H^k(M\backslash D; \mathbb{C}) \to H^{k-1}(D; \mathbb{C})$$

to preserve the integral subspaces, i.e.,

$$R(H^k(M\backslash D)) \subset H^{k-1}(D),$$

then the factor $2\pi i$ is clearly necessary in Lemma (2.15). Just think of the example $M = \mathbb{C}$, $D = \{0\}$, $\omega = 1/2\pi i \, dz/z$, and the usual Cauchy formula [GH], p. 2.

Coming back to the plane curve singularities, note that by the Conic Structure Theorem (1.5.1) we have a homotopy equivalence between $X = S_\varepsilon^3 \backslash L(C, 0)$ and $U = B_\varepsilon \backslash C$ where B_ε is a small enough open ball at the origin of $\mathbb{C}^2$. In particular, $H^1(U) = H^1(X)$ and let $[l_j^*]$ be the dual element of the 1-cycle $[l_j] \in H_1(X)$ described above with respect to the pairing

$$\langle \ \rangle: H_1(U) \times H^1(U) \to \mathbb{C}.$$

Using again the Cauchy formula and an identification as in (2.17), it is easy to see that

(2.18) $$l_j^* = \frac{1}{2\pi i} \frac{df_j}{f_j},$$

where $f_j = 0$ is a (reduced) equation for the branch C_j of C corresponding to the cycle $[l_j]$. Indeed, we clearly have

$$\langle l_k, l_j^* \rangle = \int_{l_k} l_j^* = \delta_{jk} = \begin{cases} 1 & \text{for } k = j, \\ 0 & \text{for } k \neq j. \end{cases}$$

Hence we can identify the subobject $H^1(U) \subset H^1(U; \mathbb{C})$ as being the $\mathbb{Z}$-module freely generated by l_j^* with $j = 1, \ldots, m$.

To investigate the second cohomology group $H^2(U)$ and the cup product

$$H^1(U) \times H^1(U) \to H^2(U)$$

we proceed as follows. The corresponding part in the Gysin sequence is the following (take $M = B_\varepsilon \backslash \{0\}, D = C \cap B_\varepsilon \backslash \{0\}$):

$$0 = H^2(B_\varepsilon \backslash \{0\}) \to H^2(U) \xrightarrow{R} H^1(D) \to H^3(B_\varepsilon \backslash \{0\}) = \mathbb{Z}.$$

Hence R is a *primitive embedding* of free $\mathbb{Z}$-modules (here primitive means that coker R has no torsion) and we know already that rk $H^2(U) = m - 1$, rk $H^1(D) = m$ (recall (1.4)). To simplify the notation, we denote by C (resp. C_j) the intersections $C \cap B_\varepsilon$ (resp. $C_j \cap B_\varepsilon$).

Then we have a natural identification

(2.19) $$H^1(D) = \bigoplus H^1(C_j \backslash \{0\}) \xrightarrow[\sim]{\oplus p_j^*} \mathbb{Z}^m$$

obtained as follows. For each j, let $p_j \colon (\mathbb{C}, 0) \to (C_j, 0)$ be a Puiseux parametrization of the branch C_j. Then there is an identification

$$H^1(C_j \backslash \{0\}) \xrightarrow{p_j^*} H^1(\mathbb{C} \backslash \{0\}) = \mathbb{Z},$$

the last equality being obtained by making $\omega_1 = 1/2\pi i \, dt/t$, corresponding to $1 \in \mathbb{Z}$. Here t is a local coordinate at the origin of $\mathbb{C}$ and we once again have used (2.17).

To compute $R(l_j^* \cup l_k^*) = (a_1, \ldots, a_m) \in \mathbb{Z}^m$ it is enough to describe each of the components a_j, via the above identification (2.19). Note that the rth component a_r can be identified with the Poincaré–Leray residue $R_r(\omega_{jk})$ of the 2-form

$$\omega_{jk} = \left(\frac{1}{2\pi i} \right)^2 \frac{df_j \wedge df_k}{f_j f_k}$$

along the branch $C_r \colon f_r = 0$. It follows that $a_r = 0$ for $r \neq j, k$ and that

$$a_j = -a_k = (C_j, C_k)_0 = \text{lk}(L_j, L_k).$$

Indeed, we have

$$R_j(\omega_{jk}) = \left[\frac{1}{2\pi i} \frac{df_k}{f_k} \Big|_{C_j \backslash \{0\}} \right]$$

and

$$a_j = \left[\frac{1}{2\pi i} p_j^* \left(\frac{df_k}{f_k} \right) \right] = (C_j, C_k)_0$$

using (2.12) and our conventions.

We can restate this result in the following form (compare with the corresponding result in [ABG]).

(2.20) **Corollary.**

(i) *The image of the Poincaré–Leray residue*

$$0 \to H^2(U) \xrightarrow{R} H^1(C\backslash\{0\}) \simeq \mathbb{Z}^m$$

consists exactly of the elements $(a_1, \ldots, a_m)$ *with*

$$a_1 + \cdots + a_m = 0;$$

(ii) *a basis for* $H^2(U)$ *is given by the elements*

$$\alpha_k = \left(\frac{1}{2\pi i}\right)^2 \frac{1}{(C_k, C_{k+1})_0} \frac{df_k \wedge df_{k+1}}{f_k f_{k+1}}$$

for $k = 1, \ldots, m-1$;

(iii) *the description of the linking number given in* (1.8) *is true for the case when* $J \cup K$ *is an algebraic link.*

(2.21) **Remark.** In the case $m = 2$, we can consider the complex version of the Poincaré–Leray residue map

$$R: H^2(U; \mathbb{C}) \to \mathbb{C}^2$$

obtained as above, and the *Grothendieck residue* map

$$\mathrm{Res}_0: H^2(U; \mathbb{C}) \to \mathbb{C}, \qquad \mathrm{Res}_0(\omega) = \int_\Gamma \omega,$$

where Γ is the 2-cycle $\{x \in \mathbb{C}^2; |f_1(x)| = \delta_1, |f_2(x)| = \delta_2\}$ for small enough positive numbers δ_1 and δ_2, see, for details, [GH], p. 649, but note that we again have a *different* convention about the factor $2\pi i$! Then it is easy to prove that the diagram

$$H^2(U; \mathbb{C}) \xrightarrow{\quad R \quad} \mathbb{C}^2$$

$$\mathrm{Res}_0 \searrow \qquad \nearrow \Delta$$

$$\mathbb{C}$$

with $\Delta(a) = (a, -a)$ is commutative (use [GH], p. 663).

§3. Links of Surface Singularities

In this section we relate the resolution graph of a normal surface singularity with the topology of its link (e.g., Mumford's results on the fundamental group of such a link). A brief discussion on quotient singularities is also included here, since in dimension 2 they relate nicely with Mumford's results, see (3.16). Higher dimensional quotient singularities will occur later in relation to the weighted projective spaces, see Appendix B.

Let $(X, 0) \subset (\mathbb{C}^n, 0)$ be an *irreducible isolated surface singularity*. Let $\bar{B}$ be a small closed ball at the origin of $\mathbb{C}^n$ such that the Conic Structure Theorem (1.5.1) holds for the pair $(\bar{B}, X)$. Here and in the sequal we let X denote the intersection of a representative X_1 of the germ $(X, 0)$ with the ball $\bar{B}$, i.e., X is a real four-dimensional manifold with boundary equal to the link $K = X \cap \partial\bar{B}$, except for the singular point 0!

Let $p: \tilde{X} \to X$ be a *very good resolution* for the singularity $(X, 0)$. This means that $\tilde{X}$ is a smooth complex surface (with boundary $\partial\tilde{X} \simeq \partial X$), p is a proper analytic morphism which is an isomorphism over $X\backslash\{0\}$, and $E = p^{-1}(0)$ (the *exceptional set* or *divisor* of the resolution) is a curve with only normal crossing singularities whose irreducible components $E_1, \ldots, E_r$ are smooth and $\#(E_i \cap E_j) \leq 1$ for all $i \neq j$. For the existence of such resolutions and for more details on their properties, we refer to [BPV]. (In fact, in the study of such resolutions it is usually assumed that $(X, 0)$ is a *normal* singularity. Otherwise some pathologies may occur, e.g., resolutions with a zero-dimensional exceptional set E, see, for example, (3.11) below.)

Sometimes it is useful to assume that p is a *minimal* (very good) resolution, i.e., there is no *exceptional curve E_i of the first kind* (this means that E_i is a rational curve with self-intersection $(E_i, E_i) = -1$) which can be contracted leading to a new (very good) resolution for $(X, 0)$. To a very good resolution $p: \tilde{X} \to X$ as above, we can associate the *dual graph*: its vertices correspond to the exceptional curves E_j $(j = 1, \ldots, r)$, and the two vertices E_j and E_k are joined by an edge if and only if $E_j \cap E_k \neq \varnothing$.

To include more information, we may consider the *weighted dual graph*, where in addition each vertex is weighted by the negative of the corresponding self-intersection number (E_i, E_i). It is easy to see that all these graphs are connected.

The first facts about the topology of the resolution $\tilde{X}$ are contained in the following.

(3.1) **Proposition.**

(i) *There is a deformation retract $\tilde{X} \searrow E$; in particular, these two spaces have the same homotopy type.*

(ii) *The nonzero integral homology groups of E are given by*

$$H_0(E) = \mathbb{Z}, \qquad H_1(E) = \mathbb{Z}^{2g+b}, \qquad H_2(E) = \mathbb{Z}^r,$$

where $g = \sum_{i=1,r} g(E_i)$ is the sum of the genera of the exceptional curves E_i, and b is the number of loops (i.e., closed cycles) in the dual graph of the resolution $p: \tilde{X} \to X$.

Proof. (i) The retraction $X \searrow \{0\}$ which comes from the Conic Structure Theorem (1.5.1) can be lifted in an obvious way to the surface $\tilde{X}$ to give a retraction $\tilde{X} \searrow E$.

(ii) E is connected since $(X, 0)$ was assumed to be irreducible. Then we can apply the Mayer–Vietoris exact sequence to compute the homology of E, using double induction on b and g. The key remark here is that for $b = 0$ the space E is homotopically equivalent to the bouquet $E_1 \vee \cdots \vee E_r$ of its irreducible components.   □

(3.2) **Remark.** It can be shown that the numbers g and b above are invariants of the singularity $(X, 0)$, i.e., they do not depend on the choice of the resolution $\tilde{X}$. One way to prove this is to use the existence of the minimal resolution $\tilde{X}_m \to X$ (which usually is *not* very good) and the fact that any other resolution $\tilde{X} \to X$ is obtained by blowing-up some points on the exceptional divisor of $\tilde{X}_m$. And it is easy to check that any such blowing-up affects neither g (since the new exceptional curves E_j created in this way are rational) nor b (since there are essentially two possibilities as shown in the following illustrations):

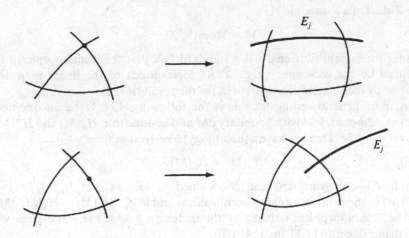

The second way to prove that g and b are invariants of the singularity $(X, 0)$ is by using the *mixed Hodge structure* on the cohomology $H^2(K)$ of the link, see Appendix C for more details.

We now turn our attention to the link K which is a connected three-dimensional compact oriented manifold. First we look at its first integral homology group. Like any finitely generated abelian group (K is compact), $H_1(K)$ has a direct sum decomposition

(3.3) $$H_1(K) = F_1(K) \oplus T_1(K),$$

where $F_1(K)$ (resp. $T_1(K)$) denotes the *free* part (resp. the *torsion* part) of $H_1(K)$.

(3.4) Proposition.

(i) $F_1(K) = \mathbb{Z}^{2g+b}$.

(ii) $T_1(K) = \text{coker}(\mathbb{Z}^r \xrightarrow{I(\tilde{X})} \mathbb{Z}^r)$ *where* $I(\tilde{X}) = ((E_i, E_j))_{1 \le i,j \le r}$ *is the* intersection matrix *of the resolution* $\tilde{X}$.

Note that since $I(\tilde{X})$ is a *negative definite* matrix [BPV], the above cokernel is indeed a finite group of order

$$|T_1(K)| = \det((E_i, E_j)).$$

Proof. The homology exact sequence of the pair $(\tilde{X}, K)$, where K is identified with $\partial \tilde{X}$, contains the sequence

$$0 \to H_3(\tilde{X}, K) \xrightarrow{\partial} H_2(K) \to H_2(\tilde{X}) \xrightarrow{j_*} H_2(\tilde{X}, K) \xrightarrow{\partial} H_1(K) \to H_1(\tilde{X}) \to 0.$$

Indeed, note that $H_3(\tilde{X}) = H_3(E) = 0$ and $H_1(\tilde{X}, K) \cong H^3(\tilde{X} \backslash K) = H^3(E) = 0$ via the Lefschetz Duality Theorem, see [Sp], p. 297. Moreover, $H_2(\tilde{X}) = H_2(E)$ and $H_2(\tilde{X}, K) \cong H^2(\tilde{X} \backslash K) = H^2(E) = H_2(E)'$, where we denote by V' the $\mathbb{Z}$-dual of a $\mathbb{Z}$-module V, i.e.,

$$V' = \text{Hom}(V, \mathbb{Z}).$$

Under these identifications, it is a standard fact that the homomorphism j_* induced by the inclusion $j: \tilde{X} \subset (\tilde{X}, K)$ corresponds to the linear map described by the intersection matrix $I(\tilde{X})$ of the resolution $\tilde{X}$.

A more general setting for this is the following. Let M be an oriented compact $2m$-manifold with boundary ∂M and assume that $H_m(M)$ and $H^m(M)$ are *torsion free*. Then we have a (homology) *intersection form*

$$(3.5) \qquad I: H_m(M) \times H_m(M) \to \mathbb{Z}$$

which is $(-1)^m$-symmetric and is obtained as follows. Let $H_m(M, \partial M) \xrightarrow{D} H^m(M)$ be the Poincaré duality isomorphism and let $j_*: H_m(M) \to H_m(M, \partial M)$ be the homomorphism induced by the inclusion $j: M \subset (M, \partial M)$. Then we can define (see also [L3], pp. 154–160)

$$(3.6) \qquad I(x, y) = \langle Dj_*(x), y \rangle \qquad \text{for} \quad x, y \in H_m(M),$$

where the pairing $\langle \ \rangle: H^m(M) \times H_m(M) \to \mathbb{Z}$ is the usual one (i.e., making the identification $H^m(M) = H_m(M)'$ then $\langle \ \rangle$ is just the *evaluation* pairing).

Let $e_1, \ldots, e_p$ be a $\mathbb{Z}$-basis for $H_m(M)$ and let $e'_1, \ldots, e'_p$ be the "dual" $\mathbb{Z}$-basis of $H_m(M, \partial M)$ defined by the property $\langle De'_j, e_k \rangle = \delta_{jk}$. Then the matrix of j_* with respect to these two bases is exactly the intersection matrix $(I(e_i, e_j))$. $\square$

(3.6') Remark. Assume that $H^{m+1}(M) = 0$ (e.g., $M \backslash \partial M$ is a Stein manifold of complex dimension m as in (1.6.8)). Then the exact sequence

$$0 = H_{m-1}(M, \partial M) \xrightarrow{\partial} H_m(\partial M) \to H_m(M) \xrightarrow{j_*} H_m(M, \partial M)$$

shows that $H_m(\partial M)$ can be identified with the radical Rad I of the intersection form I

$$\text{Rad } I = \{x \in H_m(M), I(x, y) = 0 \text{ for all } y \in H_m(M)\}.$$

(3.7) Corollary.

(i) *The link K is a $\mathbb{Q}$-homology sphere (i.e., $H_{.}(K; \mathbb{Q}) = H_{.}(S^3, \mathbb{Q})$) if and only if the dual graph of the resolution is a tree and all the exceptional curves E_i are rational (i.e., $g = b = 0$).*

(ii) *The link K is a $\mathbb{Z}$-homology sphere (i.e., $H_{.}(K) = H_{.}(S^3)$) if and only if $g = b = 0$ and the intersection matrix $I(\tilde{X})$ is unimodular, i.e.,*

$$\det(I(\tilde{X})) = \pm 1.$$

We intend to take a look at the fundamental group $\pi_1(K)$ which is a subtle topological invariant of the link K, just as in the case of curves singularities (where not the link itself but its complement in S^3 is the interesting space to look at, as we have seen in the previous two sections).

It is easy to define some elements in this group $\pi_1(K) = \pi_1(\tilde{X}\backslash E)$. Fix a base point $x \in \tilde{X}\backslash E$ and some points x_i near the curves E_i for $i = 1,\ldots,r$. Take small loops γ_i based at the points x_i and going once around the curve E_i, in a sense compatible with the orientations coming from the complex structures of $\tilde{X}$ and E_i. Then define $[c_i] = [l_i\gamma_i l_i^{-1}]$ for some paths l_i joining the point x with the points x_i.

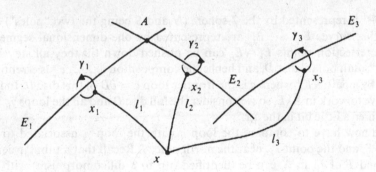

With these notations we can state the following result, due to Mumford, [Mu1], see also [L3], p. 163.

(3.8) Theorem. *Assume that the resolution graph of $\tilde{X}$ is a tree and that each of the exceptional curves E_i ($i = 1,\ldots,r$) is rational. Then $\pi_1(K)$ is presented by a set of r generators $\{v_i = [c_i]; i = 1,\ldots,r\}$ and the following relations:*

(i) *$v_k v_j = v_j v_k$ if $E_j \cap E_k \neq \varnothing$ for all (j, k);*

(ii) *$v_k^{n_k} = \prod_{(E_j, E_k) \neq 0} v_j$ for all k, where $n_k = -(E_k, E_k)$ and the factors in the product are ordered according to increasing j's, $j \neq k$.*

Instead of giving a complete formal *proof* of this result, we prefer to offer an intuitive motivation for the relations (i) and (ii) above. For (i), note that it is easy to see why $v_1 v_2 = v_2 v_1$ in our picture above. Indeed, we can move both paths γ_1 and γ_2 near the intersection point $A = E_1 \cap E_2$ and use the fact that

$$\pi_1(B \backslash (E_1 \cup E_2)) = \mathbb{Z}^2 \quad \text{is abelian,}$$

where B is a small ball centered at A.

To explain the second type of relations (ii) in (3.8) we look at the component E_k and consider all the other components $E_{j_1}, \ldots, E_{j_p}$ with $E_j \cap E_k \neq \varnothing$. Here is a drawing of the situation:

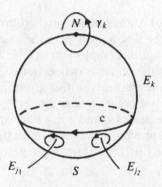

$E_k = \mathbb{P}^1$ is represented by the 2-sphere (N and S being the two "poles") and the other curves $E_{j_1}, \ldots, E_{j_p}$ are represented by one-dimensional segments. The intersection points $E_k \cap E_{j_i}$ can be pushed down till they all stay in a small "south polar" disc D, and hence the composition $v_{j_1}, \ldots, v_{j_p}$ is essentially given by a path lcl^{-1} where c is the frontier loop $c = \partial D$ of this disc D. In fact, we have to work in $\tilde{X} \backslash E$, so we consider the full disc D and all the loops γ_{j_1} and c as "lifted a little bit in the air."

We now have to compare the loop c with the loop γ_k associated to the curve E_k and the point x_k, near the north pole N. Recall that a tubular neighborhood T of E_k in $\tilde{X}$ can be identified (up to a diffeomorphism) with the total space of the line bundle $\mathcal{O}(-n_k)$, see [L3], Chap. IV. Using local coordinates (x, y) at the point S we may write the loop c as follows $c: x = \alpha$, $y = \beta \exp(2\pi i t)$, $t \in [0, 1]$, $\alpha, \beta \in \mathbb{C}^*$ with $|\alpha|, |\beta|$ small enough. The local coordinates (x_1, y_1) at the point N are related to the old ones by the next *transition formulas* [L3], p. 134,

$$x_1 = xy^{n_k}, \qquad y_1 = y^{-1}.$$

Hence our loop c becomes

$$c: x_1 = \alpha \beta^{n_k} \exp(2\pi i n_k t), \qquad y_1 = \beta^{-1} \exp(-2\pi i t).$$

If T' denotes the part of the tubular neighborhood T sitting over $E_k \backslash \mathring{D}$, it is clear that $T' \backslash E_k \simeq (E_k \backslash \overset{\circ}{D}) \times \mathbb{C}^*$ and hence

$$q_\#: \pi_1(T' \backslash E_k) \overset{\sim}{\to} \pi_1(\mathbb{C}^*) = \mathbb{Z}.$$

Here q is the projection onto the second factor and corresponds to $(x_1, y_1) \mapsto x_1$, since x_1 is a coordinate in the fibers of the line bundle $\mathcal{O}(-n_k)$! This clearly implies that $[c] = [\gamma_k]^{n_k}$ which in turn obviously implies (ii) in Theorem (3.8).

Before proceeding further, we consider the following.

(3.9) **Example.** Assume that the weighted dual graph of the resolution $\tilde{X}$ is *linear*

$$
\begin{array}{ccccc}
n_1 & n_2 & n_3 \quad n_{r-1} & & n_r \\
\bullet\!-\!\!-\!\!-\!\!\bullet\!-\!\!-\!\!-\!\!\bullet & \cdots\cdots & \bullet\!-\!\!-\!\!-\!\!\bullet
\end{array}
$$

and that all the exceptional curves E_i, $i = 1, \ldots, r$, are rational. Then according to (3.8) the fundamental group $\pi_1(K)$ is presented by r generators $v_1, \ldots, v_r$ and the relations

$$v_2 = v_1^{n_1}, \quad v_1 v_3 = v_2^{n_2}, \quad \ldots, \quad v_{r-2} v_r = v_{r-1}^{n_{r-1}}, \quad v_{r-1} = v_r^{n_r}.$$

We may apply the *inverse of the Euclidean division algorithm* to the sequence $(n_1, \ldots, n_r)$ and define

$$q_0 = 1, \quad q_1 = n_1 q_0, \quad q_2 = n_2 q_1 - q_0, \quad \ldots, \quad q_r = n_r q_{r-1} - q_{r-2}.$$

Then the relations (R) are equivalent to the relations

(R') $v_{i+1} = v_1^{q_i}$ for $i = 0, 1, \ldots, r-1$ and $v_1^{q_r} = 1$.

This clearly implies that the group $\pi_1(K)$ is finite cyclic of order q_r and that v_1 is a generator.

We recall a basic definition, see, for instance, [BS], p. 8.

(3.10) **Definition.** A complex space $(X, \mathcal{O}_X)$ is *normal* if all the fibers $\mathcal{O}_{X,x}$ of the structure sheaf $\mathcal{O}_X$ are normal rings (i.e., they are domains which are integrally closed in their fields of fractions). A singularity $(X, 0)$ is *normal* if it has a normal representative X.

(3.11) **Examples.** (i) A complete intersection singularity $(X, 0)$ is normal if and only if $\mathrm{codim}(X_{\mathrm{sing}}) \geq 2$, see [KK], p. 315.

(ii) Note that there are examples of irreducible isolated surface singularities $(X, 0) \subset (\mathbb{C}^4, 0)$ which are *not* normal. Indeed, we can take X to be the image of the following *injective* mapping $h: (\mathbb{C}^2, 0) \to (\mathbb{C}^4, 0)$

$$h(x, y) = (x^2, x^3, y, xy),$$

see [BS], p. 51, for details.

We have the following basic result due to Mumford [Mu1].

(3.12) **Theorem.** *For a normal surface singularity* $(X, 0)$ *the following statements are equivalent:*

(i) $(X, 0)$ *is a smooth germ;*
(ii) *The link* $K = L(X, 0)$ *is simply-connected.*

Proof. The implication (i) $\Rightarrow$ (ii) is clear, since $L(\mathbb{C}^2, 0) = S^3$ and $\pi_1(S^3) = 1$, for instance, [Sp], p. 58. To prove the converse, let $(\tilde{X}, E) \to (X, 0)$ be a *minimal* very good resolution, with $E = E_1 \cup \cdots \cup E_r$. Since $\pi_1(K) = 1$, it follows that all these components E_i are rational curves and that the dual graph T of the resolution is a tree.

The minimality condition refers to the number r of irreducible components of E. More precisely, it means that there is no E_i such that:

(a) $(E_i, E_i) = -1$;
(b) E_i intersects at most two other E_j's.

We do induction on r. For $r = 1$ the statement is clear by (3.7(ii)) and the Castelnuovo criterion, see [BPV], p. 78.

Assume now that $r > 1$. There are two cases to discuss.

Case 1 (The tree T is linear).

Then according to Example (3.9) the fundamental group $\pi_1(K)$ is cyclic of order $q_r > 1$ and this is a contradiction.

Case 2 (There is a component E_1 in E which meets exactly with the components $E_2, \ldots, E_m$ for some $m \geq 4$).

Let T_i be the set of the E_j's (except E_1) which are connected to E_i by a series of E_k with $k \neq 1$, for $i = 2, \ldots, m$. Then T_i are disjoint trees and $T_2, \ldots, T_m$ and $\{E_1\}$ form a partition of T. Let K_i be the link of the singularity obtained by contracting the tree T_i (which is negative definite as a subtree in T, see [BPV], p. 72).

Let $G_i = \pi_1(K_i)$ for $i = 2, \ldots, m$ and $G = \pi_1(K)/(v_1 = 1)$ where v_j corresponds to the loop around E_j as explained in the "proof" of (3.8) for $j = 1, \ldots, m$. Then Theorem (3.8) implies that

$$G = G_2 * \cdots * G_m/(v_2 \cdots v_m = 1),$$

where $*$ denotes the free product.

At this point we need a *pure group-theoretic result.*

(3.13) **Lemma.** *Let* G_i $(i = 1, \ldots, m)$ *be nontrivial groups* $(m \geq 2)$ *and let* $a_i \in G_i$ *be some elements. Then the quotient group*

$$G_1 * \cdots * G_m/(a_1 \cdots a_m = 1)$$

is nontrivial.

For the subtle proof of this result we refer to [Mu1]. In our case $\pi_1(K) = 1$ and hence there exists an index i such that $G_i = 1$. By the induction assumption it follows that the corresponding tree T_i can be contracted to a smooth point. Then, by the Zariski theorem on the factorization of birational transformations of surfaces, see [BPV], p. 79, some component E_j in T_i enjoys the properties (a) and (b) above *inside* T_i.

Because such a component does not exist in T, it is necessary that E_j is precisely the "boundary" component E_i (for simplicity of notation we even set $E_j = E_2$) and E_2 meet exactly two components (say E_2' and E_2'') in T_2

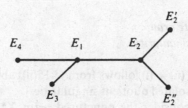

We can then apply the above argument to the curve E_2 (in place of E_1) and deduce that at least one of the curves E_1, E_2', and E_2'' (call it $\tilde{E}_2$) has self-intersection -1. But then

$$(E_2 + \tilde{E}_2, E_2 + \tilde{E}_2) = -1 + 2 - 1 = 0$$

which contradicts the fact that the intersection matrix of $\tilde{X}$ is negative-definite.

This contradiction ends the proof of Theorem (3.12).   □

Before stating a nice corollary of (3.12), we recall some basic facts on *quotient singularities*. Let $G \subset \mathrm{Gl}(n; \mathbb{C})$ be a finite group and consider the induced action of G on $\mathbb{C}^n$. Then the quotient space $\mathbb{C}^n/G$ has a natural structure of an *affine normal variety*.

(3.14) Definitions. An n-dimensional singularity $(X, 0)$ is called a *quotient singularity* if there is a finite group G such that $(X, 0) \simeq (\mathbb{C}^n/G, 0)$ (analytic isomorphism).

A linear map $R \in \mathrm{Gl}(n; \mathbb{C})$ is called a *reflection* if R has the eigenvalue 1 with multiplicity exactly $n - 1$.

A group $G \subset \mathrm{Gl}(n; \mathbb{C})$ is called a *reflection group* if it is generated by reflections.

A group $G \subset \mathrm{Gl}(n; \mathbb{C})$ is called *small* if G contains no reflections.

(Note that both these notions depend on the embedding $G \subset \mathrm{Gl}(n; \mathbb{C})$.)

One has the following basic results, see [P] for details.

(3.15) Theorem.

(i) *The quotient singularity $(\mathbb{C}^n/G, 0)$ is smooth if and only if G is a reflection group.*

(ii) *For any quotient singularity $(X, 0)$ there is a small group $G \subset \mathrm{Gl}(n; \mathbb{C})$ such that*

$$(X, 0) \simeq (\mathbb{C}^n/G, 0)$$

and then $\pi_1(X \backslash X_{\mathrm{sing}}) \simeq G$.

In particular, when $(X, 0)$ is an isolated quotient singularity, we have $\pi_1(L(X, 0)) = G$.

(3.16) **Corollary.** *For a normal surface singularity $(X, 0)$ the following statements are equivalent*:

(i) $\pi_1(L(X, 0))$ *is a finite group*;
(ii) $(X, 0)$ *is a quotient singularity.*

Proof. The implication (ii) $\Rightarrow$ (i) follows from (3.15(ii)) above and in fact holds in any dimension (for isolated quotient singularities).

To prove (i) $\Rightarrow$ (ii), consider the punctured germ $X^* = X \backslash \{0\}$ and recall that $\pi_1(X^*) = \pi_1(L(X, 0))$. Let $p: \tilde{X}^* \to X^*$ be the universal covering space and note that p extends to a ramified covering $p: (\tilde{X}, 0) \to (X, 0)$ where $(\tilde{X}, 0)$ is a normal surface singularity, see [Sn]. Since $\tilde{X} \backslash \{0\} = \tilde{X}^*$ is simply-connected, it follows by Mumford's theorem (3.12) that $(\tilde{X}, 0) = (\mathbb{C}^2, 0)$.

Let G be the covering group of the covering p. Then it is clear that $(\mathbb{C}^2/G, 0) \simeq (X, 0)$ (i.e., coordinates may be chosen on $\mathbb{C}^2$ such that G acts linearly and then this isomorphism is obvious). $\qquad \square$

(3.17) **Remark.** We warn the reader that the class of quotient singularities is quite different from the class of IHS or ICIS. Indeed, in the surface case, it is easy to show that the only singularities which are in the same time quotient singularities and ICIS are the famous rational double points A_k, D_k, E_6, E_7, and E_8, see [Df4] for a proof.

§4. Special Classes of Surface Singularities

In this section we list some of the most frequently encountered classes of isolated surface singularities. Their resolution graphs and the topology of their links are discussed in detail. Some notions and results here are not directly used in the sequel, but they have exactly the same flavor as some basic notions and results in later chapters. To give a concrete example, the pairs (α_i, β_i) in the Seifert invariant of a normal weighted homogeneous surface singularity should perhaps be compared to the transversal singularities (endowed with a finite cyclic group action) discussed in Chapter 6, §3.

(4.1) **Hirzebruch–Jung Singularities.** Consider the cyclic group of order n

$$G_{n,q} = \left\{ \begin{pmatrix} \lambda & 0 \\ 0 & \lambda^q \end{pmatrix} \in \mathrm{Gl}(2; \mathbb{C}); \lambda^n = 1 \right\}$$

and let $(X_{n,q}, 0)$ be the quotient singularity $(\mathbb{C}^2/G_{n,q}, 0)$ where we assume that $(n, q) = 1$. This final arithmetic condition is equivalent to asking that $G_{n,q}$ be a small group as defined in (3.14). Let

$$\frac{n}{q} = n_1 - \cfrac{1}{n_2 - \cfrac{\ddots}{\ - \cfrac{1}{n_r}}}$$

be the (finite) continued fraction representation of n/q such that $n_i \geq 2$ for all $i = 1, \ldots, r$.

Then it is known that the minimal (very good) resolution of the singularity $(X_{n,q}, 0)$ has a linear dual graph

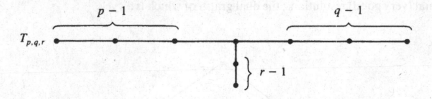

exactly as in (3.9) above, see [BPV], p. 80. Moreover, the embedding dimension of $(X_{n,q}, 0)$ is

$$\text{embdim}(X_{n,q}, 0) = \sum (n_i - 2) + 3.$$

In particular, $(X_{n,q}, 0)$ is an IHS in $\mathbb{C}^3$ if and only if $n_1 = \cdots = n_r = 2$, or equivalently $n - 1 = q = r \geq 1$. The corresponding singularity in this case is denoted by A_r and it is easy to see that an equation for it is the following:

(4.2) $\qquad\qquad A_r : x^{r+1} - yz = 0, \qquad r \geq 1.$

Note also that the link $L(X_{n,q}, 0)$ is none other than the famous *lens space* $L(n, q)$, a basic example in algebraic topology, see [R], p. 238.

(4.3) **The A-, D-, E-Singularities** (alias Rational Double Points, du Val Singularities, Simple Singularities). Consider the dual resolution graphs which are T-shaped

$$T_{p,q,r}$$

$(p, q, r \geq 1)$ and such that all the weights $n_i = -(E_i, E_i)$ are two and all the components E_i are rational. It is easy to see that the associated intersection matrix $I(\widetilde{X}) = ((E_i, E_j))$ is negative-definite if and only if

(4.4) $\qquad\qquad \dfrac{1}{p} + \dfrac{1}{q} + \dfrac{1}{r} > 1.$

Hence there are only the following possibilities (and it "happens" that in each case the dual graph determines a unique normal surface singularity by blowing-down and this singularity embeds in $\mathbb{C}^3$):

Type	Resolution Graph		Equation in $\mathbb{C}^3$
$A_k, k \geq 1$	•——•···•——•	(k points)	$x^{k+1} + y^2 + z^2 = 0$
$D_k, k \geq 4$	•——•···•——<	(k points)	$x^2 y + y^{k-1} + z^2 = 0$
E_6			$x^3 + y^4 + z^2 = 0$
E_7			$x^3 + xy^3 + z^2 = 0$
E_8			$x^3 + y^5 + z^2 = 0$

See, for details, [Df4] and [L3].

These singularities are exactly the quotient singularities $(\mathbb{C}^2/G, 0)$ for finite groups $G \subset SU(2)$ and hence their links can be identified to S^3/G. In particular, note that $\pi_1(S^3/G) = G$ and that the link $L(E_8)$ corresponding to the E_8 singularity above is precisely the *Poincaré icosahedral sphere*, i.e., a space having the same integral homology as S^3 but with a nontrivial $\pi_1(L(E_8))$.

The labels A, D, E are associated with these singularities because their resolution graphs described above coincide with the Dynkin diagrams of the simple complex Lie algebras of type A_k, D_k, E_6, E_7, and E_8, respectively. See also [B2] for a deeper connection between these singularities and simple Lie algebras. Note also that the two equations given above for the singularity A_k are clearly equivalent.

(4.5) **Cusp Singularities** (Alias Hyperbolic Singularities). These singularities are (exactly as the two previously discussed classes) determined by their minimal (very good) resolutions, the dual graph of which is a cycle

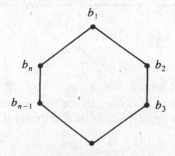

where $n \geq 3$, $b_i \geq 2$ (but not all of them are equal to 2) and the exceptional curves E_i are all rational. We let $I_n(b_1, \ldots, b_n)$ denote the normal surface singularity obtained by contracting the above graph [EW]. It is useful to extend this notation as follows:

n	Exceptional divisor
1	$-b_1 + 2$
2	$-b_1$ $-b_2$

i.e., for $n = 1$ we have a self-intersecting rational curve with self-intersection number $(-b_1 + 2)$, and for $n = 2$ we have two rational curves meeting transversally at two points. In each of these cases, the minimal very good resolution is obtained by blowing-up two successive points (resp. one point).

Moreover, we have [EW]:

$$\text{embdim } I_n(\mathbf{b}) = \max(3, \sum (b_i - 2))$$

using the notation $\mathbf{b} = (b_1, \ldots, b_n)$.

The link $L(I_n(\mathbf{b}))$ is a *torus bundle* over a circle with monodromy matrix

$$A(\mathbf{b}) = \begin{pmatrix} b_n & 1 \\ -1 & 0 \end{pmatrix} \cdots \begin{pmatrix} b_1 & 1 \\ -1 & 0 \end{pmatrix}.$$

In particular, the *Wang sequence* of the fibration

$$T^2 \to L(I_n(\mathbf{b})) \to S^1,$$

see, for instance, [M5], p. 67, implies that

$$A(\mathbf{b}) - I$$

is a presentation matrix for the torsion $\mathbb{Z}$-module $T_1(L(I_n(\mathbf{b})))$. The *dual sequence* $\mathbf{b}^*$ of a sequence $\mathbf{b}$ can be defined as follows. We may obviously write

$$\mathbf{b} = \underbrace{2, \ldots, 2}_{k_1^* - 1}, k_1 + 2, \underbrace{2, \ldots, 2}_{k_2^* - 1}, k_2 + 2, \ldots, k_g + 2$$

for integers $k_i, k_i^* \geq 1$ (up to a cyclic permutation). Then we define

$$\mathbf{b}^* = k_1^* + 2, \underbrace{2, \ldots, 2}_{k_1 - 1}, k_2^* + 2, \underbrace{2, \ldots, 2}_{k_2 - 1}, \ldots, \underbrace{2, \ldots, 2}_{k_g - 1}.$$

It is known that $L(I(\mathbf{b}^*))$ is homeomorphic to $L(I(\mathbf{b}))$.

Using the above formula for the embdim $I(\mathbf{b})$ we can easily write all the singularities which embed in $\mathbb{C}^3$, and it turns out that all of them are given by the following equation:

$$T_{p,q,r}: x^p + y^q + z^r + xyz = 0$$

with $p \le q \le r$ and $1/p + 1/q + 1/r < 1$. The correspondence between the triple (p, q, r) and the sequences $\mathbf{b}$ and $\mathbf{b}*$ is given in the following table: Here $2^a = 2, \ldots, 2$ and see, for details, [EW].

$\underbrace{\qquad\qquad}_{a \text{ times}}$

(p, q, r)	$\mathbf{b}$	$\mathbf{b}*$
$2, 3, r\, (r \ge 7)$	$2^{r-7}, 3$	$r - 4$
$2, 4, r\, (r \ge 5)$	$2^{r-5}, 4$	$2, r - 2$
$2, q, r\, (q \ge 5)$	$2^{q-5}, 3, 2^{r-5}, 3$	$q - 2, r - 2$
$3, 3, r\, (r \ge 4)$	$2^{r-4}, 5$	$2, 2, r - 1$
$3, q, r\, (q \ge 4)$	$2^{q-4}, 3, 2^{r-4}, 4$	$2, q - 1, r - 1$
$p, q, r\, (p \ge 4)$	$2^{p-4}, 3, 2^{q-4}, 3, 2^{r-4}, 3$	$p - 1, q - 1, r - 1$

Using these facts we get the following:

(4.6) **Corollary.** *The matrix*

$$\begin{pmatrix} r - 1 & 1 \\ -1 & 0 \end{pmatrix} \begin{pmatrix} q - 1 & 1 \\ -1 & 0 \end{pmatrix} \begin{pmatrix} p - 1 & 1 \\ -1 & 0 \end{pmatrix} - I$$

is a presentation matrix for $T_1(L(T_{p,q,r}))$. *In particular,*

$$|T_1(L(T_{p,q,r}))| = pqr \left(1 - \frac{1}{p} - \frac{1}{q} - \frac{1}{r} \right).$$

(4.7) **Triangle Singularities** (Alias Exceptional Singularities). Consider the dual resolution graph

where again all the exceptional curves E_i are rational. There are exactly 14 triples (p, q, r) for which the singularity obtained by contracting such a graph can be embedded in $\mathbb{C}^3$. The notation given in the next table is Arnold's notation, see, for instance, [AGV1], p. 185, where the triples (p, q, r) are called the corresponding *Dolgachev numbers*. The triples (p', q', r') are the corresponding *Gabrielov numbers* and they are used in Appendix A.

Notation	(p,q,r)	Equation	(p',q',r')
Q_{10}	$(2,3,9)$	$x^3 + y^4 + yz^2 + axy^3$	$(3,3,4)$
Q_{11}	$(2,4,7)$	$x^3 + y^2z + xz^3 + az^5$	$(3,3,5)$
Q_{12}	$(3,3,6)$	$x^3 + y^5 + yz^2 + axy^4$	$(3,3,6)$
Z_{11}	$(2,3,8)$	$x^3y + y^5 + axy^4 + z^2$	$(2,4,5)$
Z_{12}	$(2,4,6)$	$x^3y + xy^4 + ax^2y^3 + z^2$	$(2,4,6)$
Z_{13}	$(3,3,5)$	$x^3y + y^6 + axy^5 + z^2$	$(2,4,7)$
S_{11}	$(2,5,6)$	$x^4 + y^2z + xz^2 + ax^3z$	$(3,4,4)$
S_{12}	$(3,4,5)$	$x^2y + y^2z + xz^3 + az^5$	$(3,4,5)$
W_{12}	$(2,5,5)$	$x^4 + y^5 + ax^2y^3 + z^2$	$(2,5,5)$
W_{13}	$(3,4,4)$	$x^4 + xy^4 + ay^6 + z^2$	$(2,5,6)$
E_{12}	$(2,3,7)$	$x^3 + y^7 + axy^5 + z^2$	$(2,3,7)$
E_{13}	$(2,4,5)$	$x^3 + xy^5 + ay^8 + z^2$	$(2,3,8)$
E_{14}	$(3,3,4)$	$x^3 + y^8 + axy^6 + z^2$	$(2,3,9)$
U_{12}	$(4,4,4)$	$x^3 + y^3 + z^4 + axyz^2$	$(4,4,4)$

More precisely, for each triangular graph $D_{p,q,r}$ as above there are exactly *two* associated *singularities*: one of them which is *weighted homogeneous* (for $a = 0$ in the equations above) and which is usually called a *triangle singularity of type $D_{p,q,r}$*, and the other one corresponding to the value $a = 1$. Indeed, it is easy to see that for $a \neq 0$ all the singularities corresponding to a given equation above are equivalent (in spite of the fact that these equations correspond to different $\mathcal{R}$-equivalence classes, see, for instance, [D4] for the relations between $\mathcal{R}$-classes and $\mathcal{K}$-classes).

For related facts on these singularities we refer to [B5]. Using (4.6) and (3.4(ii)) we get, by a direct computation of presentation matrices, the following:

(4.8) Corollary.

$$T_1(L(T_{p,q,r})) = H_1(L(D_{p,q,r})).$$

(4.9) The Simple-Elliptic Singularities $\tilde{E}_6, \tilde{E}_7, \tilde{E}_8, \tilde{D}_5$. For these singularities the exceptional divisor E is just a smooth elliptic curve $(g(E) = 1)$ and the self-intersection $n = (E, E)$ takes the following values:

Notation	n	Equation
$\tilde{E}_8$	-1	$x^6 + y^3 + z^2 + \lambda xyz = 0$
$\tilde{E}_7$	-2	$x^4 + y^4 + z^2 + \lambda xyz = 0$
$\tilde{E}_6$	-3	$x^3 + y^3 + z^3 + \lambda xyz = 0$
$\tilde{D}_5$	-4	$\begin{cases} x^2 + y^2 + \lambda zw = 0 \\ xy + z^2 + w^2 = 0 \end{cases}$

Hence the first three of these singularities are IHS, while $\tilde{D}_5$ is an ICIS in $\mathbb{C}^4$ (the parameter λ takes values in $\mathbb{C}$ such that these equations indeed define *isolated singularities*).

Using (3.4) we get

$$H_1(L(\tilde{E}_k)) = \mathbb{Z}^2 \oplus \mathbb{Z}/(9-k)\mathbb{Z} \qquad \text{for} \quad k = 6, 7, 8,$$

and

$$H_1(L(\tilde{D}_5)) = \mathbb{Z}^2 \oplus \mathbb{Z}/4\mathbb{Z}.$$

Moreover, if L is any of these links, there is an obvious fibration (regarding L as the circle bundle of the tubular neighborhood of E in the resolution $\tilde{X}$)

$$S^1 \to L \to E.$$

Passing to fundamental groups we get

$$0 \to \mathbb{Z} \to \pi_1(L) \to \mathbb{Z}^2 \to 1$$

and this shows that $\pi_1(L)$ can be identified with the *Heisenberg group* H_n

$$\left\{ \begin{bmatrix} 1 & a & c \\ 0 & 1 & b \\ 0 & 0 & 1 \end{bmatrix} ; a, b \in \mathbb{Z}, c \in \left(\frac{1}{n}\right)\mathbb{Z} \right\},$$

where n is the corresponding intersection number. Note that *group-theoretically* the integer n can be obtained from the following easy-to-check isomorphism

$$C(H_n)/[H_n, H_n] \simeq \mathbb{Z}/n\mathbb{Z},$$

where $C(H_n)$ denotes the *center* of the group H_n and $[H_n, H_n]$ denotes its *commutator*.

(4.10) Weighted Homogeneous Surface Singularities. Consider a $\mathbb{C}^*$-action on the affine space $\mathbb{C}^n$ given by

(4.11) $$t \cdot (x_1, \ldots, x_n) = (t^{w_1} x_1, \ldots, t^{w_n} x_n),$$

where the *weights* $w_i = \text{weight}(x_i)$ are *strictly* positive integers satisfying $(w_1, \ldots, w_n) = 1$. This last condition ensures that the action is *effective* (i.e., $t \cdot x = x$ for all $x \in \mathbb{C}^n$ implies $t = 1$).

Assume that $(X, 0)$ is an isolated complete intersection singularity with $\dim X = m$ defined in $\mathbb{C}^n$ by the equations $f_1 = \cdots = f_{n-m} = 0$, where f_i is a *weighted homogeneous polynomial* of degree N_i with respect to the weights $\mathbf{w} = (w_1, \ldots, w_n)$ for all $i = 1, \ldots, n - m$. The condition on f_i means

(4.12) $$f_i(t \cdot x) = t^{N_i} f_i(x) \qquad \text{for all} \quad t \in \mathbb{C}^*, \quad x \in \mathbb{C}^n.$$

Any singularity $(Y, 0)$ which is isomorphic to the singularity $(X, 0)$ defined above is called a *weighted homogeneous ICIS of type*

$$(w_1, \ldots, w_n; N_1, \ldots, N_{n-m}) = (\mathbf{w}; \mathbf{N}).$$

We set $X^* = X \backslash \{0\}$, where X is regarded here as a global affine variety and we assume usually that X is not contained in any of the hyperplanes $x_i = 0$.

Now we restrict our attention to the surface singularities $(m = 2)$. Then the topology of the link $L(X, 0)$ and the dual graph of the minimal very good resolution for $(X, 0)$ are completely determined by the so-called *Seifert invariants* of the singularity $(X, 0)$ denoted by

(4.13) $$\{g; b; (\alpha_1, \beta_1), \ldots, (\alpha_r, \beta_r)\}$$

as explained in the excellent survey by P. Wagreich [Wa]. In what follows we recall how we may compute these Seifert invariants in terms of the weighted homogeneity type $(\mathbf{w}; \mathbf{N})$.

The basic *algebraic invariant* associated to this type is the *Poincaré series* P_A of the graded algebra

$$A = \mathbb{C}[x_1, \ldots, x_n]/(f_1, \ldots, f_{n-2}),$$

where the polynomial ring is graded by the condition $\mathrm{wt}(x_i) = w_i$. Then

(4.14) $$P_A(t) = \sum_{k \geq 0} (\dim A_k) t^k = \frac{(1 - t^{N_1}) \cdots (1 - t^{N_{n-2}})}{(1 - t^{w_1}) \cdots (1 - t^{w_n})},$$

where the first equality is just the definition of this Poincaré series and the second is a basic well-known fact (see, for instance, [D4], p. 108).

Let q be a common multiple of the weights w_i and define $\varphi(i) = \dim A_{qi}$. Then there exists a *polynomial* function $\bar{\varphi}$ such that $\bar{\varphi}(i) = \varphi(i)$ for $i \gg 0$. This function $\bar{\varphi}$ (in fact closely related to the Hilbert polynomial for A, see, for the general context, [Hn], pp. 51 and 230) occurs in the following basic result.

(4.15) **Theorem.** *The quotient variety $X^*/\mathbb{C}^* = C$ is a smooth projective curve and its genus $g(C)$ is given by*

$$g(C) = 1 - \bar{\varphi}(0).$$

In particular, for $n = 3$ and $N_1 = N$, we have

$$g(C) = \frac{1}{2}\left[\frac{N^2}{w_1 w_2 w_3} - \sum_{i < j} \frac{N(w_i, w_j)}{w_i w_j} + \sum_i \frac{(N, w_i)}{w_i} - 1 \right].$$

(4.16) **Definition.** The *virtual degree* $\deg(A)$ of a graded algebra A of dimension s is the coefficient of $(1 - t)^{-s}$ in the Laurent expansion of $P_A(t)$ at the point $t = 1$. Note that $(1 - t)^{-s}$ is in fact the first nontrivial term of this expansion and that $\deg(A) \in \mathbb{Q}$. In particular, for $n = 3$ and $N = N_1$ we have

$$\deg(A) = \frac{N}{w_1 w_2 w_3}.$$

We look next at the *isotropy groups*

$$\mathbb{C}_x^* = \{t \in \mathbb{C}^*; t \cdot x = x\} \qquad \text{for any} \quad x \in \mathbb{C}^n.$$

It is easy to see that for $x \neq 0$, this group $\mathbb{C}_x^*$ is cyclic of finite order

$$(4.17) \qquad \alpha(x) = (w_{i_1}, \ldots, w_{i_k}),$$

where x_{i_j} ($j = 1, \ldots, k$) are precisely the nonzero coordinates of the point x. Since the $\mathbb{C}^*$-action is effective, there are only a finite number of points $[a_1]$, $\ldots, [a_r] \in C$ such that the corresponding orders $\alpha_i = |\mathbb{C}_{a_i}^*| > 1$.

For any such point $a_i \in X^*$, there is an induced $\mathbb{C}_{a_i}^*$-action on a *transversal slice* S_i in X^* to the orbit $\mathbb{C}^* \cdot a_i$ of this point. Via the identification

$$\mathbb{C}_{a_i}^* = \{\lambda \in \mathbb{C}^*; \lambda^{\alpha_i} = 1\},$$

such an action is given by the formula

$$\lambda \cdot s = \lambda^{b_i} s \qquad \text{for} \quad \lambda \in \mathbb{C}_{a_i}^* \quad \text{and} \quad s \in S_i,$$

for some positive integer b_i such that $(\alpha_i, b_i) = 1$ (this is again due to the fact that our $\mathbb{C}^*$-action on X^* is effective). Define next an integer β_i by the conditions

$$(4.18) \qquad 0 < \beta_i < \alpha_i, \qquad b_i \beta_i \equiv -1 \ (\text{mod } \alpha_i).$$

In this way we have obtained the pairs (α_i, β_i), $i = 1, \ldots, r$, which are listed in the Seifert invariant (4.13). The first entry there g is just the genus $g(C)$ from (4.15), while the second entry b is a generalized Chern number and is given explicitly by the following formula:

$$(4.19) \qquad b = \deg A + \sum_{i=1,r} \beta_i/\alpha_i.$$

Next note that the link $L(X, 0)$ is an oriented compact 3-manifold with a natural S^1-action coming from the action (4.11) by looking only at those t with $|t| = 1$.

(4.20) **Theorem** (Seifert). *The invariants* $\{g; b; (\alpha_1, \beta_1), \ldots, (\alpha_r, \beta_r)\}$ *determine the link* $L(X, 0)$ *up to an orientation-preserving* S^1-*equivariant homeomorphism.*

We say that a map $p: \tilde{X} \to X$ is $\mathbb{C}^*$-equivariant if there is a $\mathbb{C}^*$-action on $\tilde{X}$ such that $p(t \cdot y) = t \cdot p(y)$ for all $t \in \mathbb{C}^*$, $y \in \tilde{X}$, where the action on the affine variety X is induced from (4.11).

The following basic result says how to derive the minimal resolution of a weighted homogeneous surface singularity.

(4.21) **Theorem** (Orlik–Wagreich [OW]). *There is a unique equivariant resolution* $p: \tilde{X} \to X$ *such that:*

(i) *The exceptional divisor* $E = p^{-1}(0) = E_1 \cup \cdots \cup E_s$ *has exactly one component (say* E_1*) which is fixed pointwise by the* $\mathbb{C}^*$*-action on* $\tilde{X}$.

(ii) *Each of the other components* E_i *($i \geq 2$) is a rational curve and has* $(E_i, E_i) \leq -2$.

(iii) E_1 *is isomorphic to* $C = X^*/\mathbb{C}^*$ *(and hence has genus g) and* $(E_1, E_1) = -b$.

(iv) *The components* E_i *meet transversally according to the following star-shaped graph*

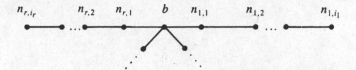

where the central curve is E_1 *and there are exactly r arms. The negatives of the intersection numbers* $(n_{k,1}, \ldots, n_{k,i_k})$ *along the kth arm are given by the following "continuous fraction" decomposition*

$$\frac{\alpha_k}{\beta_k} = n_{k,1} - \cfrac{1}{n_{k,2} - \cfrac{1}{\ddots \cfrac{}{n_{k,i_k}}}}$$

(4.22) **Example.** Consider the E_7 singularity whose equation is given in the table in (4.3). The weights $\mathbf{w}$ in this case are $(6, 4, 9)$ and the degree $N = 18$.

Using the formula given in (4.15) we get $g = g(C) = 0$, hence $E_1 = C$ is a rational curve. Next deg $A = \frac{1}{12}$ and there are exactly three points on C with nontrivial isotropy groups:

$$a_1 = (0, 1, 0) \qquad \text{with} \quad (\alpha_1, \beta_1) = (4, 3),$$

$$a_2 = (1, -1, 0) \qquad \text{with} \quad (\alpha_2, \beta_2) = (2, 1),$$

and

$$a_3 = (-1, 0, 1) \qquad \text{with} \quad (\alpha_3, \beta_3) = (3, 2).$$

Using Theorem (4.21) we get the dual resolution graph for the singularity E_7 which is already given in the table in (4.3).

We remark finally that the class of weighted homogeneous singularities is very large (e.g., it contains all our other examples given in this section with the exception of the cusp singularities $T_{p,q,r}$ and the nonweighted homogeneous triangle singularities $D_{p,q,r}$ obtained for $a = 1$ in the table in (4.7)).

CHAPTER 3

The Milnor Fibration and the Milnor Lattice

§1. The Milnor Fibration

In this section we introduce various Milnor fibrations, in particular, the global Milnor fibration associated with a weighted homogeneous polynomial. Then we discuss the basic properties of the corresponding monodromy operators. Let $\mathcal{O}_{n+1} = \mathbb{C}\{x_0, \ldots, x_n\}$ be the $\mathbb{C}$-algebra of analytic function germs at the origin 0 of $\mathbb{C}^{n+1}$ and let $(X, 0)$ be a hypersurface singularity defined by an equation $f = 0$, for some $f \in \mathcal{O}_{n+1}$ with $f(0) = 0$. Here $n \geq 0$ is a positive integer. There are *two* equivalent fibrations which, in the literature, are called *the Milnor fibration* of the function germ f (or of the hypersurface singularity $(X, 0)$).

To describe the first of these fibrations, let $\varepsilon > 0$ be small enough such that:

(i) the germ f is defined in an open neighborhood of the closed ball $\bar{B}_\varepsilon$ of radius ε, centered at the origin of $\mathbb{C}^{n+1}$;

(ii) the representative of the singularity $(X, 0)$ in the ball $\bar{B}_\varepsilon$ has a conic structure as in Theorem (5.1).

Let $S_\varepsilon^{2n+1} = \partial \bar{B}_\varepsilon$ be the boundary of this ball and let $K = L(X, 0) = X \cap S_\varepsilon^{2n+1}$ be the corresponding link. Note that $K = \emptyset$ for $n = 0$.

(1.1) **Theorem** (Milnor [M5]). *The map*

$$\varphi: S_\varepsilon^{2n+1} \setminus K \to S^1, \qquad \varphi(x) = f(x)/|f(x)|,$$

is a smooth locally trivial fibration, for any $\varepsilon > 0$ sufficiently small.

From a historical point of view, this theorem is amply motivated by the classical study of fibered knots, recall (2.1.24). Indeed, when f has an isolated singularity at the origin, then the link K is a smooth manifold. And any fiber $F_a = \varphi^{-1}(a)$ is a smooth open manifold whose closure $\bar{F}_a$ coincides with the union $F_a \cup K$. More precisely, $\bar{F}_a$ is a manifold with boundary $\partial \bar{F}_a = K$ and hence the analogy to the case of fibered knots is complete.

Perhaps the real challenge to consider the fibration φ has come from the surprising discovery by Hirzebruch [Hz2] and Brieskorn [B1] that some of the links K provide examples of exotic differentiable structures on spheres, see, for details, the last section of this chapter.

The second fibration can be described as follows. Choose an $\varepsilon > 0$ small enough and a $\delta > 0$ such that $\varepsilon \gg \delta > 0$, i.e., δ is much smaller than ε. Let

$$D_\delta^* = \{t \in \mathbb{C}; 0 < |t| < \delta\}$$

be a small open punctured disc in $\mathbb{C}$.

(1.2) **Theorem** (Lê [Lê3]).

(i) *The map*

$$\bar\psi : \bar B_\varepsilon \cap f^{-1}(D_\delta^*) \to D_\delta^*, \qquad \bar\psi(x) = f(x),$$

is a topological locally trivial fibration, for any $\varepsilon \gg \delta > 0$ small enough.

(ii) *The map*

$$\psi : B_\varepsilon \cap f^{-1}(D_\delta^*) \to D_\delta^*, \qquad \psi(x) = f(x),$$

is a smooth locally trivial fibration, for any $\varepsilon \gg \delta > 0$ small enough.

Let $S_{\delta/2}^1$ be a circle in D_δ^*, centered at 0 and having radius $\delta/2$. Then the inclusion $S_{\delta/2}^1 \to D_\delta^*$ is a homotopy equivalence. Hence all the topological information (up to homotopy) contained in the fibrations ψ and $\bar\psi$ is preserved by restricting these maps over the circle $S_{\delta/2}^1$. Moreover, by identifying $S_{\delta/2}^1$ to the unit circle S^1 via the map

$$t \mapsto 2t/\delta,$$

we will regard these new restricted fibrations as fibrations over S^1 and denote them still with ψ and $\bar\psi$.

(1.3) **Lemma.**

(i) *The fiber of the fibration ψ is a Stein complex manifold and the fiber of the fibration $\bar\psi$ is a manifold with boundary (this boundary is empty for $n = 0$).*

(ii) *Both of these fibers have the homotopy type of a CW-complex of (real) dimension n.*

Proof. The first statement for the fiber of the fibration ψ is clear since B_ε is a Stein manifold and any (closed) analytic hypersurface in a Stein space is Stein, see, for instance, [KK], p. 224. Using the fact that any manifold with boundary M is homotopy equivalent to its interior $M \setminus \partial M$, see, for instance, [Sp], p. 297 and (1.6.8), all we have still to show is that $\bar G_a = \bar\psi^{-1}(a)$ is a manifold with boundary $\partial \bar G_a = \bar G_a \cap S_\varepsilon^{2n+1}$. This fact is mentioned in the current literature only in the case when f has an isolated singularity at the origin and when the boundary $\partial \bar G_a$ can be identified to the link K.

To prove in general that $\bar G_a$ is a manifold with boundary, we identify the fiber $\bar G_a$ (for a real and positive to simplify the notations) with a part of the fiber F, following the constructions in [M5], Chap. 5. Consider the following picture.

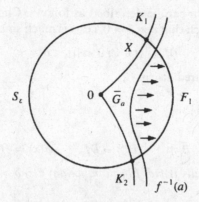

Here the link K is represented by the two points K_1 and K_2, the fiber $F_1 = \varphi^{-1}(1)$ is represented by the short arc of the circle S_ε joining K_1 and K_2.

In [M5], p. 53, a vector field like the one drawn in our picture is used to push the set $\bar{G}_a$ diffeomorphically onto the set

$$F_1^0 = \{x \in F_1; f(x) \geq a\}.$$

Note that the function $f \,|\, F_1$ takes only real positive values. Hence its critical points are the same as the critical points of the function $\log |f|$ on F_1, considered by Milnor. In [M5], p. 49, it is shown that there is a constant $\eta > 0$ (depending on f and ε) such that all these critical points lie within the compact subset $f(x) \geq \eta$ of F_1. Hence, if we choose $\delta < \eta$ we are sure that:

(i) F_1^0 is a smooth manifold with boundary

$$\partial F_1^0 = \{x \in F_1; f(x) = a\}.$$

(ii) The interior $F_1^0 \backslash \partial F_1^0 \simeq G_a = \psi^{-1}(a)$ is diffeomorphic to the whole fiber F_1.

It follows from (i) that $\bar{G}_a$ is indeed a manifold with boundary, and this completes the proof of (1.3). $\qquad\qquad\square$

Moreover, the diffeomorphism $h_a: G_a \to F_1$ whose existence is claimed in (ii) can be constructed in a way depending smoothly on the parameter a, thus giving a proof of the following basic result.

(1.4) Proposition.

(i) *The fibrations φ and ψ are fiber diffeomorphic equivalent.*
(ii) *The fibrations ψ and $\bar{\psi}$ are fiber homotopy equivalent.*

For the reader's convenience, we recall below the definition of these equivalence relations.

(1.5) Definition. Two topological (resp. smooth) fibrations $p: E \to B$ and $p': E' \to B$ are said to be *fiber homotopy equivalent* (resp. *fiber diffeomorphic*) if there is a continuous map (resp. diffeomorphism) $h: E \to E'$ such that:

(i) $p' \circ h = p$;

(ii) for all $b \in B$, h induces a map $h_b: p^{-1}(b) \to p'^{-1}(b)$ which is a homotopy equivalence (resp. a diffeomorphism).

(1.6) Exercise. Compare (1.5) with the definition given in [Sp], p. 100, using the basic results of Dold [Do].

(1.7) Definition. Any of the (equivalent) fibrations φ, ψ, or $\bar{\psi}$ (defined over S^1 or over D_δ^*) is called *the Milnor fibration* of the function germ f (or, of the singularity $(X, 0)$). Any of the corresponding fibers $\varphi^{-1}(a)$, $\psi^{-1}(a)$ or $\bar{\psi}^{-1}(a)$ is called the *Milnor fiber* of the function germ f (or, of the singularity $(X, 0)$). Sometimes, one uses the terms *open Milnor fiber* and *closed Milnor fiber* to distinguish between the fibers of ψ and $\bar{\psi}$. The closed Milnor fiber is usually denoted by $\bar{F}$.

(1.8) Remark. The Milnor fibration associated to a hypersurface singularity $(X, 0)$ does not depend on the choice of an equation $f = 0$ for $(X, 0)$. This comes essentially from the fact that K-orbits are connected, i.e., any two such equations $f_0 = 0$ and $f_1 = 0$ may be put in a continuous family f_t, $t \in [0, 1]$, such that all the singularities $(X_t, 0)$ are analytically equivalent to $(X, 0)$. See [Lê2] for details. Moreover, in the case of IHS, the Milnor fibration (equivalence class) does not change under μ-constant deformations, see [LêR].

The following example shows that the Milnor fibration can appear under various disguises.

(1.9) Example. Let $f \in \mathbb{C}[x_0, \ldots, x_n]$ be a weighted homogeneous polynomial of degree N with respect to the weights $\mathrm{wt}(x_i) = w_i$, where w_i is a positive integer for $i = 0, \ldots, n$. Consider the $\mathbb{C}^*$-action on $\mathbb{C}^{n+1}$ associated to these weights

$$(1.10) \qquad t \cdot x = (t^{w_0} x_0, t^{w_1} x_1, \ldots, t^{w_n} x_n)$$

for all $t \in \mathbb{C}^*$ and $x = (x_0, x_1, \ldots, x_n) \in \mathbb{C}^{n+1}$.

(1.11) Exercise. Show that the restriction of the polynomial mapping f given by

$$f: \mathbb{C}^{n+1} \setminus f^{-1}(0) \to \mathbb{C}^*$$

is a locally trivial fibration. *Hint.* Use the Euler relation $f(t \cdot x) = t^N f(x)$.

(1.12) Definition. Let θ be the restriction of the above fibration over the unit circle S^1. Then θ is called the (global) *affine Milnor fibration* of the weighted homogeneous polynomial f and $F = \theta^{-1}(1)$ is called its (global) *affine Milnor fiber*.

(1.13) Exercise. The affine Milnor fibration θ is fiber diffeomorphic to the Milnor fibration φ associated to the germ of f at the origin. *Hint.* Construct a diffeomorphism $h: S_\varepsilon^{2n+1} \backslash K \to f^{-1}(S^1)$ using the induced $\mathbb{R}_+$-action on $\mathbb{C}^{n+1}$, where $\mathbb{R}_+ = \{t \in \mathbb{C}^*; t \text{ real and } t > 0\}$. For $x \in S_\varepsilon \backslash K$ we consider its $\mathbb{R}_+$-orbit and note that there is exactly one $t_x \in \mathbb{R}_+$ such that $t_x \cdot x \in f^{-1}(S^1)$. So that the map $h: x \mapsto t_x \cdot x$ is indeed a diffeomorphism.

(1.14) Exercise. Consider the *weighted sphere*

$$S_\varepsilon(\mathbf{w}) = \{w \in \mathbb{C}^{n+1}; |x_0|^{2/w_0} + \cdots + |x_n|^{2/w_n} = \varepsilon^2\},$$

where $\mathbf{w} = (w_0, \ldots, w_n)$ is the set of all the weights for the weighted homogeneous polynomial f. Show that the *weighted Milnor fibration*

$$\varphi(\mathbf{w}): S_\varepsilon(\mathbf{w}) \backslash f^{-1}(0) \to S^1, x \mapsto f(x)/|f(x)|,$$

is fiber diffeomorphic to the Milnor fibration φ associated to the germ of f at the origin. *Hint.* Use the fibration θ as an intermediate step.

In the following exercises we compare the boundaries of the Milnor fiber of two homogeneous singularities.

(1.15) Exercise. Let $f_0 = x_0, \ldots, x_n$ $(n \geq 1)$ and consider the boundary of the corresponding Milnor fiber

$$B_0 = \{x \in S_\varepsilon^{2n+1}; x_0, \ldots, x_n = \delta\} \qquad \text{for} \quad \varepsilon \gg \delta > 0.$$

Show that B_0 is diffeomorphic to the product $S^{n-1} \times (S^1)^n$. *Hint.* First write B_0 in the following way

$$\left\{ B_0 = \{\bar{x} \in (\mathbb{C}^*)^n; |x_1|^2 + \cdots + |x_n|^2 + \frac{\delta^2}{|x_1|^2 \cdots |x_n|^2} = \varepsilon^2\},\right.$$

where $\bar{x} = (x_1, \ldots, x_n)$.

Consider next the identification

$$(\mathbb{C}^*)^n \to (0, \infty)^n \times (S^1)^n, \qquad \bar{x} \mapsto \left(|x_1|, \ldots, |x_n|, \frac{x_1}{|x_1|}, \ldots, \frac{x_n}{|x_n|}\right)$$

and the map $g: (0, \infty)^n \to (0, \infty)$

$$g(a) = a_1^2 + \cdots + a_n^2 + \frac{\delta^2}{a_1^2 \cdots a_n^2}.$$

Show that:

(i) g has a single critical point, namely, $a_1 = \cdots = a_n = \delta^{1/n+1}$, which is a minimum;

(ii) for $\varepsilon > g_{\min}$, the set

$$S = \{a \in (0, \infty)^n; g(a) = \varepsilon\}$$

is diffeomorphic to the sphere S^{n-1}.

(1.16) **Exercise.** Let $f_1 = x^3 + y^3 + z^3$ be our second homogeneous polynomial. Since f_1 has an isolated singularity at the origin, it follows that the boundary B_1 of its Milnor fiber coincides to the link

$$B = K = S_\varepsilon^5 \cap f_1^{-1}(0).$$

Use (2.4.9) to show that $H_1(K) = \mathbb{Z}^2 \oplus \mathbb{Z}/3\mathbb{Z}$. Hence the boundaries B_0 (for $n = 2$) and B_1 are quite distinct topologically.

There are two very useful *exact sequences* to investigate the topology of a fibration

$$F \xrightarrow{j} E \xrightarrow{p} S^1$$

over a circle. The first one is the exact homotopy sequence of a fibration, see, for instance, [Sp], p. 377.

(1.17) **Proposition.**

(i) *There is an exact sequence*

$$0 \to \pi_1(F) \xrightarrow{j_\#} \pi_1(E) \xrightarrow{p_\#} \pi_1(S^1) = \mathbb{Z} \to \pi_0(F) \to \pi_0(E) \to \pi_0(S^1).$$

(ii) *For $i \geq 2$, there are isomorphisms*

$$\pi_i(F) \simeq \pi_i(E).$$

To describe the second useful exact sequence, recall that a fibration over the circle S^1 as above is completely determined by its *monodromy homeomorphism* (or diffeomorphism when we deal with smooth fibrations) $h: F \to F$. The homomorphism induced at the homology level (any coefficients)

$$T = h_*: H_*(F) \to H_*(F)$$

is called the (homology) monodromy operator. (In this definition, homology H_* can be replaced by the reduced homology $\tilde{H}_*$.) This monodromy operator (defined up to isotropy) occurs in the following *Wang exact sequence*, see [M5], p. 67, for details,

(1.18) $$\to H_{i+1}(E) \to E_i(F) \xrightarrow{T-I} H_i(F) \to H_i(E) \to,$$

where I is the identity homomorphism. Similarly, there is a cohomology Wang sequence

(1.18') $\rightarrow H^i(E) \rightarrow H^i(F) \xrightarrow{\ T-I\ } H^i(F) \rightarrow H^{i+1}(E) \rightarrow,$

where $T = h^*$ is, this time, the cohomology monodromy operator.

(1.19) **Example.** Consider the affine Milnor fibration

$$F \rightarrow E \rightarrow S^1$$

associated to a weighted homogeneous polynomial f of type $(w_0, \ldots, w_n; N)$ as in (1.12). Let $F_t = f^{-1}(t)$ for $t \in S^1 = \mathbb{R}/\mathbb{Z}$ and consider the map $h_t \colon F_1 \rightarrow F_t$, $h_t(x) = \exp(2\pi i t/N) \cdot x$. Here $\cdot$ stands for the multiplication (1.10). Then we have

$$f(h_t(x)) = \exp(2\pi i t),$$

i.e., h_t is a one-parameter family of diffeomorphisms covering the standard generator of the fundamental group $\pi_1(S^1, 1)$. This shows that the mono-dromy operator of this affine Milnor fibration can be taken to be

$$h = h_1.$$

In particular, $h^N = I$, and hence the *complex* monodromy operator

$$T = h^* \colon H^*(F; \mathbb{C}) \rightarrow H^*(F; \mathbb{C})$$

is diagonalizable and has as eigenvalues only Nth roots of unity.

A larger class of singularities, which behave topologically exactly like the weighted homogeneous (w.h.) singularities, is the class of semiweighted homo-geneous (s.w.h.) singularities. Such a s.w.h. singularity $\overline{X} \colon \overline{f} = 0$ is the sum $\overline{f} = f + g$ of an IHS f which is w.h. of type $(w_0, \ldots, w_n; N)$ and a germ $g \in \mathcal{O}_{n+1}$, $g = \sum a_\alpha x^\alpha$ such that all the monomials $x^\alpha = x_0^{\alpha_0}, \ldots, x_n^{\alpha_n}$ with $a_\alpha \neq 0$ satisfy the condition

$$\deg(x^\alpha) = \alpha_0 w_0 + \cdots + \alpha_n w_n > N.$$

For such a s.w.h. singularity $\overline{f}$ we have an obvious equality

$$f_t(x) = \overline{f}(t \cdot x)t^{-N} = f(x) + g_t(x),$$

where g_t is a family of germs in $\mathcal{O}_{n+1}$ with $\lim_{t \to 0} g_t = 0$. The family f_t is μ-constant by [D4], p. 116, and hence we can apply the result of Lê and Ramanujam [LêR] to deduce that the Milnor fibrations of the singularities $f = f_0$ and $\overline{f} = f_1$ are equivalent.

Now let $X \colon f = 0$ be a hypersurface singularity at the origin of $\mathbb{C}^{n+1}$, $F \rightarrow S_\varepsilon^{2n+1} \setminus K \xrightarrow{\varphi} S^1$ its Milnor fibration, and $T = h_{\mathbb{C}}^*$ the corresponding com-plex monodromy operator acting on $H^p(F; \mathbb{C})$. We list below some of the basic properties of this operator.

(1.20) Theorem.

(i) *All the eigenvalues of T are roots of unity. In other words, T is quasi unipotent, i.e, there are positive integers u and v such that*

$$(T^u - I)^v = 0.$$

Moreover, we can take $v = p + 1$.

(ii) *Let $L(h) = \sum (-1)^i \operatorname{Tr}(h_{\mathbb{C}}^* | H^i(F; \mathbb{C}))$ be the Lefschetz number of the map h. Then $L(h) = 0$ if 0 is a singular point for f (i.e., if $df(0) = 0$).*

For a recent proof of assertion (i) (usually known as the *Monodromy Theorem*) we refer to the paper [Fr] and to the references listed there. As for a proof of (ii), we refer to [AC2]. Note that (1.20(ii)) has the following consequence.

(1.21) Corollary. *If 0 is a singular point for f, then the associated Milnor fiber F cannot be homologically contractible, i.e., $H^*(F; \mathbb{C}) \neq H^*(point; \mathbb{C})$.*

(1.22) Exercise. Prove (1.20(ii)) in the case when f is a weighted homogeneous polynomial using the classical Lefschetz Fixed Point Theorem, see, for instance, [Sp], p. 195. *Hint.* Use the description of the monodromy operator given in (1.19). Use also the cylindric structure at infinity of algebraic sets (1.6.9) to replace the affine Milnor fiber with a compact polyhedron.

Another basic property of the Milnor fiber is given in the following result.

(1.23) Proposition. *The Milnor fiber is a parallelizable manifold.*

Proof. Let F be the fiber of the Milnor fibration ψ. Since F is a smooth hypersurface in the open ball B_ε, it follows that:

(i) the normal bundle N_F to F is trivial;
(ii) the sum $T_F \oplus N_F$ of the tangent bundle to F with the normal bundle to F is trivial. $\square$

To deduce that the tangent bundle T_F is trivial we have to apply the following simple fact, see, for instance, [KeMi].

Fact. Let F be a CW-complex of dimension n and let $E \to F$ be a (real) vector bundle of rank $k > n$ which is stably trivial (i.e., $E \oplus \theta^s = \theta^{k+s}$ where θ^m denotes the trivial vector bundle over F of rank m). Then E is trivial.

Before ending this section, we remark that the concepts of the Milnor fibration, the Milnor fiber, and the monodromy operator have been extended by several authors to the more general situation of function germs $f: (Y, 0) \to (\mathbb{C}, 0)$ defined on an analytic space germ $(Y, 0)$. Usually it is assumed that $(Y, 0)$

is an isolated complete intersection singularity (abbreviated in the sequel to ICIS) and that the special fiber $(X, 0) = (f^{-1}(0), 0)$ is again an ICIS. For examples, the reader can see Hamm [H1], [Lê3], [Lg], [D1], and Theorem (2.12) at the end of the next section.

§2. The Connectivity of the Link, of the Milnor Fiber, and of Its Boundary

In this section we consider a hypersurface singularity $X: f = 0$ at the origin of $\mathbb{C}^{n+1}$ and denote by K its link, by F its (open) Milnor fiber, and by $\partial \overline{F}$ the boundary of its closed Milnor fiber. The main theme is to discuss the connectivity of these three topological spaces. The first result says that the connectivity of the link K is indendent of how bad the singularity $(X, 0)$ is.

(2.1) **Proposition** (Milnor [M5]). *The link K is $(n - 2)$-connected.*

We recall that a space T is called s-connected (for some integer $s \geq 0$) if T is path-connected and $\pi_i(T, t) = 0$ for all i with $1 \leq i \leq s$ and some base point $t \in T$. By convention (-1)-connected means nonempty. Unlike in the case of the link, the connectivity of the Milnor fiber depends on the dimension of the singular subspace $(X_{\text{sing}}, 0)$ of the germ $(X, 0)$. Note that $(X, 0)$ was *not* supposed to be reduced (i.e., f may have multiple factors) and hence $\dim(X_{\text{sing}}, 0) \leq \dim(X, 0)$.

(2.2) **Theorem.** *Let $s = \dim(X_{\text{sing}}, 0)$. Then the Milnor fiber F is $(n - s - 1)$-connected.*

The case $s = 0$ (when $(X, 0)$ is an IHS) was treated by Milnor in [M5], while the general case is due to Kato–Matsumoto [KM].

When $s = n$, we have the following more precise result.

(2.3) **Proposition.** *Let $f = f_1^{m_1}, \ldots, f_k^{m_k}$ be the decomposition of the germ f as a product of distinct irreducible factors in the factorial ring $\mathcal{O}_{n+1}$. Let $m =$ g.c.d.$(m_1, \ldots, m_k)$. Then the Milnor fiber F has exactly m connected components and for any base point $a \in F$ we have an exact sequence*

$$0 \to [\pi_1(S_\varepsilon^{2n+1} \setminus K, a), \pi_1(S_\varepsilon^{2n+1} \setminus K, a)] \to \pi_1(F, a) \to \mathbb{Z}^{k-1} \to 0.$$

Proof. As the case $n = 0$ is clear, we assume $n \geq 1$.

Using (1.17(i)) we get

$$0 \to \pi_1(F, a) \to \pi_1(S_\varepsilon^{2n+1} \setminus K, a) \xrightarrow{\varphi_*} \pi_1(S^1) \to \pi_0(F) \to 0.$$

Since $\pi_1(S^1) = \mathbb{Z}$ is abelian, the morphism φ_* factorizes as

$$\pi_1(S_\varepsilon^{2n+1} \setminus K, a) \xrightarrow{H} H_1(S_\varepsilon^{2n+1} \setminus K) \xrightarrow{\overline{\varphi_*}} \pi_1(S^1),$$

where H = the Hurewicz morphism. And the kernel ker H is precisely the commutator subgroup

$$[\pi_1(S_\varepsilon^{2n+1} \setminus K, a), \pi_1(S_\varepsilon^{2n+1} \setminus K, a)].$$

Now we describe explicitly a basis for the homology group $H_1(S_\varepsilon^{2n+1} \setminus K)$, as was done in Chapter 2, §2, in the case $n = 1$. Let $Z = X_{\mathrm{red}} = Z_1 \cup \cdots \cup Z_k$ be the reduced hypersurface corresponding to X. Here Z_i: $f_i = 0$ are the irreducible components of Z for $i = i, \ldots, k$.

Then by the Conic Structure Theorem (1.5.1) we have

$$H_1(S_\varepsilon \setminus K) = H_1(B_\varepsilon \setminus Z).$$

Let $M = B_\varepsilon \setminus Z_{\mathrm{sing}}$, $D = Z \setminus Z_{\mathrm{sing}}$, and apply the homology Gysin sequence (2.2.13). It is obvious that D has k connected components and the isomorphism

$$\bar{\theta}: H_0(D) \to H_1(M \setminus D)$$

shows that a basis $e_i = \theta(a_i)$, $i = 1, \ldots, k$, is obtained by taking small circles e_i in a transversal to the smooth part of $Z_i \setminus Z_{\mathrm{sing}}$. It is easy to see that

$$\bar{\varphi}_*(e_i) = m_i$$

under the usual identification $\pi_1(S^1) \simeq \mathbb{Z}$. Hence im $\varphi_* = $ im $\bar{\varphi}_* = m\mathbb{Z}$ and this implies that F has m connected components. It is also clear that ker $H \subset \pi_1(F, a)$ and that

$$\pi_1(F, a)/\mathrm{ker}\, H \simeq \mathrm{ker}\, \bar{\varphi}_* \simeq \mathbb{Z}^{k-1}.$$

This ends the proof of (2.3). $\square$

Consider now the boundary $\partial \bar{F}$ of the closed Milnor fiber $\bar{F}$. The following result gives us information about the connectivity of $\partial \bar{F}$, compare with (1.6.12).

(2.4) **Proposition.** *The inclusion map $j = \partial \bar{F} \to \bar{F}$ is an $(n-1)$-equivalence.*

Proof. We can regard the Milnor fiber $\bar{F}$ as part of a larger open Milnor fiber F^1 of the form $B_{\varepsilon^1} \cap f^{-1}(\delta)$ for some $\varepsilon^1 > \varepsilon$. Consider the squared-distance function from the points in F^1 to the origin, i.e.,

$$d: F^1 \to \mathbb{R}_+, \qquad d(x) = |x|^2,$$

and assume that this is a Morse function. Then using the classical Morse theory for the function $-d$ and the fact that every critical point of $-d$ has index $\geq n$, see [M3], §7, it follows that the manifold F^1 is obtained from $F^1 \setminus B_\varepsilon$ by attaching cells of dimension $\geq n$. Since $F^1 \setminus B_\varepsilon$ is clearly homotopy equivalent to the boundary $\partial \bar{F}$, it follows that the pair $(\bar{F}, \partial \bar{F})$ is $(n-1)$-connected, see [Sp], p. 402.

This ends the proof in this case.

If the function d is not a Morse function, we replace it by the function

$$d^1\colon F^1 \to \mathbb{R}_+, \qquad d(x) = \|x - x_0\|^2,$$

where x_0 is a generic point close to the origin and such that d^1 is a Morse function. The existence of such points x_0, arbitrarily close to 0, follows from [M3], §6. Moreover, the boundary $\partial\overline{F}$ can be identified with the set $\{x \in F^1;\ d^1(x) = \varepsilon^2\}$. Indeed, the intersection $S_\varepsilon^{2n+1} \cap F^1$ is transversal and hence it is stable under small movements of the sphere S_ε^{2n+1} (e.g., when we move its center from 0 to the nearby point x_0). The argument can be then completed exactly as in the case $x_0 = 0$.   □

(2.5) **Example.** Let $f = x_0 x_1, \ldots, x_n$. Then the Milnor fiber F is clearly homotopy equivalent to $(S^1)^n$. Using (1.15), it follows that the boundary $\partial\overline{F}$ is diffeomorphic to $S^{n-1} \times (S^1)^n$. Hence the inclusion

$$j\colon \partial\overline{F} \to \overline{F}$$

cannot be an n-equivalence, i.e., our result (2.4) is sharp.

(2.6) **Exercise.** Show that the morphism induced by the inclusion

$$j_*\colon H_i(\partial\overline{F}) \to H_i(\overline{F})$$

is an isomorphism for $i < n - 1$ and an epimorphism for $i = n - 1$ (without using (2.4). *Hint.* By the Lefschetz Duality Theorem (see [Sp], p. 298) we have

$$H_i(\overline{F}, \partial\overline{F}) \cong H^{2n-i}(\overline{F} \setminus \partial\overline{F}).$$

Then use the fact that $\overline{F} \setminus \partial\overline{F}$ is a Stein manifold.

As Example (2.5) shows, the homology groups $H_i(F)$ of the Milnor fiber associated to a hypersurface singularity can be nonzero in all the dimensions $0 \le i \le n$.

The following result describes the situation for an isolated singularity.

(2.7) **Theorem** (Milnor [M5]). *Assume that $(X, 0)$ is an IHS. Then the associated Milnor fiber F has the homotopy type of a bouquet of n-spheres. The number of spheres in this bouquet is given by the formula*

$$\mu(f, 0) = \mu(X, 0) = \dim_{\mathbb{C}} \frac{\mathcal{O}_{n+1}}{J_f},$$

where $J_f = (\partial f/\partial x_0, \ldots, \partial f/\partial x_n)$ is the Jacobian ideal of the singularity f.

This number is called the *Milnor number* of the function germ f (or of the singularity $(X, 0)$).

(2.8) **Example.** When f is a weighted homogeneous polynomial of type $(w_0, \ldots, w_n; N)$ having an isolated singularity at the origin, then

$$\mu(f, 0) = \prod_{i=0, n} \frac{N - w_i}{w_i}.$$

For a proof see, for instance, [D4], p. 112.

A special case of this situation is the case of a zero-dimensional IHS. Any such singularity is analytically equivalent to the simple singularity

$$A_\mu : x^{\mu+1} = 0 \qquad \text{for some} \quad \mu \geq 1.$$

The corresponding Milnor fiber F_μ consists of $(\mu + 1)$-points and hence can be regarded as a bouquet of μ spheres S^0. Since $\mu = \mu(x^{\mu+1}, 0)$, by a simple computation, this zero-dimensional case can be regarded as a special case of (2.7).

Note however that

$$H_0(F_\mu) \simeq \mathbb{Z}^{\mu+1},$$

while for the Milnor fiber F of an n-dimensional IHS f with $n > 0$ we have

$$H_0(F) \simeq \mathbb{Z}^\mu \qquad \text{where} \quad \mu = \mu(f).$$

To unify the treatment, it is usual to consider the *reduced* homology

$$\tilde{H}_0(F_\mu) \simeq \tilde{H}_n(F) \simeq \mathbb{Z}^\mu$$

of the Milnor fiber as a basic topological invariant (or, the reduced cohomology). However, since $H_n(F) \neq \tilde{H}_n(F)$, only for $n = 0$, some authors (including ourselves) sometimes omit the notation for the reduced homology.

The following result provides us with a subtle and efficient way of deciding whether a Milnor fiber is simply-connected.

(2.9) **Theorem** (Lê–Saito [LêS]). *Assume that there is a subgerm of analytic spaces $(A, 0) \subset (X, 0)$ such that*:

(i) $\text{codim}_X A \geq 2$;
(ii) *any singular point p in $X \backslash A$ is locally equivalent to the normal crossing of two hyperplanes (i.e., (X, p) has a local equation of type $u_0 u_1 = 0$, for a local coordinate system $(u_0, u_1, \ldots, u_n)$ on $\mathbb{C}^{n+1}$ at point p).*

Then

$$\pi_1(S_\varepsilon^{2n+1} \backslash K) \simeq \mathbb{Z}^c,$$

where c is the number of irreducible components of the singularity $(X, 0)$. In particular, when $(X, 0)$ is irreducible, the associated Milnor fiber F is simply-connected.

(Conditions (i) and (ii) are usually stated as: "X has only·normal crossing singularities in codimension 1.")

The local result (2.9) can sometimes be used to get global results, as the following example shows.

(2.10) **Example.** Identify $\mathbb{C}^n$ (resp. $\mathbb{C}^m$) with the space of monic polynomials

$$p_a(x) = x^n + a_1 x^{n-1} + \cdots + a_n, \qquad a_i \in \mathbb{C},$$

(resp. $q_b(x) = x^m + b_1 x^{m-1} + \cdots + b_m, b_j \in \mathbb{C}$). Let $R \subset \mathbb{C}^n \times \mathbb{C}^m$ be the resultant hypersurface, i.e.,

$$R = \{(a, b) \in \mathbb{C}^{n+m}; \text{ the equations } p_a(x) = 0, q_b(x) = 0 \text{ have a common root}\}.$$

Then R is the zero set of a weighted homogeneous polynomial (a well-known determinant in the coefficients a_i and b_j) and $(R, 0)$ satisfies the assumptions in (2.9). Using (1.17(i)) for the associated affine Milnor fibration we get

$$\pi_1(\mathbb{C}^{n+m} \setminus R) \simeq \mathbb{Z}.$$

For more details, see Choudary [Ch].

Finally, we state some connectivity results in the relative case. Consider an analytic germ $Y: g_1 = \cdots = g_s = 0$ at the origin of $\mathbb{C}^m$ and an analytic map germ $f = (f_1, \ldots, f_p): (Y, 0) \rightarrow (\mathbb{C}^p, 0)$ such that:

(i) $Y \setminus X$ is smooth, where $(X, 0) = (f^{-1}(0), 0) \subset (Y, 0)$;
(ii) 0 is not an isolated point in X.

For $\varepsilon > 0$ small enough we set

$$K_Y = Y \cap S_\varepsilon^{2m-1}, \qquad K_X = S_\varepsilon^{2m-1}.$$

(2.11) **Definition.** The singularity $(X, 0)$ is a *complete intersection relative to* $(Y, 0)$ if for any irreducible component Y_α of Y and any irreducible component X_β of $X \cap Y_\alpha$ we have

$$\dim Y_\alpha - \dim X_\beta = p.$$

Note that this definition is a natural generalization of Definition (1.6.1) to the case when the ambient space $(Y, 0)$ is itself singular.

(2.12) **Theorem** (Hamm [H1]). *Let n be the dimension of $(X, 0)$ and assume that $(X, 0)$ is a complete intersection relative to $(Y, 0)$.*

(i) *Then the pair (K_Y, K_X) is $(n-1)$-connected. In particular, when $(Y, 0)$ is a smooth germ, the link K_X is $(n-2)$-connected.*
(ii) *Assume that $p = 1$ and write $f = f_1$. Then the map*

$$\varphi: K_Y \setminus K_X \rightarrow S^1, \qquad \varphi(x) = f(x)/|f(x)|,$$

   is a smooth fibration.
(iii) *Assume that $p = 1$ and that $(Y, 0)$ and $(X, 0)$ are both ICIS. Then the fiber F of the fibration φ above has the homotopy type of a bouquet of n-spheres.*

(2.13) **Example.** Consider the germ

$$(X, 0) = (H_1, 0) \cup (H_2, 0) \subset (\mathbb{C}^4, 0),$$

which is the union of the planes

$$H_1: x_1 = x_2 = 0 \qquad \text{and} \qquad H_2: x_3 = x_4 = 0.$$

We have claimed in (1.6.2) that $(X, 0)$ is *not* a complete intersection.

To prove this, consider the link K_X which is a disjoint union of two spheres S^3. If $(X, 0)$ had been a complete intersection, then we infer that K_X would have been a connected space by (2.12). This contradiction proves our claim.

(2.14) **Exercise.** Consider the nondegenerate quadratic singularity

$$A_1: f = x_0^2 + \cdots + x_n^2, \qquad n \geq 1.$$

 (i) Show that the corresponding link K can be identified to the Stiefel manifold $V_2(\mathbb{R}^{n+1})$ of orthonormal 2-frames in $\mathbb{R}^{n+1}$ (for a definition of these Stiefel manifolds, we refer to [Hu], p. 13).
(ii) Show that the corresponding open Milnor fiber F can be identified to the total space of the tangent bundle of the sphere S^n.

§3. Vanishing Cycles and the Intersection Form

In this section we introduce the vanishing cycles and the intersection form on the homology of the Milnor fiber of an IHS. Special attention is devoted to the basic Thom–Sebastiani construction. Let $X: f = 0$ be an IHS at the origin of $\mathbb{C}^{n+1}$. Let $G: (X, 0) \to (S, 0)$ be a *miniversal deformation* for the singularity $(X, 0) = (G^{-1}(0), 0)$, see, for details, [Lg], Chap. 6.

Then S can be taken to be a small open ball at the origin in $\mathbb{C}^\tau$, where

$$(3.1) \qquad \tau = \tau(f, 0) = \tau(X, 0) = \dim_{\mathbb{C}} \frac{\mathcal{O}_{n+1}}{(f) + J_f}$$

is the *Tjurina number* of the singularity $(X, 0)$. The subgerm

$$(D, 0) = \{s \in (S, 0); G^{-1}(s) \text{ is singular}\}$$

is called the *discriminant* of the deformation G. It is known that $(D, 0)$ is an irreducible hypersurface singularity such that

$$(3.2) \qquad \text{mult}(D, 0) = \mu(X, 0),$$

see, for instance, [Lg], pp. 63–64.

Hence a generic complex line L in the base space $\mathbb{C}^\tau$ close to the origin meets the discriminant D in $\mu = \mu(X, 0)$ points $s_1, \ldots, s_\mu$ (all of them smooth on D). It is known that the fibers $X_i = G^{-1}(s_i)$ corresponding to these points have just one singularity and this is of type A_1, see [Lg], p. 64.

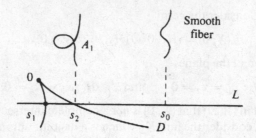

Let $\bar{D}_\delta$ be a closed disc contained in the intersection $L \cap S$ and such that s_1, ..., $s_\mu \in D_\delta$. Let $s_0 \in \partial\bar{D}_\delta$ be a fixed point. Consider the restrictions

$$\tilde{f} = G \,|\, G^{-1}(L \cap S),$$

$$\tilde{\psi} = G \,|\, G^{-1}(\partial\bar{D}_\delta).$$

Then $U = G^{-1}(L \cap S)$ can be regarded as an open neighborhood of the origin in $\mathbb{C}^{n+1}$, and the function $\tilde{f}: U \to L \simeq \mathbb{C}$ can be regarded as a small deformation of the function f. Moreover, $\tilde{f}$ has only singularities of type A_1 with distinct critical values (corresponding exactly to the points $s_1, \ldots, s_\mu$).

Such a deformation $\tilde{f}$ is called a "morsification" of the singularity f, since the complex A_1-singularities are the analog of the real nondegenerate singularities which appear in Morse Theory [M3]. The following result establishes a basic relation to the results discussed in the previous two sections, see [Sil], Chap. 6.

(3.3) **Lemma.** *The map $\tilde{\psi}$ is a fibration fiber diffeomorphically equivalent to the Milnor fibration ψ associated to f.*

In particular, we consider $F = \tilde{\psi}^{-1}(s_0)$ to be the Milnor fiber of the singularity f. Consider now a system of μ paths $\alpha_1, \ldots, \alpha_\mu$ starting from s_0 and ending at $s_1, \ldots, s_\mu$, respectively, and such that:

(i) each path α_i has no self-intersection points;
(ii) two distinct paths α_j and α_k meet only at their common origin $\alpha_j(0) = \alpha_k(0) = s_0$; and
(iii) the points s_i and the paths α_i are numbered in the order they start from the point s_0, counting clockwise (see the next illustration and [AGV2], p. 14).

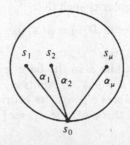

Let us now look carefully at the topology of the fibers

$$F_t = G^{-1}(\alpha_i(t)), \qquad t \in [0, 1],$$

along one of these paths. We can assume that $\alpha_i(t) = s_i$ only for $t = 1$, and then all the fibers F_t for $t \in [0, 1)$ are diffeomorphic to the Milnor fiber $F_0 = F$. The "limit" fiber F_1 has a singular point of type A_1, which appears by contracting a cycle Δ_i^t in F_t to a point $p \in F_1$. Moreover, a local investigation of the singularity A_1 shows that this cycle Δ_i^t is represented by an embedded sphere S^n in the fiber F_t (use 2.14(ii)).

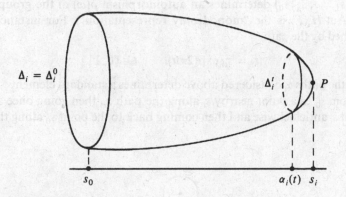

In this way we obtain a collection of cycles $\Delta_i = \Delta_i^0$ for $i = 1, \ldots, \mu$ in $H_n(F)$, the integral n-homology of the Milnor fiber F.

(3.4) **Definition.** The set $(\Delta_1, \ldots, \Delta_\mu)$ of cycles in $\tilde{H}_n(F)$, with the numbering obtained as described above, is called a *distinguished basis* of vanishing cycles for $\tilde{H}_n(F)$.

This definition is motivated by the following:

(3.5) **Proposition.** $\tilde{H}_n(F)$ *is a free abelian group and the cycles* $\Delta_1, \ldots, \Delta_\mu$ *form a basis of this group.*

The group $\tilde{H}_n(F)$ has an *intersection form* $\langle \ , \ \rangle$ which is defined as explained in (2.3.6), taking $M = \bar{F}$, the closed Milnor fiber, and identifying $\tilde{H}_n(F)$ with $\tilde{H}_n(\bar{F})$ (recall that the inclusion $F \subset \bar{F}$ is a homotopy equivalence).

(3.6) **Definition.** The pair $(\tilde{H}_n(F), \langle \ , \ \rangle)$ is called the *Milnor lattice* of the singularity $(X, 0)$. The matrix

$$I_\Delta = (\langle \Delta_i, \Delta_j \rangle)_{1 \le i, j \le \mu}$$

is called the *intersection matrix* of the singularity $(X, 0)$ with respect to the distinguished basis $\Delta = (\Delta_1, \ldots, \Delta_\mu)$.

For general facts concerning lattices we refer to Appendix A.

Note that the elements on the diagonal of the intersection matrix I_Δ are determined by the following general formula:

$$(3.7) \qquad \langle \Delta_i, \Delta_i \rangle = \begin{cases} 0 & \text{for } n \text{ odd,} \\ (-1)^{n(n-1)/2} 2 & \text{for } n \text{ even.} \end{cases}$$

We can determine the *integral* monodromy operator

$$T = h_* \colon \tilde{H}_n(F) \to \tilde{H}_n(F)$$

from the intersection matrix I_Δ as follows. Any element $\omega \in \pi = \pi_1(\bar{D}_\delta \setminus \{s_1, \ldots, s_\mu\} s_0)$ determines an automorphism $\rho(\omega)$ of the group $\tilde{H}_n(F)$ ($\rho \colon \pi \to \operatorname{Aut} \tilde{H}_n(F)$ is the "monodromy representation"). For instance, T is determined by the path

$$\gamma(t) = s_0 \exp(2\pi i t), \qquad t \in [0, 1].$$

Each of the paths α_i considered above determines a standard element $\gamma_i \in \pi$ by going from s_0 to a point nearby s_i along the path α_i, then going once around the point s_i anticlockwise and then coming back to the point s_0 along the path α_i.

(3.8)

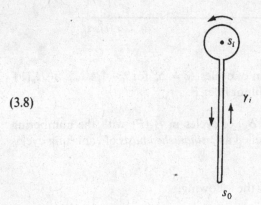

Moreover, it is clear that

$$\gamma = \gamma_\mu \cdots \gamma_1 \quad \text{in } \pi,$$

using the following:

(3.9) **Convention** (About the Composition of Paths). If α, β are two paths in a space M, then the "product" path $\alpha\beta$ is defined by going first along the path α and then along the path β. In particular, it is necessary that $\beta(0) = \alpha(1)$. Note that this convention is the same as in [Sp], p. 46.

Using the above formula for γ we get

$$(3.10) \qquad T = T_1 \ldots T_\mu,$$

where $T_i = \rho(\gamma_i)$ is called a *Picard–Lefschetz transformation* associated to the vanishing cycle Δ_i and is given by the following explicit formula (see, for instance, [Lg], p. 42, or [AGV2], p. 26)

$$(3.11) \qquad T_i(x) = x - (-1)^{n(n-1)/2} \langle x, \Delta_i \rangle \Delta_i.$$

Hence all the automorphisms T_i and the monodromy operator T can be computed from the intersection matrix I_Δ using (3.11) and (3.10).

Another way in which the intersection form I_Δ determines the monodromy operator T is via the *Seifert matrix*. In some sense, what follows now can be regarded as inspired by the results on Seifert matrices for knots in Chapter 2, §1. First, we extend Definition (2.17) of the linking number of two knots in S^3 to a more general situation.

(3.12) Definition. Let a and b be disjoint n-dimensional oriented cycles in the sphere S^{2n+1}. Let A be an $(n+1)$-dimensional oriented chain in S_ε^{2n+1} such that $\partial A = a$ and the intersection $A \cap b$ is transversal. Then the linking number $\mathrm{lk}(a, b)$ of the oriented cycles a and b is the integer in $\mathbb{Z} = H_0(S_\varepsilon^{2n+1})$ corresponding to the oriented 0-cycle $A \cap b$.

Let $\varphi: S_\varepsilon^{2n+1} \setminus K \to S^1$ be the Milnor fibration of the singularity $(X, 0)$. Let $h_t: F_0 \to F_t$ be a family of homeomorphisms, where $F_t = \varphi^{-1}(\exp(2\pi i t))$, $t \in [0, 1]$, such that h_0 is the identity on the Milnor fiber $F = F_0$.

(3.13) Definition. The *Seifert bilinear form*

$$S: \tilde{H}_n(F) \times \tilde{H}_n(F) \to \mathbb{Z}$$

is defined by the formula

$$S(a, b) = \mathrm{lk}(a, (h_{1/2})_* b).$$

The matrix $V_\Delta = (S(\Delta_i, \Delta_j))_{1 \le i, j \le \mu}$ is called the *Seifert matrix* with respect to the distinguished basis $\Delta = (\Delta_1, \ldots, \Delta_\mu)$.

(3.14) Proposition. *The Seifert matrix is unimodular, i.e.,*

$$\det V_\Delta = \pm 1.$$

Proof. Consider the following isomorphisms

$$\tilde{H}_n(F_{1/2}) \xrightarrow[\sim]{j_*} \tilde{H}_n(\bar{F}_{1/2}) \xleftarrow[\sim]{\partial} H_{n+1}(S_\varepsilon^{2n+1}, \bar{F}_{1/2})$$

$$\xrightarrow[\sim]{A} \tilde{H}^n(S_\varepsilon^{2n+1} \setminus \bar{F}_{1/2}) \xrightarrow[\sim]{i^*} \tilde{H}^n(F_0).$$

Here $j: F_{1/2} \to \bar{F}_{1/2}$ and $i: F_0 \to S_\varepsilon^{2n+1} \setminus \bar{F}_{1/2}$ are both inclusions and homotopy equivalences, while A comes from the Alexander Duality, see [Sp], p. 296. It is not difficult to check that under the composed isomorphism $\tilde{H}_n(F_{1/2}) \simeq \tilde{H}^n(F_0)$, the linking pairing

$$\tilde{H}_n(F_0) \times \tilde{H}_n(F_{1/2}) \xrightarrow{\mathrm{lk}} \mathbb{Z}$$

corresponds to the *evaluation pairing*

$$\tilde{H}_n(F_0) \times \tilde{H}^n(F_0) \overset{ev}{\to} \mathbb{Z}.$$

Since $\tilde{H}^n(F_0)$ can be identified to $\mathrm{Hom}(\tilde{H}_n(F_0), \mathbb{Z})$, it follows that ev is a unimodular pairing and this ends our proof.   □

The intersection form $\langle \ , \ \rangle$ and the Seifert form S on $\tilde{H}_n(F)$ are related by the following formula, see [AGV2], p. 41,

(3.15) $\langle a, b \rangle = -S(a, b) - (-1)^n S(b, a).$

This is a complete analog of the formula (2.1.17) from Knot Theory.

Moreover, we have the following basic result:

(3.16) **Theorem** ([AGV2], p. 47).

 (i) *The Seifert matrix V_Δ is upper triangular;*
 (ii) $I_\Delta = -V_\Delta - (-1)^n V_\Delta^\tau;$ *and*
 (iii) $T_\Delta = (-1)^{n+1} V_\Delta^{-1} V_\Delta^\tau;$

where the symbol M^τ denotes the transpose of a matrix M and T_Δ is the matrix of the monodromy operator T with respect to the distinguished basis $\Delta = (\Delta_1, \ldots, \Delta_\mu)$.

Note that (3.15) is equivalent to (3.16(ii)).

(3.17) **Corollary.** *For an algebraic knot $K \subset S^3_\varepsilon$ we have*:

 (i) *the Seifert matrix V_Δ is unimodular; and*
 (ii) $T_\Delta = V_\Delta^{-1} V_\Delta^\tau.$

This result says that (2.1.21) and (2.1.26) (resp. (2.1.23) and (2.1.28)) are equivalent for algebraic knots. To see this, we have to use only the following simple fact.

(3.18) **Exercise.** Let $\mathbb{Z}^m \overset{A}{\to} \mathbb{Z}^m \to G \to 0$ be a presentation for the abelian group G. Then $\mathbb{Z}^m \overset{A^\tau}{\to} \mathbb{Z}^m$ is again a presentation for G, i.e.,

$$\mathrm{coker}\, A^\tau \simeq \mathrm{coker}\, A = G.$$

Hint. Up to base changes in domain and codomain, the matrix A can be chosen to be diagonal and then $A^\tau = A$.

It is a common procedure in mathematics to build more complicated objects by summing simpler ones, e.g., any finite abelian group is a direct sum of finite cyclic groups. The construction which embodies this idea in the framework of the Singularity Theorem is the *Thom–Sebastiani construction*.

Let $f \colon (\mathbb{C}^{n+1}, 0) \to (\mathbb{C}, 0)$ and $g \colon (\mathbb{C}^{m+1}, 0) \to (\mathbb{C}, 0)$ be two singularities and consider their sum

$$f + g: (\mathbb{C}^{n+m+2}, 0) \to (\mathbb{C}, 0),$$

$$(f + g)(x, y) = f(x) + g(y) \qquad \text{for all} \quad x \in \mathbb{C}^{n+1}, \quad y \in \mathbb{C}^{m+1}.$$

Let F_f, F_g, F_{f+g} be the corresponding Milnor fibers and h_f, h_g, h_{f+g} the associated monodromy homeomomomorphisms. When f and g are weighted homogeneous polynomials, then $f + g$ is also a weighted homogeneous polynomial and we can take F_f, F_g, F_{f+g}, h_f, h_g, and h_{f+g} to be the corresponding global affine objects as well. With these notations, we have the following basic result, see [ST], [Sk], [O1].

(3.19) Theorem. *There is a homotopy equivalence*

$$j: F_f * F_g \to F_{f+g}$$

such that the diagram

$$
\begin{array}{ccc}
F_f * F_g & \xrightarrow{\ j\ } & F_{f+g} \\
{\scriptstyle h_f * h_g}\big\downarrow & & \big\downarrow{\scriptstyle h_{f+g}} \\
F_f * F_g & \xrightarrow{\ j\ } & F_{j+g}
\end{array}
$$

is commutative up to homotopy.

To understand fully this result and its consequences, we recall some definitions. For two topological spaces X and Y, we can define the space $X * Y$ (called the *join* of X and Y) starting with the product $X \times [0, 1] \times Y$ and making the following identifications:

(i) $(x, 0, y) \sim (x', 0, y)$ for all $x, x' \in X$; $y \in Y$; and
(ii) $(x, 1, y) \sim (x, 1, y')$ for all $x \in X$; $y, y' \in Y$.

Let $[x, t, y]$ denote the equivalence class in $X * Y$ of the triple $(x, t, y) \in X \times [0, 1] \times Y$. Informally, $X * Y$ is the disjoint union of all the segments $[x, y]$ joining a point $x \in X$ to a point $y \in Y$.

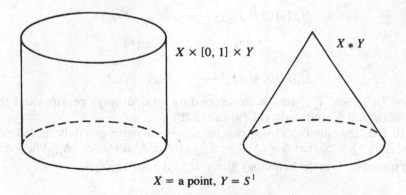

$$X \times [0, 1] \times Y \qquad\qquad X * Y$$

$$X = \text{a point}, \ Y = S^1$$

Use the surjection $X \times [0, 1] \times Y \to X * Y$ to put a topology on $X * Y$, the so-called weak topology.

For maps $a: X \to X$ and $b: Y \to Y$, we get a map $a * b: X * Y \to X * Y$ by setting

$$(a * b)([x, t, y]) = [a(x), t, b(y)].$$

(3.19') **Remark.** The homotopy equivalence j in (3.19) can be taken to be injective. Such a construction of the homotopy equivalence j in the case of an IHS is given in [AGV2], pp. 54–55. To give the reader an idea about this explicit construction of the map j, consider the case when both f and g are homogeneous polynomials of degree d. Let F_f, F_g, and F_{f+g} be the corresponding affine Milnor fibers and define

$$j: F_f * F_g \to F_{f+g}, \qquad j([x, t, y]) = (t^{1/d}x, (1-t)^{1/d}y).$$

Then it is clear that the map j is well-defined, injective, and satisfies the relation

$$h_{f+g} \circ j = j \circ (h_f * h_g).$$

To see that j is a homotopy equivalence, use the homotopy equivalence $\alpha: F_{f+g} \to F_f * F_g$ constructed by Oka in [O1]. The reader will have no difficulty in checking that $\alpha \circ j = 1$ and $j \circ \alpha \sim 1$. This will show that our map j is indeed a homotopy equivalence.

It turns out that the reduced homology groups of a join $X * Y$ can be expressed in terms of the reduced homology groups of the factors X and Y (integer coefficients are used throughout).

(3.20) **Lemma** (Milnor [M1]).

$$\tilde{H}_{r+1}(X * Y) \overset{\theta}{\underset{\sim}{\to}} \bigoplus_{i+j=r} \tilde{H}_i(X) \otimes \tilde{H}_j(Y) \oplus \bigoplus_{i+j=r-1} \mathrm{Tor}(\tilde{H}_i(X), \tilde{H}_j(Y)).$$

Moreover, this isomorphism θ is compatible with the homomorphisms induced by $a * b$, a, and b, respectively, at the homology level.

(3.21) **Corollary** (Sebastiani–Thom). *Assume that f and g are both IHS. Then $f + g$ is also an IHS and the following diagram is commutative:*

$$
\begin{array}{ccc}
\tilde{H}_n(F_f) \otimes \tilde{H}_m(F_g) & \overset{\bar{\theta}}{\underset{\sim}{\longrightarrow}} & \tilde{H}_{n+m+1}(F_{f+g}) \\
\downarrow{\scriptstyle T_f \otimes T_g} & & \downarrow{\scriptstyle T_{f+g}} \\
\tilde{H}_n(F_f) \otimes \tilde{H}_m(F_g) & \overset{\bar{\theta}}{\underset{\sim}{\longrightarrow}} & \tilde{H}_{nm+1}(F_{f+g}).
\end{array}
$$

Here T_f, T_g, and T_{f+g} are the corresponding monodromy operators and the isomorphism $\bar{\theta}$ is induced by (3.19) and (3.20).

In fact, the isomorphism $\bar{\theta}$ can be described more precisely as follows, see [AGV2], p. 55. Let $\Delta = (\Delta_1, \ldots, \Delta_{\mu(f)})$ (resp. $\Delta' = (\Delta'_1, \ldots, \Delta'_{\mu(g)})$ be a distinguished basis in the homology group $\tilde{H}_n(F_f)$ (resp. $\tilde{H}_m(F_g)$).

(3.22) **Theorem** (Gabrielov [AGV2], p. 59).

(i) *The cycles $\Delta_{ij} = \bar{\theta}(\Delta_i \otimes \Delta'_j)$ form a distinguished basis in $H_{n+m+1}(F_{f+g})$ when they are ordered lexicographically, i.e., the cycle $\Delta_{i_1 j_1}$ precedes the cycle $\Delta_{i_2 j_2}$ if $i_1 < i_2$ or $i_1 = i_2$ and $j_1 < j_2$.*

(ii) *The intersection numbers of the cycles Δ_{ij} are determined by the following formulas:*

$$\langle \Delta_{ij_1}, \Delta_{ij_2} \rangle = \text{sgn}(j_2 - j_1)^{n+1}(-1)^{(n+1)(m+1)+[n(n+1)/2]}\langle \Delta'_{j_1}, \Delta'_{j_2} \rangle$$

$$\text{for } \quad j_1 \neq j_2;$$

$$\langle \Delta_{i_1 j}, \Delta_{i_2 j} \rangle = \text{sgn}(i_2 - i_1)^{m+1}(-1)^{(n+1)(m+1)+[m(m+1)/2]}\langle \Delta_{i_1}, \Delta_{i_2} \rangle$$

$$\text{for } \quad i_1 \neq i_2;$$

$$\langle \Delta_{i_1 j_1}, \Delta_{i_2 j_2} \rangle = 0 \quad \text{for } \quad (i_2 - i_1)(j_2 - j_1) < 0;$$

$$\langle \Delta_{i_1 j_1}, \Delta_{i_2 j_2} \rangle = \text{sgn}(i_2 - i_1)(-1)^{(n+1)(m+1)}\langle \Delta_{i_1}, \Delta_{i_2} \rangle \langle \Delta'_{j_1}, \Delta'_{j_2} \rangle$$

$$\text{for } \quad (i_2 - i_1)(j_2 - j_1) > 0.$$

We consider now a special case of the Thom–Sebastiani construction. Let $f: (\mathbb{C}^{n+1}, 0) \to (\mathbb{C}, 0)$ be an IHS as above and consider the function germ

$$f_m: \mathbb{C}^{n+m+1}, 0) \to (\mathbb{C}, 0), \qquad f_m(x, y) = f(x) + y_1^2 + \cdots + y_m^2$$

$$\text{where} \quad y = (y_1, \ldots, y_m).$$

This new singularity f_m is called a *stabilization* of the singularity f, see [AGV2], p. 60. Using the formula for the Milnor number in (2.7) we get

$$\mu(f_m, 0) = \mu(f, 0),$$

and we denote for simplicity this number by μ.

By the process described in (3.22), to each distinguished basis $\Delta = (\Delta_1, \ldots, \Delta_\mu)$ for $\tilde{H}_n(F_f)$ corresponds a distinguished basis $\bar{\Delta} = (\bar{\Delta}_1, \ldots, \bar{\Delta}_\mu)$ for $\tilde{H}_{n+m}(F_{f_m})$ such that

(3.22') $\quad \langle \bar{\Delta}_i, \bar{\Delta}_j \rangle = \text{sgn}(j - i)^m(-1)^{(n+1)m+[m(m-1)/2]}\langle \Delta_i, \Delta_j \rangle \qquad \text{for } \quad i \neq j.$

From this formula it is clear that the sequence of the corresponding Milnor lattices $(\tilde{H}_{n+m}(F_{f_m}), \langle \ , \ \rangle)$ for $m \in \mathbb{N}$ has period 4. In other words, if we identify a singularity f with its various stabilizations f_m for $m \in \mathbb{N}$, each singularity has four distinct associated Milnor lattices. However, the lattices for m and $m + 2$ differ only by a sign, so that essentially we get

(3.23)
$$\begin{cases} \text{(i) a symmetric Milnor lattice} \\ \quad L_f^s = (\tilde{H}_{n+m}(F_{f_m}), \langle \ , \ \rangle) \qquad \text{for } \quad n + m \equiv 2 \pmod 4. \\ \text{(ii) a skew-symmetric Milnor lattice} \\ \quad L_f^{ss} = (\tilde{H}_{n+m}(F_{f_m}), \langle \ , \ \rangle) \qquad \text{for } \quad n + m \equiv 1 \pmod 4. \end{cases}$$

Note that in the symmetric lattice L_f^s we have $\langle \Delta, \Delta \rangle = -2$ for any vanishing cycle Δ.

It is usual to describe the intersection form in L_f^s pictorially by a graph, called a *Dynkin diagram* for the singularity f, as follows:

(i) the vertices of this graph are in one-to-one correspondence with a distinguished basis $(\Delta_1, \ldots, \Delta_\mu)$ in $\tilde{H}_{n+m}(F_{f_m})$ for $n + m \equiv 2 \pmod 4$;
(ii) the vertices corresponding to Δ_i and Δ_j are joined by an edge with multiplicity $\langle \Delta_i, \Delta_j \rangle$.

Several examples of Dynkin diagrams are given in Appendix A.

(3.24) Remark. There is a close relation between another special type of the Thom–Sebastiani construction and *cyclic coverings of a sphere* S_ε^{2n+1} branched along a singularity link $K \subset S_\varepsilon^{2n+1}$. Recall that such coverings (for $n = 1$) were considered in our review of classical Knot Theory in Chapter 2. Indeed, let $X: f = 0$ be an IHS at the origin of $\mathbb{C}^{n+1}$ and let $K = X \cap S_\varepsilon^{2n+3}$ be the corresponding link. Consider the new IHS $X_p: f(x) + t^p = 0$ at the origin of $\mathbb{C}^{n+2}$ for some integer $p \geq 2$. Let $K_p = X_p \cap S^{2n+3}$ be the associated link. Then it is not difficult to show that K_p is diffeomorphic to the cyclic covering of order p of the sphere S_ε^{2n+1} ramified along the link K, see, for details, [DK], [M6], p. 177, and [St].

We end this section with a property of Milnor lattices in relation to miniversal deformations. Let $(X, 0)$ be an IHS and let $G: (X, 0) \to (S, 0)$ be a miniversal deformation for $(X, 0)$, as considered at the beginning of this section. For a point $s \in D$ in the discriminant D of this deformation, the fiber $X_s = G^{-1}(s)$ has some isolated singularities, say $a_1, \ldots, a_k$.

Let L_i be the Milnor lattice corresponding to the singularity (X_s, a_i) for $i = 1, \ldots, k$, and let L be the Milnor lattice of the singularity $(X, 0)$. For a proof of the following result we refer to [Lg], p. 121.

(3.25) Proposition. *There is a primitive embedding of lattices*

$$\psi: L_1 \oplus \cdots \oplus L_k \to L$$

such that $\operatorname{coker} \psi$ *is isomorphic as a group to* $H_n(X_s)$.

(3.26) Remark. Most of the results stated in this section for IHS hold in the more general case of an ICIS $(X, 0)$. Let $f_1 = \cdots = f_c = 0$ be local equations for $(X, 0)$ at the origin of $(\mathbb{C}^{n+c}, 0)$ where $n = \dim X \geq 1$. Let $f_1^t, \ldots, f_c^t$ be small deformations of the equations $f_1, \ldots, f_c$, such that the Stein variety

(i)
$$F = \{x \in B_\varepsilon; f_1^t(x) = \cdots = f_c^t(x) = 0\}$$

is smooth where $B_\varepsilon = \{x \in \mathbb{C}^{n+c}; |x| < \varepsilon\}$. Then F can be identified with the Milnor fiber of the singularity $(X, 0)$ and the corresponding lattice $(H_n(F), \langle \ , \ \rangle)$ is called the Milnor lattice of the singularity $(X, 0)$.

If $s = \text{mult}(D, 0)$ is the multiplicity of the discriminant hypersurface of the miniversal deformation of $(X, 0)$, then in analogy to (3.5) we have the following:

(ii) The free $\mathbb{Z}$-module $H_n(F)$ has a system of generators formed by vanishing cycles Δ_i satisfying (3.7). In particular, $s \geq \text{rk } H_n(F)$ and, unlike the hypersurface case, this inequality may be strict.

Another fact which we need is that:

(iii) Proposition (3.25) holds unchanged in this more general situation.

On the negative side, note that there is no good notion of stabilization in the ICIS case as in (3.23). We refer for more details to the books by Looijenga [Lg] and Ebeling [E3].

§4. Homology Spheres, Exotic Spheres, and the Casson Invariant

In this section we investigate which links of IHS are $\mathbb{Q}$- or $\mathbb{Z}$-homology spheres. When such a link is homeomorphic to a sphere, it can give rise to an exotic sphere, i.e., produce a nonstandard differentiable structure on the sphere. We briefly survey basic facts on these exotic spheres and mention two other related applications of singularities to differential topology.

Let $X: f = 0$ be an IHS at the origin of $\mathbb{C}^{n+1}$ and let K be its link ($n \geq 2$). Then K is a compact oriented $(2n - 1)$-manifold which is $(n - 2)$-connected by (2.1). Hence the only "interesting" integer (co)homology for K is in dimensions $(n - 1)$ and n. Recall the isomorphism

$$(4.1) \qquad \tilde{H}^{j-1}(K) \overset{\partial}{\underset{\sim}{\to}} H^j(S, K) \overset{A}{\underset{\sim}{\to}} H_{2n+1-j}(S\backslash K)$$

for $j < 2n + 1$, where A comes from the Alexander duality and $S = S_\varepsilon^{2n+1}$ to simplify the notation. The Wang sequence (1.18) contains the sequence

$$(4.2) \qquad 0 \to H_{n+1}(S\backslash K) \to H_n(F) \overset{T-I}{\longrightarrow} H_n(F) \to H_n(S\backslash K) \to 0.$$

Since $H_n(F)$ is a free $\mathbb{Z}$-module of rank μ, we can regard the morphism $T - I$ as given by a $\mu \times \mu$ matrix with entries in $\mathbb{Z}$ and denoted again by $T - I$. Then (4.2) can be restated as

(4.2') **Corollary.** *The matrix* $T - I$ *is a presentation matrix for the* $\mathbb{Z}$-*module* $H_n(S\backslash K) \simeq H^n(K)$.

Since we have assumed $n \geq 2$, the Milnor fiber F of our singularity $(X, 0)$ is simply-connected. It follows that the inclusion $F \to S\backslash K$ is, *up to homotopy equivalence*, the infinite cyclic (and universal) covering of the link complement $S\backslash K$. With this identification, the monodromy homeomorphism $h: F \to F$ corresponds precisely to a generator of the group of covering transformations.

(4.3) Exercise. Is there a contradiction between the fact that the covering transformation group is infinite cyclic and the fact that the monodromy homeomorphism h can have finite order as seen in (1.19)?

Now consider $H_n(F)$ as an A-module where $A = \mathbb{Z}[t, t^{-1}]$ is the Laurent polynomial ring from (2.1.12). The multiplication is defined by $t \cdot x = T(x)$ with $T = h_* =$ the monodromy operator. It is easy to see that there is an exact sequence of A-modules

$$A^\mu \xrightarrow{\ tI-T\ } A^\mu \xrightarrow{\ \rho\ } H_n(F) \to 0,$$

where $A^\mu = \mathbb{Z}^\mu \otimes A = H_n(F) \otimes A$ and T corresponds to $T \otimes 1_A$ (choose $\rho: A^\mu \to \mathbb{Z}^\mu$ to be the map which associates to a μ-tuple of Laurent polynomials their μ-tuple of constant terms, i.e, the coefficients of t^0).

(4.4) Corollary. *The matrix $tI - T$ is an Alexander matrix for the A-module $H_n(F)$.*

In particular, the characteristic polynomial of the monodromy operator

$$\Delta(t) = \det(tI - T)$$

is the Alexander polynomial of the A-module $H_n(F)$.

Proof. Just recall Definition (2.1.19). □

In fact (4.4) is the analog of the result (2.1.26) for fibered knots and was included in this book only to show that the topology of the higher dimensional links has many things in common with the topology of (fibered) knots.

However, from now on in this section we concentrate on some interesting phenomena which occur only for higher dimensional links.

(4.5) Definition. Let X, Y be two topological spaces and let R be a ring:

(i) We say that the spaces X and Y have the same *R-homology type* if there is an R-module isomorphism $H_*(X; R) \simeq H_*(Y; R)$.

(ii) We say that the spaces X and Y have the same *R-cohomology type* if there is an isomorphism $H^*(X; R) \simeq H^*(Y; R)$ of R-algebras, the multiplication being given by the cup product.

Usually, in this definition, $R = \mathbb{Z}, \mathbb{C}$ and the isomorphisms are *not* induced by a map $h: X \to Y$. When a space X has the same R-homology (resp. R-cohomology)type as a sphere S^n, then we say that X is an *R-homology* (resp. *R-cohomology*) *n-sphere*. When $R = \mathbb{Z}$, the ring $\mathbb{Z}$ is usually omitted.

(4.6) Exercise (Compare with [BD]). Show that for any ring R the following statements are equivalent:

(i) X is an R-homology n-sphere;
(ii) X is an R-cohomology n-sphere.

Hence we need only speak about R-homology spheres.

(4.7) **Proposition.** (A) *The following statements are equivalent*:

(i) K *is a* $\mathbb{Q}$-*homology sphere*;
(ii) $\Delta(1) \neq 0$, *i.e.,* 1 *is not an eigenvalue of the monodromy operator*;
(iii) *the Milnor lattice* $(H_n(F), \langle\ ,\ \rangle)$ *is nondegenerate*.

(B) *The following statements are equivalent*:

(i) K *is a* $\mathbb{Z}$-*homology sphere*;
(ii) $\Delta(1) = \pm 1$;
(iii) *the Milnor lattice* $(H_n(F), \langle\ ,\ \rangle)$ *is unimodular*.

Moreover, if these last three statements hold and $n \geq 3$, then K is homeomorphic to the sphere S_ε^{2n-1}, and the embedding $K \subset S_\varepsilon^{2n+1}$ is not equivalent to the trivial equatorial embedding $S_\varepsilon^{2n-1} \subset S_\varepsilon^{2n+1}$ except for the smooth case $df(0) \neq 0$.

Proof. The equivalence (i) $\Leftrightarrow$ (ii) in both parts (A) and (B) follows from (4.1), (4.2), and (4.6). The equivalence (ii) $\Leftrightarrow$ (iii) in both parts (A) and (B) follows from (3.16).

Indeed, using (3.16) and its notation we get

$$1 - T_\Delta = -V_\Delta^{-1} \cdot I_\Delta,$$

where 1 is the identity matrix. Hence the matrix $1 - T_\Delta$ (corresponding to the operator $1 - T$) differs from the intersection matrix I_Δ by the unimodular matrix $(-V_\Delta^{-1})$, see (3.14). Alternatively, we can use the general remark (2.3.6′).

Assume now that K is a $\mathbb{Z}$-homology sphere and $n \geq 3$. Then K is simply-connected by (2.1) and has dimension ≥ 5. Hence we can apply the generalized Poincaré conjecture, which is true in these dimensions by the work of Smale and Stallings, as quoted in [M5], p. 65.

The trivial equatorial embedding $S_\varepsilon^{2n+1} \subset S_\varepsilon^{2n+1}$ can be regarded as the link of the smooth "singularity" $f(x) = x_0$. This "singularity" has obviously a contractible Milnor fiber, and hence by applying (1.17) we get

$$\pi_i(S_\varepsilon^{2n+1} \setminus S_\varepsilon^{2n-1}) = 0 \qquad \text{for} \quad i \geq 2.$$

Alternatively, we can get this result using the following.

(4.8) **Exercise** (Compare to (2.1.2(i))). The space $S_\varepsilon^{2n+1} \setminus S_\varepsilon^{2n-1}$ has the homotopy type of a circle S^1.

Coming back to our link K and using (1.17) we get

$$\pi_i(S_\varepsilon^{2n+1} \setminus K) = \pi_i(F) \qquad \text{for } i \geq 2.$$

Since F is simply-connected, we cannot have $\pi_i(F) = 0$ for all $i \geq 2$, since then F would have the homotopy type of a point. Indeed, recall that F has the homotopy type of a CW-complex. The inclusion of any point $x \hookrightarrow F$ is a weak homotopy equivalence if $\pi_i(F) = 0$ for all $i \geq 0$. Any weak homotopy equivalence between CW-complexes is a homotopy equivalence, see [Sp], p. 405. But according to (1.21) this happens only when f is nonsingular at the origin.

(4.9) **Remark.** According to Mumford's Theorem (2.3.12) the link of an IHS in dimension $n = 2$ cannot be homeomorophic to the sphere S^3, except for the smooth germ.

On the other hand, the following discussion shows that there are plenty of $\mathbb{Z}$-homology spheres in any dimension $n \geq 2$. Consider a Brieskorn–Pham singularity

$$X(\mathbf{a}): f = x_0^{a_0} + \cdots + x_n^{a_n} = 0,$$

where $n \geq 2$, $a_i \geq 2$, are some integers and $\mathbf{a} = (a_0, \ldots, a_n)$.

Let $K(\mathbf{a})$ be the corresponding link. In order to give necessary and sufficient conditions for this link to be an R-homology sphere ($R = \mathbb{Z}, \mathbb{Q}$), we associate to the set of integers $\mathbf{a}$ a graph $G(\mathbf{a})$ as follows. The graph $G(\mathbf{a})$ has $(n + 1)$ vertices denoted by the letters $a_0, \ldots, a_n$, and two vertices a_i and a_j are connected by an edge if and only if $(a_i, a_j) > 1$. Note that in the case where some of the integers a_i are even, all of them belong to the same connected component of the graph $G(\mathbf{a})$ which we denote by C_{ev}. Note that C_{ev} may contain some odd vertices a_j as well.

Consider the following condition on C_{ev}:

Condition $(\mathscr{C})$. The component C_{ev} contains an odd number of vertices and $(a_i, a_j) = 2$ for any two distinct points a_i, a_j in C_{ev}.

(4.10) **Theorem** (Brieskorn [B1]). (A) *The following statements are equivalent:*

(i) *the link $K(\mathbf{a})$ is a $\mathbb{Q}$-homology sphere;*
(ii) *the graph $G(\mathbf{a})$ contains at least one isolated point or the component C_{ev} satisfies the condition $(\mathscr{C})$.*

(B) *The following statements are equivalent:*

(i) *the link $K(\mathbf{a})$ is a $\mathbb{Z}$-homology sphere;*
(ii) *the graph $G(\mathbf{a})$ contains at least two isolated points or an isolated point which is odd and the component C_{ev} satisfies the condition $(\mathscr{C})$.*

Proof. We prove here only part (A) and refer the reader to the paper [B1] for the more involved proof of part (B). To apply (4.7), we have to compute the monodromy operator T of the singularity $X(\mathbf{a})$. By the Sebastiani–Thom result (3.21) we have

(4.11) $$T = T_{a_0} \otimes \cdots \otimes T_{a_n},$$

where $T_a: \tilde{H}_0(F_a) \to \tilde{H}_0(F_a)$ is the monodromy operator of the zero-dimensional singularity $x^a = 0$. Here F_a is the associated affine Milnor fiber, i.e.,

$$F_a = \{\lambda \in \mathbb{C}; \ \lambda^a = 1\}.$$

A basis for the group $\tilde{H}_0(F_a)$ is given by the elements

$$\Delta_j = t_a^{j-1} - t_a^j \quad \text{for} \quad j = 1, \ldots, a - 1,$$

where $t_a = \exp(2\pi i/a)$.

Since the monodromy homeomorphism $h: F_a \to F_a$ is given by multiplication with t_a as in (1.19), the matrix of the monodromy operator $T_a = h_*$, with respect to the basis $\Delta_1, \ldots, \Delta_{a-1}$ above, is the following

(4.12) $$T_a = \begin{pmatrix} 0 & 0 & 0 & -1 \\ 1 & 0 & 0 & -1 \\ 0 & 1 & 0 & -1 \\ \vdots & & \ddots & \\ 0 & 0 & 1 & -1 \end{pmatrix}$$

It follows that

$$\Delta_a(t) = \det(tI - T_a) = t^{a-1} + \cdots + 1.$$

Hence the eigenvalues of the monodromy operator T_a are exactly the a-roots of unity different from 1, namely, $t_a^j, j = 1, \ldots, a - 1$.

Using (4.11) it follows that the eigenvalues of the monodromy operator T are exactly the products

$$t_{a_0}^{j_0} \cdots t_{a_n}^{j_n} \quad \text{for} \quad 1 \le j_k \le a_k - 1.$$

Using (4.7(A)), we see that $K(\mathbf{a})$ is a $\mathbb{Q}$-homology sphere if and only if the equation

(E) $$\frac{j_0}{a_0} + \cdots + \frac{j_n}{a_n} = m$$

has no solutions with $m \in \mathbb{Z}, 1 \le j_k \le a_k - 1$. Let $C_1, \ldots, C_r$ be the connected components of the graph $G(\mathbf{a})$.

It is easy to see that any solution for the equation (E) above consists in fact, of r solutions, one for each of the following equations

$$(\text{E}_k) \qquad \sum_{a_s \in C_k} \frac{j_s}{a_s} = m_s,$$

where $m_s \in \mathbb{Z}$, $1 \leq j_s \leq a_s - 1$, $k = 1, \ldots, r$. $\qquad\qquad\qquad\qquad\square$

Hence the equation (E) has no solutions if and only if at least one of these "partial" equations (E_k) has no solutions. To end the proof of part (A) we have only to apply the following.

(4.13) Lemma. *The equation* (E_k) *has no solutions exactly in the following two cases:*

(i) *the component* C_k *is an isolated point; or*
(ii) *the component* C_k *is the component* C_{ev} *and it satisfies the condition* $(\mathscr{C})$.

Proof. The proof is by induction on $c = |C_k|$, the number of points in the connected graph C_k.

The case $c = 1$, i.e., C_k is an isolated point, is clear.
The case $c = 2$, i.e., $C_k = \{a_\alpha, a_\beta\}$ is again obvious: just take

$$j_\alpha = a_\alpha/g, \qquad j_\beta = a_\beta - a_\beta/g \qquad \text{where} \quad g = (a_\alpha, a_\beta).$$

Assume now that the statement (4.13) is true for all connected graphs with at most $c - 1$ points.

Case 1. $(a_\alpha, a_\beta) = 2$ for all $a_\alpha, a_\beta \in C_k$.

Then it is easy to check that the equation (E_k) has no solution exactly when this component C_k (which is the even component C_{ev}) satisfies the condition $(\mathscr{C})$.

Case 2. $(a_\alpha, a_\beta) = g > 2$ for some $a_\alpha, a_\beta \in C_k$.

Take out the point a_α and consider the connected components of the resulting graph $C_k \backslash \{a_\alpha\} = \overline{C}_k$.

There are two possibilities.

Case 2a. There is an isolated point a_γ in $\overline{C}_k$ such that $(a_\alpha, a_\gamma) = g > 2$.

Then the graph $C_k \backslash \{a_\gamma\}$ is connected and we can use the induction hypothesis. Hence there are two possiblities.

(i) The equation corresponding to the graph $C_k \backslash \{a_\gamma\}$ has a solution (j_s), $s \neq \gamma$. Then take

$$j'_\gamma = a_\gamma(1 - \varepsilon/g), \qquad j'_\alpha = \varepsilon a_\alpha/g + j_\alpha,$$

and $j'_\beta = j_\beta$ for $\beta \neq \alpha, \gamma, \varepsilon = 1, 2$.

For at least one value of ε, the number $j'_\alpha a_\alpha^{-1}$ is not an integer and we get in this way a solution for equation (E_k).

(ii) The equation corresponding to the graph $C_k \backslash \{a_\gamma\}$ has no solution. Then $C_k \backslash \{a_\gamma\}$ should be the even component and satisfy the condition $(\mathscr{C})$.

In this case we can take

$$j_\alpha = a_\alpha/g, \qquad j_\gamma = a_\gamma - a_\gamma/g, \qquad j_\beta = a_\beta/2,$$

for $\beta \neq \alpha, \gamma$. This is clearly a solution for equation (E_k).

Case 2b. If all the equations associated with the connected components of the graph $\overline{C}_k$ have solutions, then we can get a solution for equation (E_k) as in case (i) above. In view of the above discussion, only one of the following two cases can cause problems:

(iii) the graph $\overline{C}_k$ has an isolated point a_γ such that $(a_\alpha, a_\gamma) = 2$; or
(iv) the even component $(\overline{C}_k)_{ev}$ satisfies the condition $(\mathscr{C})$.

In this latter case, we let a_γ denote a point in $(\overline{C}_k)_{ev}$ such that

$$g = (a_\alpha, a_\gamma) \geq 2.$$

A solution for equation (E_k) is obtained by taking

$$j_\alpha = a_\alpha/2, \qquad j_\gamma = a_\gamma/2 \qquad \text{in case (iii)},$$

or

$$j_\alpha = a_\alpha/g, \qquad j_\gamma = a_\gamma - a_\gamma/g, \qquad j_\beta = a_\beta/2,$$

for all the other points $a_\beta \in (\overline{C}_k)_{ev}$ in case (iv).
This ends the proof of (4.13). $\square$

(4.14) Example. Consider the singularity

$$f_i = x_0^{6i-1} + x_1^3 + x_2^2 + \cdots + x_n^2,$$

where $i \geq 1$ and $n = 2m$ is an even number ≥ 4.
Then the associated link K_i is homeomorphic to the $(4m - 1)$-sphere.

We discuss now some basic facts about exotic spheres and their relations with singularities. We refer to Hirzebruch [Hz2] and Kervaire–Milnor [KeMi] for more information. A k-dimensional compact oriented differentiable manifold K is called a k-*sphere* if it is homeomorphic to the standard k-sphere S^k. Two k-spheres K_1 and K_2 are *equivalent* if there is an orientation-preserving diffeomorphism $h: K_1 \rightarrow K_2$. A k-sphere not equivalent to the standard sphere S^k is called an *exotic sphere*.

The equivalence classes of k-spheres form a finite abelian *group* θ_k for $k \geq 5$, the operation being the connected sum. This group θ_k contains the *subgroup* bP_{k+1} of these k-spheres which bound a parallelizable manifold.

The group bP_{4m} for $m \geq 2$ is known to be cyclic of order

$$|bP_{4m}| = 2^{2m-2}(2^{2m-1} - 1) \text{ numerator of } \left(\frac{4B_m}{m}\right),$$

where B_m is the mth Bernoulli number. As an example, we have

$$|bP_8| = 28 \qquad \text{and} \qquad |bP_{12}| = 992.$$

(4.15) **Proposition.** *If a $(4m - 1)$-sphere K bounds a parallelizable manifold F, then up to a suitable choice of the generator g_m for the cyclic group bP_{4m} we have*

$$K = \frac{\tau(F)}{8} g_m,$$

where $\tau(F)$ is the index of the intersection form of F.

Coming back now to singularities, note that the link K of an IHS $(X, 0)$ bounds the parallelizable manifold F where F is the corresponding Milnor fiber (1.23). Note also that when the singularity $(X, 0)$ is weighted homogeneous (in particular, when $X = X(\mathbf{a})$ is a Brieskorn–Pham singularity), there are effective methods to compute the index $\tau(F)$ which is exactly the index of the corresponding Milnor lattice, see Appendix C.

To be more specific, consider the singularity f_i from Example (4.14), its link K_i, and the corresponding Milnor fiber F_i. Then using (C26) in Appendix C it follows that

(4.16) $$\tau(F_i) = (-1)^m 8i.$$

Hence by taking the links K_i for $i = 1, \ldots, |bP_{4m}|$ we get all the classes in the group bP_{4m}. It is a striking fact that we can write simple explicit equations for all the exotic spheres in bP_{4m}!

Another spectacular application of singularities to differential topology is the construction of codimension 1 smooth foliations on every odd-dimensional sphere. We refer the interested reader to Lawson [Lw] and Durfee [Df1] for details. Here we just mention that the construction is based on the surface singularity $T_{7,3,2}$ and on the Thom–Sebastiani Theorem (3.19).

There is still another, more recent, development in topology which leads to a formula surprisingly similar to the formulas in (4.15). To each oriented $\mathbb{Z}$-homology 3-sphere M we can associate an integer $\lambda(M)$ called the Casson invariant of M, satisfying the following properties, see [AM].

(4.17) **Proposition/Definition.**

   (i) $\lambda(-M) = -\lambda(M)$ *where* $-M$ *denotes the manifold M with reversed orientation;*
   (ii) *if $\lambda(M) \neq 0$, then the fundamental group $\pi_1(M)$ has a nontrivial representation in* $\mathrm{SU}(2, \mathbb{C})$;
   (iii) $\lambda(M_1 \# M_2) = \lambda(M_1) + \lambda(M_2)$ *where $\#$ denotes connected sum.*

This invariant $\lambda(M)$ is relevant to the study of the knots $K \subset M$ and of the 3-manifolds obtained by Dehn surgery on K and, on the other hand, it counts the nontrivial conjugacy classes of representations of the fundamental group $\pi_1(M)$ in $SU(2, \mathbb{C})$.

We have the following explicit computation for the Casson invariant of homology 3-spheres which are links of Brieskorn–Pham singularities.

(4.18) **Proposition** (Neumann–Wahl [NW]). *Assume that the link* $K = K(a_0, a_1, a_2)$ *is a* $\mathbb{Z}$-*homology sphere. Then*

$$\lambda(K) = \frac{\tau(F)}{8},$$

where $\tau(F)$ *is the index of the associated Milnor fiber* F.

As a concrete example, consider the links K_i from Example (4.14) where this time we take $n = 2$. Using (4.16) we get

$$\lambda(K_i) = -i.$$

In particular, any integer is the Casson invariant of some oriented homology 3-sphere.

(4.19) **Exercise.** Show that the monodromy operator T of the cusp $A_2: x^3 + y^2 = 0$ is given by

$$T = \begin{pmatrix} 0 & 1 \\ -1 & 1 \end{pmatrix}$$

Hint. Use (4.11) and (4.12).

(4.20) **Open Problem.** Compute the integral monodromy operator associated with a weighted homogeneous IHS in terms of weight and degree.

Note that the *complex* monodromy operator of such a singularity is well known by [MO] or [B3] as well as the integral monodromy operator for *several special cases*, e.g., the Brieskorn–Pham singularities considered above or the singularities in [OR].

(4.21) **Open Probem.** List the topological invariants of an IHS X which can be computed from the characteristic polynomial $\Delta(t)$ of its monodromy operator. Special cases:

(i) If the corresponding Milnor lattice L_X is nondegenerate, then

$$\det L_X = |\Delta(1)|$$

see (A3 (ii)). Does there exist a similar formula for the reduced Milnor lattice $\bar{L}_X$ holding in general?

(ii) If X is a plane curve singularity consisting of two branches X_1 and X_2, then

$$(X_1, X_2)_0 = |\Delta'(1)|,$$

see Akbulut–McCarthy [AM], p. 144. (Recall also the various definitions of the intersection number $(X_1, X_2)_0$ in (2.2.12).) Does this formula have a higher dimensional analog?

CHAPTER 4

Fundamental Groups of Hypersurface Complements

§1. Some General Results

We have seen in Chapter 2 that a basic idea in studying a link $L \subset S^3$ is to investigate the topology of its complement $S^3 \setminus L$. In particular, the fundamental group $\pi_1(S^3 \setminus L)$ of this space played a crucial role. Note that since most of the spaces of interest to us are path-connected, we usually pay no attention to base points.

Before beginning our considerations, we explain why hypersurfaces are the only relevant subvarieties in the projective space $\mathbb{P}^n$ for this problem. Let $W \subset \mathbb{P}^n$ be a quasi-projective subvariety (i.e., W is not necessarily closed) and let $\overline{W}$ be its closure. Decompose W and $\overline{W}$ into disjoint unions of smooth algebraic sets (e.g., using (1.1.12)) such that $W_1, \ldots, W_s$ are the top dimensional strata for both W and $\overline{W}$.

(1.1) **Proposition.**

(i) $\pi_1(\mathbb{P}^n \setminus W) = 0$ if $\dim W < n - 1$;

(ii) $\pi_1(\mathbb{P}^n \setminus W) = \pi_1(\mathbb{P}^n \setminus \overline{W}) = \pi_1(\mathbb{P}^n \setminus (W_1 \cup \cdots \cup W_s))$ if $\dim W = n - 1$.

Proof. Any element in $\pi_1(\mathbb{P}^n \setminus W)$ is represented by a smooth map $\alpha: S^1 \to \mathbb{P}^n \setminus W$, and any relation among the elements in $\pi_1(\mathbb{P}^n \setminus W)$ is represented by a smooth map $\beta: D^2 \to \mathbb{P}^n \setminus W$ of the 2-disc D^2 into $\mathbb{P}^n$. This follows from the fact that any continuous map between smooth manifolds can be approximated, without changing the homotopy class, by a smooth map. See, for instance, [BT], p. 213, or [Hr], p. 124.

Moreover we need work only with maps $S^1 \to \mathbb{P}^n$ and $D^2 \to \mathbb{P}^n$ which are transversal to all the strata in the decompositions of W and $\overline{W}$ constructed above. To see this, use the transversality theorem, as presented, for instance, in [Hr], p. 74. Hence all the strata of complex codimension strictly greater than one play no role. □

(1.2) **Remark.** The reader is warned not to apply a similar argument when the ambient space is singular. Consider, for instance, the cone

$$C = \{x \in \mathbb{C}^3 : x_0^3 + x_1^3 + x_2^3 = 0\}$$

which is contractible to its vertex. The pointed cone $C \backslash \{0\}$ is homotopy equivalent to the link $K = C \cap S_\epsilon^5$ and hence there is an epimorphism

$$\pi_1(C \backslash \{0\}) = \pi_1(K) \to H_1(K) = \mathbb{Z}^2 \oplus (\mathbb{Z}/3\mathbb{Z})$$

by (2.4.9). Hence the fundamental group $\pi_1(C) = 0$ was drastically changed by removing the point 0, which is a subset of codimension 2 in C.

Proposition (1.1) tells us that it is enought to study fundamental groups of the form $\pi_1(\mathbb{P}^n \backslash \overline{W})$ where $\overline{W}$ is a (closed) hypersurface in $\mathbb{P}^n$. To fix notations for the whole chapter, let V be a reduced hypersurface of degree d in $\mathbb{P}^n$ given by a homogeneous equation $f = 0$. If $f = f_1 \dots f_k$ is the decomposition of the polynomial f into irreducible factors, it follows that $V_i \colon f_i = 0$ for $i = 1$, $\dots$, k are precisely the irreducible components of the hypersurface V. Let $U = \mathbb{P}^n \backslash V$ denote the complement of V and let homology and cohomology groups in this chapter be with integer coefficients $\mathbb{Z}$ if not stated otherwise.

(1.3) **Proposition.** *Assume that the hypersurface V has k irreducible components V_i with degrees $\deg V_i = d_i$ for $i = 1, \dots, k$. Then*

$$H_1(U) = \mathbb{Z}^{k-1} \oplus (\mathbb{Z}/(d_1, \dots, d_k)\mathbb{Z}),$$

where $(d_1, \dots, d_k)$ denotes the greatest common divisor of the integers $d_1, \dots, d_k$.

Proof. By Lefschetz duality, [Sp], p. 297, we get

$$H_1(U) \stackrel{L}{\cong} H^{2n-1}(\mathbb{P}^n, V).$$

The exact sequence of the pair $(\mathbb{P}^n, V)$ gives the exact sequence

$$H^{2n-2}(\mathbb{P}^n) \stackrel{j^*}{\to} H^{2n-2}(V) \to H^{2n-1}(\mathbb{P}^n, V) \to 0,$$

where j is the inclusion $V \to \mathbb{P}^n$.

The cohomology $H^*(\mathbb{P}^n)$ is well known, see, for instance, [Gn], p. 90, or Chapter 5, §1, in this book. In particular,

$$H^{2n-2}(\mathbb{P}^n) = \mathbb{Z}.$$

Let $S(V)$ denote the singular locus of V. Since the real dimension of $S(V)$ (e.g., regarded as polyhedron) is at most $2n - 4$ it follows that

$$H^{2n-2}(V) \cong H^{2n-2}(V, S(V)) \stackrel{L}{\cong} H_0(V \backslash S(V)) \cong \mathbb{Z}^k.$$

Indeed, $V \backslash S(V)$ has exactly k path-connected components, one for each irreducible component V_i. We have in fact, the following commutative diagram:

$$
\begin{array}{ccc}
 & & H^{2n-2}(V) \\
 & \nearrow^{j^*} & \downarrow \oplus j_i^* \\
H^{2n-2}(\mathbb{P}^n) & \xrightarrow[\oplus \bar{j}_i^*]{} & \oplus\, H^{2n-2}(V_i)
\end{array}
$$

where $j_i: V_i \to V$ are the corresponding inclusions for $i = 1, \ldots, k$.

If $\bar{j}: V_i \to \mathbb{P}^n$ is the composition $j \circ j_i$, then it is well known that the induced morphism

$$\bar{j}_i^*: H^{2n-2}(\mathbb{P}^n) = \mathbb{Z} \to \mathbb{Z} = H^{2n-2}(V_i)$$

is just the multiplication by the degree d_i. This is another way of saying that a ' generic line L in $\mathbb{P}^n$ intersects the hypersurface V_i in exactly d_i points, see [L2] for details, or refer to our results in Chapter 5, §2, in this book. $\square$

(1.4) **Corollary.** *If* $X \subset \mathbb{C}^n$ *is an affine hypersurface with* k *irreducible components, then*

$$H_1(\mathbb{C}^n \backslash X) \cong \mathbb{Z}^k.$$

Proof. Apply (1.3) by taking $V = \overline{X} \cup H_\infty$, where $\overline{X}$ is the closure of X in the projective space $\mathbb{P}^n$ and $H_\infty = \mathbb{P}^n \backslash \mathbb{C}^n$ is the hyperplane at infinity. $\square$

There is a *local* problem similar to the problem we started to investigate.

Let $(X, 0)$ be a hypersurface singularity at the origin of $\mathbb{C}^n$ and let $\varepsilon > 0$ be small enough such that $(X, 0)$ has a conic structure inside the closed ball $\overline{B}_\varepsilon$, see (1.5.1). Let $K = X \cap S_\varepsilon$ be the corresponding link where $S_\varepsilon = \partial \overline{B}_\varepsilon$. Then we can consider the (embedded) *local fundamental group* of the hypersurface singularity $(X, 0)$ defined by

$$\pi_1^{\text{loc}}(X, 0) = \pi_1(B_\varepsilon \backslash X) = \pi_1(S_\varepsilon \backslash K)$$

exactly as in (1.1.3(ii)).

For $n = 2$, K is an algebraic knot as defined in (2.2.1) and we can use the methods of knot theory to compute this local fundamental group. In general, the link K should be regarded as a fibered link via the corresponding Milnor fibration

$$F \to S_\varepsilon \backslash K \xrightarrow{\varphi} S^1.$$

(1.5) **Exercise.** Assume that the analytic germ $(X, 0)$ has k irreducible components. Show that

$$H_1(S_\varepsilon \backslash K) = \mathbb{Z}^k.$$

Hint. Use the same argument as in the proof of (1.3) or recall the proof of (3.2.3).

Consider now the case when the singularity $(X, 0)$ is defined by a weighted homogeneous polynomial g. Then the local Milnor fibration considered above is equivalent by (3.1.13) to the global Milnor fibration

$$g^{-1}(1) \to \mathbb{C}^n \backslash X \xrightarrow{g} \mathbb{C}^*,$$

where $X = g^{-1}(0)$ is the associated affine hypersurface. It follows that

(1.6) $$\pi_1(\mathbb{C}^n \backslash X) = \pi_1(S_\varepsilon \backslash K),$$

a fact already used in (3.2.10).

(1.7) **Exercise.** Show that $\pi_1(\mathrm{GL}(n, \mathbb{C})) = \mathbb{Z}$. *Hint.* Write the general linear group $\mathrm{GL}(n, \mathbb{C})$ as a complement

$$\mathbb{C}^{n^2} \backslash D,$$

where D is the hypersurface $\{A : \det A = 0\}$. Then use the fact that $(D, 0)$ is a *normal* hypersurface singularity, see [D4], p. 45, and hence we can use (3.2.2) or (3.2.9).

Let us come back to our projective hypersurface $V : f = 0$ in $\mathbb{P}^n$ and to its complement U. Let $F = f^{-1}(1)$ be the associated affine Milnor fiber and let $h : F \to F$ be the monodromy homeomorphism $h(x) = t \cdot x$ for $t = \exp(2\pi i/d)$ and $x \in F$ as in (3.1.19). Let $\langle h \rangle = \{h^s; s = 0, \ldots, d - 1\} = \mathbb{Z}/d\mathbb{Z}$ be the cyclic group generated by this monodromy homeomorphism. There is an obvious identification

(1.8) $$F/\langle h \rangle = U.$$

Since the group $\langle h \rangle$ generated by h acts freely on the Milnor fiber F, we get an exact sequence

(1.9) $$1 \to \pi_1(F) \to \pi_1(U) \xrightarrow{p} \mathbb{Z}/d\mathbb{Z} \to 0,$$

where the last homomorphism p is surjective by (3.2.3) and the fact that we use a *reduced* equation for V.

(1.10) **Corollary.** *When the hypersurface V is irreducible, then*

$$\pi_1(F) = [\pi_1(U), \pi_1(U)],$$

the commutator subgroup of $\pi_1(U)$.

Proof. Use (1.9) and (1.3), together with the standard fact

$$H_1(U) = \pi_1(U)/[\pi_1(U), \pi_1(U)],$$

see [Gn], p. 48.   □

Let $X = f^{-1}(0)$ and note that there is a natural identification

(1.11) $$\mathbb{C}^{n+1} \backslash X/\mathbb{C}^* = U.$$

This gives the following exact seqence

$$(1.12) \qquad \pi_1(\mathbb{C}^*) \to \pi_1(\mathbb{C}^{n+1} \setminus X) \to \pi_1(U) \to 1.$$

Using this fact, the isomorphism (1.6), and the deep result by Lê–Saito (3.29) we get the following.

(1.13) **Theorem.** *If the hypersurface V has only normal crossing singularities in codimension 1, then the fundamental group $\pi_1(U)$ is abelian. When V is in addition irreducible, the Milnor fiber F is simply-connected.*

This result is an old conjecture due to Zariski [Z1] and was first proved by different methods by Fulton [F] and Deligne [De2]. For more details and another proof, see (3.18) below.

Consider for a moment the more general case when f is a weighted homogeneous polynomial of type $(w_0, \ldots, w_n; d)$ and V is the corresponding hypersurface in the weighted projective space $\mathbb{P}(w)$, $w = (w_0, \ldots, w_n)$ as in Appendix B. Then the group $\langle h \rangle$ acts on the Milnor fiber F, but this action may be no longer free and hence the exact sequence (1.9) is no longer valid. However, in many cases we can compute the fundamental group $\pi_1(\mathbb{P}(w) \setminus V)$ using the following result of Armstrong [Ar].

(1.14) **Proposition.** *Let X be a simply-connected space and let G be a finite group acting on X. Let N be the normal subgroup in G generated by elements $g \in G$ such that there is a point $x \in X$ with $g \cdot x = x$. Then*

$$\pi_1(X/G) = G/N.$$

To apply this result to our case we need some notation. Consider the set $I = \{0, 1, \ldots, n\}$ and the family of subsets $I(f) \subset P(I)$ in I defined by

$$J \in I(f) \Leftrightarrow \text{there is a point } x \in F \text{ such that } x_j \neq 0 \text{ exactly for } j \in J.$$

Let $J_1, \ldots, J_m$ be the minimal elements in the set $I(f)$, ordered by inclusion. Define

$$k_i = \min\{k \in \mathbb{N}; \, kw_j \text{ is divisible by } d \text{ for all } j \in J_i\},$$

$$k = \text{g.c.d.}(k_1, \ldots, k_m).$$

(1.15) **Corollary.** *If the Milnor fiber F is simply-connected, then*

$$\pi_1(\mathbb{P}(w) \setminus V) = \mathbb{Z}/k\mathbb{Z}.$$

In particular, this is true when

$$\dim\{x \in \mathbb{C}^{n+1}; \, df(x) = 0\} < n - 1,$$

e.g., for all quasismooth hypersurfaces $V \subset \mathbb{P}(w)$ with $\dim V \geq 1$.

(1.16) **Exercise.** Let $f = x_0^{a_0} + x_1^{a_1} + \cdots + x_n^{a_n}$ $(a_i \geq 2)$ be a Brieskorn–Pham polynomial as in (3.4.10). Show that

$$\pi_1(\mathbb{P}(\mathbf{w}) \setminus V) = \mathbb{Z}/k\mathbb{Z},$$

where $k = \text{g.c.d.}(a_0, \ldots, a_n)$ and the weights $\mathbf{w}$ are defined by

$$w_i = \text{l.c.m.}(a_0, \ldots, a_n)/a_i$$

for $i = 0, \ldots, n$.

We come back now to the hypersurface V in the usual projective space $\mathbb{P}^n$ and state a basic result allowing us to restrict our attention to the case of plane curves (i.e., $n = 2$).

(1.17) **Zariski Theorem of Lefschetz Type.** *Let V be a hypersurface in $\mathbb{P}^n$ and let $\mathscr{S}$ be a Whitney stratification for V. For any 2-plane $E \subset \mathbb{P}^n$ which is transversal to all the strata in $\mathscr{S}$ we have an isomorphism*

$$\pi_1(E \setminus (E \cap V) \xrightarrow{\sim} \pi_1(\mathbb{P}^n \setminus V)$$

induced by the inclusion.

This is a special case of (1.6.5) and for details of the proof we refer to [Ch2] or [HLe]. Note that $C = E \cap V$ is indeed a plane curve in $E \simeq \mathbb{P}^2$. In the following sections we will study in detail such fundamental groups $\pi_1(\mathbb{P}^2 \setminus C)$.

We have seen in Chapter 2 that the Alexander polynomial of a knot $K \subset S^3$ is a powerful invariant in understanding the topology of the complement $S^3 \setminus K$. Following Libgober [Li1], [Li3], we define a sequence of Alexander polynomials for any hypersurface $V \subset \mathbb{P}^n$.

(1.18) **Definition.** Let M be a connected, locally contractible space having the same homotopy type as a CW-complex of dimension m. Let $\varphi : \pi_1(M) \to \mathbb{Z}$ be an epimorphism and let $\tilde{M} \to M$ be the covering space associated to the normal subgroup $\ker \varphi$ in $\pi_1(M)$, see [Sp], p. 82. The group of covering transformations for the covering $\tilde{M} \to M$ is infinite cyclic: let T denote one of its two generators. We assume that the cohomology with $\mathbb{C}$-coefficients $H^k(\tilde{M}; \mathbb{C})$ is finite dimensional and then define

$$\Delta^k(M, \varphi)(t) = \det(t \cdot I - T^* | H^k(\tilde{M}; \mathbb{C}))$$

for $k = 0, \ldots, m$.

Since T^* is defined over $\mathbb{Z}$, it follows that $\Delta^k(M, \varphi) \in \mathbb{Z}[t]$. We set by convention $\det(0 \to 0) = 1$. To define the Alexander polynomials of the hypersurface $V : f = 0$ in $\mathbb{P}^n$, we take in the above definition

$$M = M_V = \mathbb{C}^{n+1} \setminus f^{-1}(0)$$

and

$$\varphi = \varphi_V : \pi_1(M_V) \xrightarrow{f_*} \pi_1(\mathbb{C}^*) = \mathbb{Z}.$$

The resulting cyclic covering $\tilde{M}_V \to M_V$ can be explicitly described as follows. Let $\exp: \mathbb{C} \to \mathbb{C}^*$ be the standard universal covering of $\mathbb{C}^*$. Then the covering $\tilde{M}_V \to M_V$ can be regarded as the pull-back of this covering $\exp$ via the map $f: M_V \to \mathbb{C}^*$. It follows that

$$\tilde{M}_V \simeq F \times \mathbb{C}$$

and the action of the covering transformation T on $\tilde{M}_V$ corresponds to the action of $h \times \mathrm{id}_{\mathbb{C}}$ on $F \times \mathbb{C}$, where $h: F \to F$ is the monodromy homeomorphism (if the generator T for $\mathrm{Deck}(\tilde{M}_V \mid M_V)$ was properly chosen).

Let $s = \dim S(V)$ be the dimension of the singular locus of the hypersurface V. In order to avoid having many Alexander polynomials identically equal to 1, we make a shift in the indices and give the following.

(1.19) **Definition.** The Alexander polynomials of a hypersurface $V \subset \mathbb{P}^n$ with $s = \dim S(V)$ are defined by the following formula

$$\Delta_V^k(t) = \Delta^{n-s-2+k}(M_V, \varphi_V) = \det(t \cdot I - h^* \mid H^{n-s-2+k}(F; \mathbb{C}))$$

for $k = 1, \ldots, s + 2$.

(1.20) **Remark.** The expression given in (2.1.26) for the Alexander polynomial of a fibered knot $K \subset S^3$ can be regarded as a historical motivation for this definition.

Moreover, when V is an irreducible hypersurface with $G = \pi_1(U)$, it follows from (1.10) that $H = \pi_1(F) = [G, G]$. Since

$$H_1(F) = H/[H, H],$$

and since we can use homology instead of cohomology in (1.18) and (1.19), we see again a clear analogy of (1.19) with the "group-theoretic" definition of the Alexander polynomial for a knot as presented in Chapter 2, §1.

These $(s + 2)$ Alexander polynomials Δ_V^k associated to the hypersurface V are related by the following formula.

(1.21) **Proposition.**

$$\prod_{k=1, s+2} \Delta_V^k(t)^{(-1)^{n-s+k}} = \frac{(t^d - 1)^{\chi(F)/d}}{t - 1},$$

where $d = \deg(V)$ and $\chi(F)$ denotes the Euler characteristic of the Milnor fiber F.

Proof. We start with a formula in linear algebra

(1.22) $$\det(I - tA) = \exp\left(-\sum_{i \geq 1} \frac{\mathrm{tr}(A^i)t^i}{i}\right),$$

where A is any $n \times n$ complex matrix (or any linear endomorphism of an n-dimensional $\mathbb{C}$-vector space).

Using the Jordan canonical form of such a matrix A, it is enough to check the formula (1.22) for a Jordan block

$$A_\lambda = \begin{bmatrix} \lambda & 1 & 0 & \dots & 0 \\ 0 & \lambda & 1 & \dots & 0 \\ & & \ddots & \ddots & 1 \\ 0 & 0 & 0 & \dots & \lambda \end{bmatrix}$$

For A_λ, the formula (1.22) follows from the series expansion

$$\log(1 - x) = - \sum_{i \geq 1} \frac{x^i}{i}.$$

Consider now the monodromy homeomorphism

$$h: F \to F,$$

the *Lefschetz numbers* of its iterates h^k

$$\Lambda(h^k) = \sum_{i=0,n} (-1)^i \operatorname{Tr}((h^k)^* | H^i(F; \mathbb{C})),$$

and the *zeta function* of h

$$Z(h)(t) = \prod_{i \geq 0} \det(1 - th^* | H^i(F))^{(-1)^{i+1}}.$$

Using (1.22) we get

$$Z(h)(t) = \exp\left(\sum_{k \geq 1} \Lambda(h^k) t^k / k \right).$$

When $k \equiv 0 \pmod d$, then h^k is the identity of F and hence $\Lambda(h^k) = \chi(F)$. We leave the reader to show that

$$\Lambda(h^k) = 0 \quad \text{for} \quad k \not\equiv 0 \pmod d,$$

using Exercise (3.1.22) and its hint.

Putting all this together we get

$$Z(h)(t) = (1 - t^d)^{-\chi(F)/d},$$

a nice formula already obtained by Oka [O2]. Moreover,

$$\det\left(1 - \frac{1}{t} h^* | H^i(F; \mathbb{C})\right) = t^{-b_i(F)} \det(tI - h^* | H^i(F; \mathbb{C})),$$

where $b_i(F) = \dim H^i(F; \mathbb{C})$ is the ith *Betti number* of F. It follows that

$$\prod_{i \geq 0} \det(tI - h^* | H^i(F; \mathbb{C}))^{(-1)^{i+1}} = (1 - t^d)^{-\chi(F)/d},$$

and this formula is clearly equivalent to the formula in (1.21).   □

As a result of (1.21), it is enough to compute only the first $(s + 1)$ Alexander polynomials $\Delta_V^1, \dots, \Delta_V^{s+1}$ and the Euler characteristic $\chi(F)$.

(1.23) **Example.** Assume that V is a smooth hypersurface. Then $s = -1$ since $S(V) = \varnothing$. The only Alexander polynomial is given in this case by the following formula:

$$\Delta_V^1(t) = \det(tI - h^* | H^n(F; \mathbb{C})) = \left(\frac{(t^d - 1)^{\chi(F)/d}}{t-1} \right)^{(-1)^n},$$

where $\chi(F) = 1 + (-1)^n (d-1)^{n+1}$ by (3.2.8).

(1.24) **Lefschetz Theorem for Alexander Polynomials.** *Let V be a hypersurface in $\mathbb{P}^n$ and let $s = \dim S(V)$. Then*

$$\Delta_{V \cap H}^k = \Delta_V^k \qquad for \quad k \leq s,$$

and the polynomial Δ_V^{s+1} divides the polynomial $\Delta_{V \cap H}^{s+1}$, for any generic hyperplane H in $\mathbb{P}^n$.

Proof. It is enough to remark that for a generic hyperplane $\bar{H}$ in $\mathbb{C}^{n+1}$ passing through the origin, the inclusion

$$j: F \cap \bar{H} \to F$$

is an $(n-1)$-equivalence. This follows from (1.6.5) by taking $X = \bar{F}$, the closure of the Milnor fiber F in $\mathbb{P}^{n+1}$, and $Z = \mathbb{P}^{n+1} \setminus \mathbb{C}^{n+1}$, the hyperplane at infinity. It is clear that $X \cap Z$ is precisely V.   □

(1.25) **Corollary.** *Let $V \subset \mathbb{P}^n$ be a hypersurface with $s = \dim S(V) \geq 1$. Then*

$$\Delta_V^1 = \Delta_{V \cap H_1 \cap \cdots \cap H_s}^1$$

for generic hyperplanes $H_1, \ldots, H_s$ in $\mathbb{P}^n$.

Since $V \cap H_1 \cap \cdots \cap H_s$ is a hypersurface with isolated singularities, it follows that the computation of the first Alexander polynomial Δ_V^1 can be always reduced to this special case.

(1.26) **Example.** Consider the following irreducible surfaces

$$V: x^{d-1}z + xt^{d-1} + y^d + xyt^{d-2} = 0$$

and

$$W: x^{d-1}z + y^d + xyt^{d-2} = 0,$$

for $d \geq 3$. Then V has a single singular point, namely, $a = (0:0:1:0)$, while W is singular along the line $L: x = y = 0$.

Let F_V and F_W be the associated Milnor fibers. Then:

(i) $H_0(F_V) = H_0(F_W) = \mathbb{Z}$, $H_1(F_V) = H_1(F_W) = 0$,
     $H_2(F_V) = H_2(F_W) = \mathbb{Z}^{d-1}$ and $H_3(F_V) = H_3(F_W) = 0$;
(ii) $\Delta_V^2(t) = \Delta_W^2(t) = (t^d - 1)/(t-1)$;
(iii) $H_.(V) = H_.(W) = H_.(\mathbb{P}^2)$, i.e., both surfaces V and W have the same integral homology as the projective plane $\mathbb{P}^2$.

Proof. To prove (i) and (ii), we compute the integral homology groups $H_.(F_W)$ and the corresponding Alexander polynomials Δ_W^i. The same computations for the surface V are completely analogous.

The hyperplane section $D_W = F_W \cap \{x = 0\}$ consists of d disjoint copies of the affine plane $\mathbb{C}^2$. Moreover, there is a fibration

$$\mathbb{C}^2 \to F_W \setminus D_W \xrightarrow{p} \mathbb{C}^*$$

induced by the projection on the x-coordinate. It follows that

$$p_\# : \pi_i(F_W \setminus D_W) \to \pi_i(\mathbb{C}^*)$$

is an isomorphism for all $i \geq 0$ and hence p is a homotopy equivalence [Sp], p. 405.

The homology Gysin sequence from (2.2.13) implies

$$H_3(F_W) = H_1(D_W) = 0.$$

For a generic plane $H \subset \mathbb{P}^3$, the intersection $W \cap H$ is an irreducible nodal curve. Applying (1.13) we get

$$H_1(F_W) = 0.$$

Coming back to the Gysin sequence, we get the exact sequence

$$0 \to H_2(F_W) \to H_0(D_W) \to H_1(F_W \setminus D_W) \to 0.$$

By the above remarks, $H_0(D_W) = \mathbb{Z}^d$ and $H_1(F_W \setminus D_W) = \mathbb{Z}$. It follows that

$$H_2(F_W) = \mathbb{Z}^{d-1}.$$

The formula for Δ_W^2 follows from (1.21). To prove (iii), note that

$$V \cap H = W \cap H = \mathbb{P}^1$$

and

$$V \setminus H = W \setminus H \simeq \mathbb{C}^2,$$

where H is the plane given by $x = 0$ in $\mathbb{P}^3$. The cohomology exact sequences for $(V, V \cap H), (W, W \cap H)$, and Alexander duality end the proof.   □

(1.27) **Remark.** (i) Using the Whitehead theorem on the homotopy type of a simple four-dimensional CW-complex, see, for instance, [BaK], p. 84, it follows that the surfaces V and W are homotopy equivalent. (To determine the cup product $H^2 \times H^2 \to H^4$, use (5.2.10).) It is an interesting question to determine whether the complements $\mathbb{P}^3 \setminus V$ and $\mathbb{P}^3 \setminus W$ are homotopy equivalent too.

(ii) The proof in (1.26) is based on the fact that both the surfaces V, W and their Milnor fibers F_V, F_W have nice decompositions in very simple pieces. This idea is similar in spirit to Durfee's paper [Df 8].

§2. Presentations of Groups and Monodromy Relations

Let us discuss first some elementary group-theoretic notions. For any set S, we denote by $F(S)$ the *free group* generated by S, which is unique up to isomorphism, see [La], p. 34. When we say that a group G has a *presentation*

$$G = \langle (x_i) \rangle_{i \in I} : (r_j)_{j \in J} \rangle \cdot$$

with generators x_i and relations r_j, we mean the following.

Let X be the set consisting of the elements x_i, $i \in I$. The relations r_j are elements in the free group $F(X)$, i.e., they are "words" in the letters x_i and x_i^{-1}. Let R be the normal subgroup in $F(X)$ generated by all these elements $r_j, j \in J$. Then the quotient group $F(X)/R$ is isomorphic to the given group G, see [La], pp. 36–37. Usually the relations r_j are written in the more intuitive form

$$r_j = 1, \qquad j \in J,$$

since these equalities hold in $F(X)/R = G$.

(2.1) **Examples.** (i) In the notation above, the *free group* $F(X)$ is generated by the elements x_i, $i \in I$, with no relations, i.e., $J = \varnothing$.

(ii) The *finite cyclic group* $\mathbb{Z}/n\mathbb{Z}$ of order n has the following presentation

$$\mathbb{Z}/n\mathbb{Z} = \langle x : x^n = 1 \rangle.$$

(iii) The binary k-dihedral group $\tilde{D}_k$ is an important type of finite subgroup in $SL(2, \mathbb{C})$ and has the following presentation, see [L3], p. 51,

$$\tilde{D}_k = \langle a, b : a^2 = b^k = (ab)^2 \rangle.$$

$\tilde{D}_k$ is a finite group of order $4k$ and satisfies

$$[\tilde{D}_k, \tilde{D}_k] = \mathbb{Z}/k\mathbb{Z},$$

$$\tilde{D}_k/[\tilde{D}_k, \tilde{D}_k] = \begin{cases} \mathbb{Z}/4\mathbb{Z} & \text{for } k \text{ odd,} \\ (\mathbb{Z}/2\mathbb{Z})^2 & \text{for } k \text{ even,} \end{cases}$$

see [L3], pp. 53 and 64.

(iv) Suppose, given two groups,

$$G^k = \langle (x_i^k)_{i \in I^k}; (r_j^k)_{j \in J^k} \rangle \qquad \text{where} \quad k = 1, 2.$$

Then the *free product* group $G^1 * G^2$ is the product of these groups in the category of groups (and not the product in the category of abelian groups, which is denoted by $G^1 \times G^2$ and exists only when both G^1 and G^2 are abelian).

In terms of presentation, we have

$$G^1 * G^2 = \langle (x_i^1)_{i \in I^1}, (x_i^2)_{i \in I^2} : (r_j^1)_{j \in J^1}, (r_j^2)_{j \in J^2} \rangle,$$

i.e., we put together all the generators and all the relations.

For instance, the group

$$(\mathbb{Z}/2\mathbb{Z}) * (\mathbb{Z}/3\mathbb{Z}) = \langle a, b \colon a^2 = b^3 = 1 \rangle$$

is known to be isomorphic to the group

$$\mathrm{PSL}(2, \mathbb{Z}) = \mathrm{SL}(2, \mathbb{Z})/(\pm I),$$

e.g., we can identify a with $\begin{pmatrix} 0 & 1 \\ -1 & 0 \end{pmatrix}$ and b with $\begin{pmatrix} 1 & 1 \\ -1 & 0 \end{pmatrix}$, see [CM], p. 85. In particular, it is clear from this isomorphism that the group $(\mathbb{Z}/2\mathbb{Z}) * (\mathbb{Z}/3\mathbb{Z})$ is an infinite noncommutative group.

We present now a *"topological" construction* for the group

$$(\mathbb{Z}/p\mathbb{Z}) * (\mathbb{Z}/q\mathbb{Z}) \qquad \text{for} \quad (p, q) = 1.$$

First, some general constructions. Let S be a connected, locally contractible topological space and let $N \colon S \to S$ be a homeomorphism of finite order d. Let $p \colon \tilde{S} \to S$ be the universal covering space of S. Then the group $G = \pi_1(S)$ acts on $\tilde{S}$ as the group of covering transformations. Let $\tilde{N} \colon \tilde{S} \to \tilde{S}$ be a lifting of the homeomorphism N and let $\tilde{G}$ be the subgroup in the group of all homeomorphisms of $\tilde{S}$ generated by G and $\tilde{N}$. Since two liftings of N differ by an element in G, it is clear that this group $\tilde{G}$ is well defined, i.e., depends only on S and N.

For this reason, we use the notation $\tilde{G}(S, N)$ when we want to be more accurate. It is also clear that G is a normal subgroup in $\tilde{G}$ and that $\tilde{G}/G \simeq \mathbb{Z}/d\mathbb{Z}$. In other words, $\tilde{G}$ is an extension of the group $G = \pi_1(S)$ by the group $\mathbb{Z}/d\mathbb{Z}$, i.e,

$$(2.2) \qquad\qquad 0 \to \pi_1(S) \to \tilde{G} \to \mathbb{Z}/d\mathbb{Z} \to 0.$$

Some simple properties of the action of the group $\tilde{G}$ on the space $\tilde{S}$ are contained in the following.

(2.3) Exercise. Show that:

(i) $\tilde{S}/\tilde{G} = S/\langle N \rangle$ where $\langle N \rangle$ is the finite cyclic group of order d generated by the homeomorphism N.

(ii) Let $\tilde{s} \in \tilde{S}$ and $s = p(\tilde{s}) \in S$ and let $\tilde{G}_{\tilde{s}}$ and $\langle N \rangle_s$ be the corresponding isotropy groups. Then

$$\tilde{G}_{\tilde{s}} = \langle N \rangle_s.$$

(iii) If the group $\langle N \rangle$ acts freely on S, then the group $\tilde{G}$ acts freely on $\tilde{S}$ and

$$\pi_1(S/\langle N \rangle) = \tilde{G}.$$

(2.4) Example. Let $S = F$ be the Milnor fiber associated to the hypersurface V in $\mathbb{P}^n$ and let $N = h$, the monodromy homeomorphism. Using (2.3(iii)) we deduce

$$\pi_1(U) = \tilde{G}(F, h).$$

In fact, the extension (2.2) in this case is nothing other than the extension (1.9). The next example is quite different, since the actions involved are no longer free.

(2.5) **Example.** Let $F_a = \{\lambda \in \mathbb{C}; \lambda^a = 1\}$ and $h_a \colon F_a \to F_a$

$$h_a(x) = \exp\left(\frac{2\pi i}{a}\right) \cdot x$$

be the Milnor fiber and the monodromy homeomorphism of the singularity x^a for some integer $a > 1$. Take two positive integers p, q such that $(p, q) = 1$ and consider the join space

$$S = F_p * F_q \quad \text{and the join map} \quad N = h_p * h_q.$$

It is clear that the quotient $S/\langle N \rangle$ can be identified with a segment $[\alpha, \beta]$, with $\alpha \in F_p$, $\beta \in F_q$, i.e., the segment $[\alpha, \beta]$ is part of the join S.

The universal covering space $\widetilde{S}$ in this case can be regarded as a tree, since it is a 1-complex which is simply-connected. Let $[\tilde{\alpha}, \tilde{\beta}]$ be a segment in this tree $\widetilde{S}$, which is a lift of the segment $[\alpha, \beta]$. Using (2.3) (i) and (ii), we obtain:

(i)   The segment $[\tilde{\alpha}, \tilde{\beta}]$ is a fundamental domain of $\widetilde{S}$ mod $\widetilde{G}$, see [Se], p. 48.

(ii) $\widetilde{G}_{\tilde{\alpha}} = \langle N \rangle_\alpha = \mathbb{Z}/q\mathbb{Z}$, $\widetilde{G}_{\tilde{\beta}} = \langle N \rangle_\beta = \mathbb{Z}/p\mathbb{Z}$.

(iii) The segment $[\tilde{\alpha}, \tilde{\beta}]$ is invariant only by the identity element in $\widetilde{G}$.

Using now a basic result on groups acting on graphs, see [Se], p. 48, we get an isomorphism

$$\widetilde{G}(F_p * F_q, h_p * h_q) = (\mathbb{Z}/p\mathbb{Z}) * (\mathbb{Z}/q\mathbb{Z}).$$

We now discuss briefly two important examples of *braid groups*. Let M be a connected manifold and for $n \geq 2$ consider the configuration space

$$M^{(n)} = \{(x_1, \ldots, x_n) \in M^n; x_i \neq x_j \text{ for } i \neq j\}.$$

The full symmetric group Σ_n on n letters acts on $M^{(n)}$ in an obvious way and let

$$\widetilde{M}^{(n)} = M^{(n)}/\Sigma_n$$

be the corresponding quotient.

(2.6) **Definition.** The fundamental group $\pi_1(M^{(n)})$ (resp. $\pi_1(\widetilde{M}^{(n)})$) is called the *pure braid group* (resp. the *full braid group*) on n strings of the manifold M.

(2.7) **Example** $(M = \mathbb{C})$. Consider the canonical projection

$$\sigma = (\sigma_1, \ldots, \sigma_n) \colon \mathbb{C}^n \to \mathbb{C}^n/\Sigma_n = \mathbb{C}^n,$$

where σ_k is the kth symmetric function in $x_1, \ldots, x_n$. We can identify the points in the base space $\mathbb{C}^n/\Sigma_n$ with monic polynomials

$$p = x^n + a_1 x^{n-1} + \cdots + a_n.$$

Then $\tilde{\mathbb{C}}^{(n)}$ can be identified with the space of all such polynomials p having no multiple roots, i.e.,

$$\tilde{\mathbb{C}}^{(n)} = \mathbb{C}^n \backslash \Delta,$$

where Δ is the *discriminant hypersurface*. The group

$$B_n(\mathbb{C}) = \pi_1(\tilde{\mathbb{C}}^{(n)})$$

is called the *classical braid group of Artin* on n strings. Our discussion above implies

(2.8) $$B_n(\mathbb{C}) = \pi_1(\mathbb{C}^n \backslash \Delta).$$

From a purely algebraic point of view, we have the following presentation for the group $B_n(\mathbb{C})$, see [Bi], p. 18.

(2.9) **Theorem** (Artin). *The group $B_n(\mathbb{C})$ admits a presentation with generators $g_1, \ldots, g_{n-1}$ and defining relations*

$$g_i g_j = g_j g_i \quad \text{for} \quad |i - j| \geq 2, \quad 1 \leq i, j \leq n - 1,$$

$$g_i g_{i+1} g_i = g_{i+1} g_i g_{i+1} \quad \text{for} \quad 1 \leq i \leq n - 2.$$

(2.10) **Example** ($M = \mathbb{P}^1$). The set of unordered n-points in $\mathbb{P}^1$ corresponds to the set of homogeneous polynomials

$$\tilde{p} = a_0 x^n + a_1 x^{n-1} y + \cdots + a_n y^n$$

of degree n in two variables x, y (modulo multiplicative nonzero constants). These polynomials $\tilde{p}$ form a projective space $\mathbb{P}^n$ and we have a *projective discriminant hypersurface* $\tilde{\Delta} \subset \mathbb{P}^n$, consisting of those polynomials $\tilde{p}$ with multiple roots. Hence the full braid group $B_n(\mathbb{P}^1)$ of the projective line $\mathbb{P}^1 \simeq S^2$ is given by

(2.11) $$B_n(\mathbb{P}^1) = \pi_1(\mathbb{P}^n \backslash \tilde{\Delta}).$$

We have the following analog of (2.9).

(2.12) **Theorem.** *The braid group $B_n(\mathbb{P}^1)$ admits a presentation with generators $g_1, \ldots, g_{n-1}$ and defining relations*

$$g_i g_j = g_j g_i \quad \text{for} \quad |i - j| \geq 2, \quad 1 \leq i, j \leq n - 1$$

$$g_i g_{i+1} g_i = g_{i+1} g_i g_{i+1} \quad \text{for} \quad 1 \leq i \leq n - 2,$$

$$g_1 \cdots g_{n-2} g_{n-1}^2 g_{n-2} \cdots g_1 = 1.$$

As a trivial case, consider the case $n = 2$. Then the discriminant $\tilde{\Delta}$ has the equation

$$a_1^2 - 4a_0 a_2 = 0,$$

i.e., $\tilde{\Delta}$ is a smooth conic on $\mathbb{P}^2$. By (1.3) and (1.13) we get

$$\pi_1(\mathbb{P}^2 \backslash \tilde{\Delta}) = \mathbb{Z}/2\mathbb{Z}.$$

This clearly agrees with the above presentation for $B_n(\mathbb{P}^1)$.

(2.13) **Remark.** The complement $\mathbb{C}^n \backslash \Delta$ offers a nice example of a $K(\pi, 1)$-space, see [Sp], p. 424, and [B4]. Indeed, we have

$$\pi_i(\mathbb{C}^n \backslash \Delta) = \pi_i \tilde{\mathbb{C}}^{(n)} = \pi_i(\mathbb{C}^{(n)})$$

for $i > 1$.

The natural projection

$$\mathbb{C}^{(n)} \to \mathbb{C}^{(n-1)}, \qquad (x_1, \ldots, x_n) \mapsto (x_1, \ldots, x_{n-1}),$$

is a fibration with typical fiber

$$F = \mathbb{C} \backslash \{(n-1)\text{-points}\}.$$

Since $\pi_i(F) = 0$ for $i > 1$, it follows by induction on n that

$$\pi_i(\mathbb{C}^{(n)}) = 0 \qquad \text{for} \quad i > 1,$$

i.e., $\mathbb{C}^n \backslash \Delta$ is indeed a $K(\pi, 1)$-space.

(2.14) **Exercise.** Is $\mathbb{P}^n \backslash \tilde{\Delta}$ a $K(\pi, 1)$-space? *Hint.* Use the fact that $\pi_2(\mathbb{P}^1) = \pi_2(S^2) = \mathbb{Z}$.

(2.15) **Exercise.** Let $A = \{a_1, \ldots, a_n\}$ be a set consisting of n distinct points in $\mathbb{C}$. Show that $\pi_1(\mathbb{C} \backslash A)$ is a free group on n generators. *Hint.* Consider a system of n loops $\gamma_1, \ldots, \gamma_n$ (as defined, for instance, in (3.3.8)) going once around each of the points a_i.

Two possible illustrations are the following, in which the loops are numbered using the convention introduced after (3.3.3):

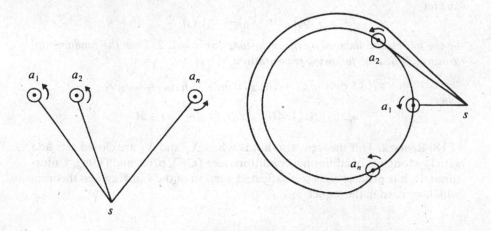

In the first illustration the points a_i are all situated on a line, while in the second picture they are situated on a circle. Note that in both cases the composition

$$\gamma_n \gamma_{n-1} \cdots \gamma_1$$

is a loop γ going once anticlockwise around all the points a_i. Show that

$$\pi_1(\mathbb{C} \setminus A) = F(\gamma_1, \ldots, \gamma_n).$$

We call such a set $\gamma_1, \ldots, \gamma_n$ of loops in $\pi_1(\mathbb{C} \setminus A)$ a *set of generating loops*.

(2.16) **Exercise.** Let $A = \{a_1, \ldots, a_n\}$ be a set consisting of n distinct points in $\mathbb{P}^1$. Show that $\pi_1(\mathbb{P}^1 \setminus A)$ is a free group on $(n-1)$ generators. *Hint.* There are two possible ways to solve this exercise:

(i) Note that $\mathbb{P}^1 \setminus \{a_n\} = \mathbb{C}$.
(ii) Suppose that all the points a_i are in the "finite part" $\mathbb{C}$ of the projective line $\mathbb{P}^1$. Let $\gamma_1, \ldots, \gamma_n$ be a set of generating loops in $\pi_1(\mathbb{C} \setminus A)$ as in (2.15). Then the loop $\gamma = \gamma_n \gamma_{n-1} \cdots \gamma_1$ is trivial in $\pi_1(\mathbb{P} \setminus A)$. Hence $\pi_1(\mathbb{P} \setminus A)$ has a presentation

$$\langle \gamma_1, \ldots, \gamma_n : \gamma_n \gamma_{n-1} \cdots \gamma_1 = 1 \rangle$$

from which it is clear that $\pi_1(\mathbb{P} \setminus A) = F(\gamma_1, \ldots, \gamma_{n-1})$.

We recall now a very useful tool for doing computations of fundamental groups (for a proof, see, for instance, [CF]).

(2.17) **Theorem** (van Kampen). *Let X be a topological space with an open covering $X = X_1 \cup X_2$ such that $X_0 = X_1 \cap X_2$, X_1, and X_2 are all nonempty path-connected spaces. Suppose there are given presentations*

$$\pi_1(X_1) = \langle (x_i)_{i \in I} : (r_j)_{j \in J} \rangle,$$

$$\pi_1(X_2) = \langle (y_k)_{k \in K} : (s_l)_{l \in L} \rangle,$$

$$\pi_1(X_0) = \langle (z_m)_{m \in M} : (t_n)_{n \in N} \rangle,$$

and let

$$(i_a)_\# : \pi_1(X_0) \to \pi_1(X_a)$$

be the morphisms induced by the inclusions for $a = 1, 2$. Then the fundamental group of X has the following presentation:

$$\pi_1(X) = \langle (x_i)_{i \in I}, (y_k)_{k \in K} : (r_j)_{j \in J}, (s_l)_{l \in L}, (u_m)_{m \in M} \rangle,$$

where

$$u_m : (i_1)_\#(z_m) = (i_2)_\#(z_m) \qquad for \quad m \in M.$$

(2.18) **Remark.** This theorem also holds when X_1 and X_2 are closed sets and satisfy a long list of additional conditions, see [CF], p. 65, and [Om]. Unfortunately, it is precisely this sophisticated version of the van Kampen theorem which we need in the sequel.

Before going further, we introduce an important class of groups, following Oka [O5]. Let p, q be two positive integers and consider the group

(2.19) $G(p, q) = \langle \beta, (a_i)_{i \in \mathbb{Z}}: \beta = a_{p-1} \cdots a_0, R_1, R_2 \rangle$

where

$$R_1: a_i = a_{i+q} \qquad \text{for any } i \in \mathbb{Z},$$
$$R_2: a_{i+p} = \beta a_i \beta^{-1} \qquad \text{for any } i \in \mathbb{Z}.$$

Note that this presentation has infinitely many generators and relations, but it is clear that the group $G(p, q)$ is finitely generated (e.g., by the elements $a_0, \ldots, a_{p-1}$).

(2.20) **Exercise.** Show that when $(p, q) = 1$, the group $G(p, q)$ has the following simpler presentation:

$$G(p, q) = \langle \alpha, \beta: \alpha^p = \beta^q \rangle.$$

Hint. Prove the following claims:

(a) $\beta = a_j \cdots a_{j-p+1}$ for any $j \in \mathbb{Z}$;
(b) $\beta^q = a_{pq-1} a_{pq-2} \cdots a_1 a_0 = \alpha^p$ where $\alpha = a_{q-1} \cdots a_0$;
(c) $a_r = \beta^m a_0 \beta^{-m}$ if $r = mp + nq$;
(d) $a_0 = \beta^l \alpha^k$ if $1 = lp + kq$.

The topological significance of the abstract group $G(p, q)$ is explained by the following result.

(2.21) **Proposition** (Oka [O5]). *Consider the affine plane curve*

$$C: x^p - y^q = 0$$

for some positive integers $p, q \geq 1$. Then

$$\pi_1(\mathbb{C}^2 \backslash C) = G(p, q).$$

Proof. Let $\varphi: \mathbb{C}^2 \backslash C \to \mathbb{C}, (x, y) \mapsto y$ be the second projection. Then φ is a locally trivial fibration over $\mathbb{C}^*$ with fiber $F = \varphi^{-1}(1)$. We consider the fibers $\varphi^{-1}(t)$ as subsets in $\mathbb{C}$ by projection onto the x-coordinate. Take generators $a_0, \ldots, a_{p-1}$ of $\pi_1(F)$ as shown in the figure below.

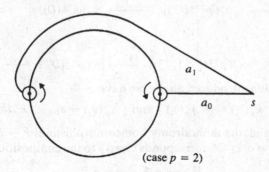

(case $p = 2$)

If D is the closed disc $\{y \in \mathbb{C}; |y| \leq 1\}$, then $\varphi^{-1}(D)$ is a deformation retract of the complement $\mathbb{C}^2 \setminus C$. Let D^+ and D^- be the upper and lower closed half-discs in D, respectively. Note that the map $y \mapsto |y|$ induces deformation retracts of the half-discs D^+ and D^- onto the segment $[0, 1]$. These deformation retracts can be lifted to produce deformation retracts r^+ and r^- of the sets $\varphi^{-1}(D^+)$ and $\varphi^{-1}(D^-)$, respectively, onto $\varphi^{-1}([0, 1])$. Note also that $\varphi^{-1}([0, 1])$ can be deformed onto $F = \varphi^{-1}(1)$. We apply van Kampen's theorem (2.17) twice (in fact, the version of it involving a closed covering as in (2.18)).

First consider the closed covering

$$\varphi^{-1}([-1, 1]) = \varphi^{-1}([-1, 0]) \cup \varphi^{-1}([0, 1]).$$

Let $F^- = \varphi^{-1}(-1)$ and note that $\varphi^{-1}(0) = \mathbb{C}^*$. This covering leads to the following commutative diagram

$$
\begin{array}{ccccc}
\pi_1(F) & \xrightarrow{\quad\sim\quad} & \pi_1(\varphi^{-1}([0, 1])) & \longrightarrow & \pi_1(\varphi^{-1}([-1, 1])) \\
& & \Big\uparrow{\scriptstyle j'_\#} & & \Big\uparrow \\
\pi_1(\mathbb{C}^*) & \xrightarrow{\quad j''_\#\quad} & \pi_1(\varphi^{-1}([-1, 0])) & \xleftarrow{\quad\sim\quad} & \pi_1(F^-).
\end{array}
$$

Here $\pi_1(F) = F(a_0, \ldots, a_{p-1})$ as in (2.15) and $\pi_1(F^-) = F(b_0, \ldots, b_{p-1})$ for a set of generating loops $b_0, \ldots, b_{p-1}$ in F^- chosen similarly to the loops $a_0, \ldots, a_{p-1}$ in F.

The generator σ of $\pi_1(\mathbb{C}^*) = \mathbb{Z}$ can be taken to be a large circle in $\mathbb{C}^*$ going anticlockwise and hence

$$j'_\#(\sigma) = a_{p-1} \cdots a_0,$$

$$j''_\#(\sigma) = b_{p-1} \cdots b_0.$$

Hence, by van Kampen's theorem (2.17) we have

$$\pi_1(\varphi^{-1}([-1, 1])) = \langle a_0, \ldots, a_{p-1}, b_0, \ldots, b_{p-1} : a_{p-1} \cdots a_0 = b_{p-1} \cdots b_0 \rangle.$$

Now apply van Kampen's theorem to the closed covering

$$\varphi^{-1}(D) = \varphi^{-1}(D^+) \cup \varphi^{-1}(D^-).$$

To do this, we consider the following commutative diagram:

$$
\begin{array}{ccccc}
\pi_1(F) & \xrightarrow{\quad\sim\quad} & \pi_1(\varphi^{-1}(D^+)) & \longrightarrow & \pi_1(\varphi^{-1}(D)) \\
& & \Big\uparrow{\scriptstyle k'_\#} & & \Big\uparrow \\
\pi_1(\varphi^{-1}([-1, 1])) & \xrightarrow{\quad k''_\#\quad} & \pi_1(\varphi^{-1}(D^-)) & \xleftarrow{\quad\sim\quad} & \pi_1(F)
\end{array}
$$

Using the notation introduced above, we have

$$k'_\#(a_i) = a_i, \qquad k'_\#(b_i) = (r^+)_\#(b_i) \quad \text{and} \quad k''_\#(a_i) = a_i, \qquad k''_\#(b_i) = (r^-)^*(b_i).$$

On the other hand, the monodromy homeomorphism $h: F \to F$ of the fibration induced by φ over $\mathbb{C}^*$ corresponds clearly to the composition

$$F \xrightarrow{\;(r^+)^{-1}\;} F_- \xrightarrow{\;r^-\;} F.$$

Since $r^+: F_- \to F$ is a homeomorphism, it follows that the elements $(r^+)_\#(b_i)$ for $i = 0, \ldots, p - 1$ generate the group $\pi_1(F)$. Using van Kampen's theorem (2.17) it follows that

$$\pi_1(\varphi^{-1}(D)) = \langle a_0, \ldots, a_{p-1} : h_\#(a_i) = a_i \rangle.$$

The relations $h_\#(a_i) = a_i$ are called the *monodromy relations* associated to the projection φ of the pair $(\mathbb{C}^2, C)$ at the origin on $\mathbb{C}$. In our case, it is easy to see that

$$h_\#(a_j) = \begin{cases} \beta^m a_{r+j} \beta^{-m} & \text{for } j = 0, \ldots, p - r - 1, \\ \beta^{m+1} a_{p-r+j} \beta^{-m-1} & \text{for } j = p - r, \ldots, p - 1, \end{cases}$$

where the integers m and r are defined by the equation $q = mp + r, 0 \le r < p$, and

$$\beta = a_{p-1} \cdots a_0.$$

As an example we draw the loops $h_\#(a_j)$ in the case $p = 2, q = 3$.

(Note that the monodromy homeomorphism $h: F \to F$ is the exactly rotation with angle $2\pi q/p$.)

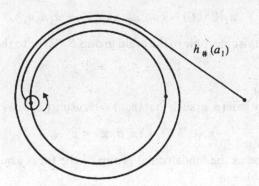

$h_\#(a_1)$

Hence $h_\#(a_0) = a_1 a_0 a_1 (a_1 a_0)^{-1}$.

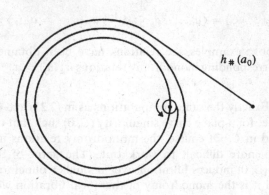

$h_\#(a_0)$

Hence $h_\#(a_1) = (a_1 a_0)^2 a_0 (a_1 a_0)^{-2}$.

To end the proof of (2.21), we can add new generators and new relations to the above presentation of the group

$$\pi_1(\mathbb{C}^2, C) = \pi_1(\varphi^{-1}(D)),$$

namely,

$$a_{kp+j} = \beta^k a_j \beta^{-k} \qquad \text{for} \quad k \in \mathbb{Z}$$

and $0 \le j < p$. Then the monodromy relation $h_*(a_j) = a_j$ becomes

$$a_j = a_{j+q}$$

and this clearly ends the proof. $\square$

(2.22) **Special Cases.**

(i) $(C, 0)$ is a node A_1, i.e., $p = q = 2$. Then

$$\pi_1(\mathbb{C}^2 \backslash C) = \langle a_0, a_1 : a_0 a_1 = a_1 a_0 \rangle = \mathbb{Z}^2.$$

(ii) $(C, 0)$ is a cusp A_2, i.e., $p = 2$, $q = 3$ (recall the illustration in the proof above). Then

$$\pi_1(\mathbb{C}^2 \backslash C) = \langle a_0, a_1 : a_0 a_1 a_0 = a_1 a_0 a_1 \rangle.$$

This group is thus isomorphic to the braid group $B_2(\mathbb{C})$ or to the group

$$\pi_{2,3} = \langle \alpha, \beta : \alpha^2 = \beta^3 \rangle$$

of the trefoil knot.

In fact, for any pair (p, q) such that $(p, q) = 1$, using (2.20) we get

$$\pi_1(\mathbb{C}^2 \backslash C) = \langle \alpha, \beta : \alpha^p = \beta^q \rangle,$$

which is the same as the fundamental group of the torus knot of type (p, q) discussed in (2.1.6).

(iii) $(C, 0)$ is smooth and the line $y = 0$ is an inflectional tangent of order p, i.e., $q = 1$. Then

$$\pi_1(\mathbb{C}^2 \backslash C) = \langle a_0, \ldots, a_{p-1} : a_0 = a_1 = \cdots = a_{p-1} \rangle \simeq \mathbb{Z}.$$

In all the above examples, the relations have been obtained by writing explicitly the corresponding monodromy relations $h_*(a_j) = a_j$.

(2.23) **Remark.** Exactly the same computations as in (2.21) and (2.22) work in the local case, i.e., for a plane curve singularity $(C, 0)$ such that the line $\varphi^{-1}(0)$ is not contained in C. Of course, the monodromy relations in such a case might be much more difficult to work out. (The germ of the projection $\varphi : (\mathbb{C}^2 \backslash C, 0) \to (\mathbb{C}, 0)$ induces fibration over a smaller punctured disc at the origin of $\mathbb{C}$ and it is the monodromy of this local fibration which is meant here.)

§3. The van Kampen–Zariski Theorem

In this section we discuss a general method for finding a presentation of the fundamental group $G = \pi_1(\mathbb{P}^2 \setminus C)$ of a given (reduced) plane curve C.

First we consider the easiest part, namely, finding a set of generators for this group G. A special case of the Zariski theorem of Lefschetz type (1.6.5) is the following.

(3.1) **Proposition.** *For any hypersurface $V \subset \mathbb{P}^n$ and any line L in $\mathbb{P}^n$ intersecting V transversally and avoiding the singular part $S(V)$, there is an epimorphism*

$$\pi_1(L \setminus (V \cap L)) \to \pi_1(\mathbb{P}^n \setminus V)$$

induced by the inclusion.

Note that for such a line L, the intersection $V \cap L$ consists exactly of $d = \deg(V)$ points and hence by (2.16) the group $\pi_1(L \setminus (V \cap L))$ is a free group on $(d - 1)$-generators.

To following result describes the *behavior of the fundamental group with respect to degenerations* of curves.

(3.2) **Corollary.** *Let C_t, $t \in [0, \varepsilon]$, be a smooth family of plane curves in $\mathbb{P}^2$ such that:*

(i) *the family C_t for $t \in (0, \varepsilon]$ is equisingular;*
(ii) *the limit curve $C_0 = \lim_{t \to 0} C_t$ is a reduced curve.*

Then there is an epimorphism

$$\pi_1(\mathbb{P}^2 \setminus C_0) \to \pi_1(\mathbb{P}^2 \setminus C_\varepsilon).$$

Proof. We recall that a familly C_t, $t \in (0, \varepsilon]$, of plane curves is *equisingular* if the singular points of the curve C_t can be indexed as $a_1(t), \ldots, a_p(t)$ in such a way that all the families of singularities $(C_t, a_i(t))$ are μ-constant. Then, by our discussion in Chapter 1, §3, it follows that the topological type of the pair $(\mathbb{P}^2, C_t)$ is independent of t for $t \in (0, \varepsilon]$.

Consider now a "tubular neighborhood" T of the curve C_0 in $\mathbb{P}^2$. In other words, T is a small open neighborhood of the curve C_0, which retracts to C_0, see [Df6] and our discussion in Chapter 5, §2, below.

Choose a t such that $\varepsilon \gg t > 0$ and such that the curve C_t is contained in the neighborhood T. Let L be a generic line with respect to both curves C_0 and C_t. Then the intersection $L \cap C_0$ (resp. $L \cap C_t$) consists of d points $p_1^0, \ldots, p_d^0$ (resp. $p_1^t, \ldots, p_p^t$) where $d = \deg C_t = \deg C_0$. We can arrange that the intersection $L \cap T$ consists of d disjoint small discs $D_1, \ldots, D_d$, each of them containing a pair of points p_i^0 and p_i^t for some $i = 1, \ldots, d$.

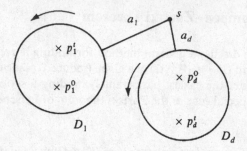

Consider a system of paths $a_1, \ldots, a_d$ as drawn in the figure above, i.e., each of these paths is a path in the line L, starting from a base point $s \in \mathbb{P}^1 \setminus (C_0 \cup C_t)$ and going around one of the discs D_i for each $i = 1, \ldots, d$. It is clear that the classes of these loops generate both groups $\pi_1(\mathbb{P}^2 \setminus C_0)$ and $\pi_1(\mathbb{P}^2 \setminus C_t)$.

Consider the morphisms induced by the corresponding inclusions

$$\pi_1(\mathbb{P}^2 \setminus C_0) \xleftarrow[\; j_{\#} \;]{\sim} \pi_1(\mathbb{P}^2 \setminus T) \xrightarrow{\; k_{\#} \;} \pi_1(\mathbb{P}^2 \setminus C_t)$$

and note that $j_{\#}$ is an isomorphism since $\mathbb{P}^2 \setminus T$ is a deformation retract of $\mathbb{P}^2 \setminus C_0$. Since $\pi_1(\mathbb{P}^2 \setminus C_t) = \pi_1(\mathbb{P}^2 \setminus C_\varepsilon)$, it follows that $k_{\#} \circ j_{\#}^{-1}$ is the epimorphism that we are looking for. $\qquad \square$

(3.3) **Exercise.** Show that the assumption C_0 reduced in (3.2) is crucial. *Hint.* Consider the family of conics in $\mathbb{P}^2$

$$Q_t: x(x + ty) = 0, \qquad t \in [0, 1).$$

For $t \neq 0$, we have $\pi_1(\mathbb{P}^2 \setminus Q_t) = \mathbb{Z}$, while $\pi_1(\mathbb{P}^2 \setminus Q_0) = 0$.

We also remark that the epimorphism in (3.1) does *not* exist for some special lines.

(3.4) **Example.** Consider the smooth conic

$$C: x^2 + yz = 0 \quad \text{in } \mathbb{P}^2$$

and the line $L: y = 0$, which is a tangent to C at the point $a = (0:0:1)$.

Since $C \cap L = \{a\}$, it follows that the group $\pi_1(L \setminus C \cap L) = \pi_1(\mathbb{C}) = 0$ has no epimorphism onto the group

$$\pi_1(\mathbb{P}^2 \setminus C) = \mathbb{Z}/2\mathbb{Z}.$$

There is, however, a *larger class* of lines L for which (3.1) holds than stated there. Let $p \in \mathbb{P}^2$ be a point and let $\mathbb{P}^1 \subset \mathbb{P}^2$ be a line which does not contain the point p. Consider the projection

$$\varphi: \mathbb{P}^2 \setminus \{p\} \to \mathbb{P}^1$$

with center p. Namely, for each point $x \in \mathbb{P}^2 \setminus \{p\}$, $\varphi(x)$ is the unique intersection point of the line $\overline{px}$ with the fixed line $\mathbb{P}^1$. We can use the points $a \in \mathbb{P}^1$ to parametrize the set of lines in $\mathbb{P}^2$ passing through the point p: the line $\varphi^{-1}(a) \cup \{p\}$ is denoted in the sequel by L_a.

We say that a line L_a through p is *exceptional* with respect to a curve C if it satisfies at least one of the following conditions:

(i) L_a is tangent to the curve C;
(ii) L_a passes through a singular point q of the curve C, $q \neq p$.

Here we say that a line L is tangent to the curve C at a point a if we have

$$(L, C)_a > \mathrm{mult}_a(C).$$

In particular, note that in (i) above the tangent point of the line L_a with the curve C may be the fixed point p. Let B be the union of all these exceptional lines through p with respect to the curve C and let $a_1, \ldots, a_s$ be the corresponding points in $\mathbb{P}^1$.

(3.5) **Lemma.** *With the above notations, the restriction*

$$\varphi \colon \mathbb{P}^2 \backslash (C \cup B) \to \mathbb{P}^1 \backslash \{a_1, \ldots, a_s\}$$

is a locally trivial fibration.

Proof. Let $Z = B_p(\mathbb{P}^2)$ be the blow-up of the projective plane $\mathbb{P}^2$ at the point p, see, for instance, [D4], p. 170. Let $\pi \colon Z \to \mathbb{P}^2$ be the canonical projection and let $E = \pi^{-1}(p)$ be the exceptional divisor. Let $\overline{\varphi} = \varphi \circ \pi \colon Z \backslash E \to \mathbb{P}^1$ and note that $\overline{\varphi}$ extends in an obvious way to a map

$$\tilde{\varphi} \colon Z \to \mathbb{P}^1.$$

Let $\tilde{C}, \tilde{B} \subset Z$ be the proper transforms of the curves C and B, respectively, and set

$$Z_0 = Z \backslash \tilde{B}, \qquad Z_\infty = Z_0 \cap E, \qquad \tilde{C}_0 = \tilde{C} \cap Z_0.$$

Then the restriction

$$\tilde{\varphi} \colon Z_0 \to \mathbb{P}^1 \backslash \{a_1, \ldots, a_s\}$$

is a proper submersion and induces a locally trivial fibration of the pair $(Z_0, \tilde{C}_0 \cup Z_\infty)$ over $\mathbb{P}^1 \backslash \{a_1, \ldots, a_s\}$ by (1.3.1). Since

$$\mathbb{P}^2 \backslash (B \cup C) = Z_0 \backslash (\tilde{C}_0 \cup Z_\infty),$$

this ends the proof of (3.5). □

(3.6) **Corollary.** *Asssume that no exceptional line in B is a component of the curve C. The for any line L_a through p which is not exceptional, there is an epimorphism*

$$\pi_1(L_a \backslash C) \to \pi_1(\mathbb{P}^2 \backslash C)$$

induced by the inclusion.

In other words, given a curve C and a point p in the projective plane $\mathbb{P}^2$, the result (3.1) holds for any line through the point p which is generic with respect to the curve C (even when $p \in C$).

Proof. The space $F = L_a \backslash C$ is exactly the fiber of the fibration φ considered in (3.5). The exact homotopy sequence of a fibration implies

$$\pi_1(F) \xrightarrow{\ j_* \ } \pi_1(\mathbb{P}^2 \backslash (C \cup B)) \xrightarrow{\ \varphi_* \ } \pi_1(\mathbb{P}^1 \backslash A) \to 1,$$

where $A = \{a_1, \ldots, a_s\}$ and j is the corresponding inclusion.

Hence a system of generators for the group $\pi_1(\mathbb{P}^2 \backslash (C \cup B)$ can be obtained as follows:

(i) take any generators $g_1, \ldots, g_m$ for $\pi_1(F)$ and consider the elements $\bar{g}_i = j_*(g_i)$ for $i = 1, \ldots, m$;
(ii) consider a system of generating loops $\alpha_1, \ldots, \alpha_s$ for $\pi_1(\mathbb{P}^1 \backslash A)$ as in (2.16) and lift then to some loops $\bar{\alpha}_1, \ldots, \bar{\alpha}_s$ in $\pi_1(\mathbb{P}^2 \backslash (C \cup B))$.

The morphism

$$\pi_1(\mathbb{P}^2 \backslash (C \cup B)) \xrightarrow{\ k_* \ } \pi_1(\mathbb{P}^2 \backslash C)$$

induced by the inclusion is an epimorphism, by a transversality argument similar to the one in the proof of (1.1). Moreover, $k_*(\bar{\alpha}_j) = 0$ for any loop $\bar{\alpha}_j$, since this loop goes once around the line L_{a_j} and becomes trivial when this line is added back to the space $\mathbb{P}^2 \backslash (C \cup B)$. It follows that the elements

$$k_*(\bar{g}_i) \qquad \text{for} \quad i = 1, \ldots, m,$$

generate the group $\pi_1(\mathbb{P}^2 \backslash C)$, i.e., the morphism

$$k_* \circ j_*$$

is an epimorphism. $\qquad\qquad\qquad\qquad\qquad\qquad\qquad\qquad\qquad\qquad\qquad\quad$ $\square$

(3.7) **Exercise.** Let the curve C consist of d lines, all passing through a point p. Compute the group $\pi_1(\mathbb{P}^2 \backslash C)$ and compare the result to (3.6).

(3.8) **Corollary.** *Let C be an irreducible curve of degree d such that there is a point $p \in C$ with*

$$\text{mult}_p(C) = d - 1.$$

Then the group $\pi_1(\mathbb{P}^2 \backslash C)$ is abelian.

Proof. Use (3.6) and note in this case that $L_a \backslash C \simeq \mathbb{C}^*$. The result (3.8) can be informally stated as follows: if an irreducible curve C has a "very bad" singularity (i.e., a singularity having the highest possible multiplicity), then the corresponding group $\pi_1(\mathbb{P}^2 \backslash C)$ is abelian. $\qquad\qquad\qquad\qquad\quad$ $\square$

The next result shows that the same conclusion holds when the irreducible curve C has just one singular point which is not very degenerate, in the following precise sense. First we give a local definition.

(3.9) Definition. Let $X: g = 0$ be an IHS at the origin of $\mathbb{C}^{n+1}$. Let m denote the maximal ideal in the ring $\mathcal{O}_{n+1}$ of germs of analytic functions at the origin. The *μ-constant determinacy order* of the singularity $(X, 0)$, denoted by $\mu\text{-det}(X, 0)$, is the smallest positive integer s such that the family

$$g_t = g + th, \qquad t \in [0, \varepsilon),$$

is μ-constant for any germ $h \in m^s$ and $\varepsilon > 0$ small enough (here ε may depend on h).

(3.10) Example. Let $X: g = 0$ be a plane curve singularity which is an *ordinary r-fold* point, see [Hn], p. 38, for a definition. Using the formula for the Milnor number of a semiweighted homogeneous singularity, recall (3.1.19), we deduce that $\mu\text{-det}(X, 0) = r$.

(3.11) Proposition. *Assume that the irreducible curve C of degree d has just one singular point p and that*

$$\mu\text{-det}(C, p) < d.$$

Then the fundamental group $\pi_1(\mathbb{P}^2 \setminus C)$ is abelian.

Proof. Let $m = \text{mult}_p(C)$ and let $L_1, \ldots, L_s$ be all the exceptional lines through p. Let $\mathbb{P}^1$ be a line in $\mathbb{P}^2$ such that:

(i) $\#(\mathbb{P}^1 \cap C) = d$, i.e., $\mathbb{P}^1$ is transverse to C;
(ii) $\mathbb{P}^1$ does not pass through any of the intersection points in $L_i \cap C$, $i = 1, \ldots, s$. Choose a linear system of coordinates $(x : y : z)$ on $\mathbb{P}^2$ such that:

(α) $p = (0 : 0 : 1)$;
(β) $z = 0$ is an equation for $\mathbb{P}^1$.

In the affine coordinate chart $\mathbb{C}^2$ given by $z = 1$ the curve C is given by an equation

$$f = f_m(x, y) + \cdots + f_{d-1}(x, y) + f_d(x, y) = 0,$$

where f_k denotes the homogeneous component of degree k in this equation.
   Moreover, we have the following:

(3.12) $\begin{cases} \text{(i) } f_m \neq 0; \\ \text{(ii) } f_d \text{ has no multiple factors;} \\ \text{(iii) } (f_d, f_{d-1}) = 1. \text{ In particular, } f_{d-1} \neq 0. \end{cases}$

Indeed, (i) and (ii) are obvious and to prove (iii) we can proceed as follows. Consider the line

$$L = \{(tx, ty), t \in \mathbb{C}\} \qquad \text{for some} \quad (x, y) \in \mathbb{C}^2 \setminus \{0\}.$$

If the equations $f_{d-1} = 0$ and $f_d = 0$ had (x, y) as a common root, then the corresponding line L would have a multiple intersection point with C on the line at infinity $z = 0$. But this is a contradiction to our choice of this line at infinity $\mathbb{P}^1$.

Consider now the family of curves $\bar{C}_t$ in $\mathbb{P}^2$ having the following affine equation

$$\bar{C}_t \colon f_t = t(f_m + \cdots + f_{d-2}) + f_{d-1} + f_d = 0$$

for $t \in \mathbb{C}$. Using (3.12(iii)), we find out that there is a finite set $B \subset \mathbb{C}$ such that:

  (i) $\bar{C}_t$ is an irreducible curve for $t \in \mathbb{C} \backslash B$;
 (ii) $\bar{C}_t$ has exactly one singular point, namely, p for $t \in \mathbb{C} \backslash B$; and
(iii) $\{0, 1\} \subset \mathbb{C} \backslash B$.

(*Hint.* In fact, we first prove (iii) and then use standard semicontinuity arguments from algebraic geometry.)

Consider now the function

$$\mu \colon \mathbb{C} \backslash B \to \mathbb{Z}, \qquad \mu(t) = \mu(\bar{C}_t, p),$$

the Milnor number of the singularity $(\bar{C}_t, p)$. Since μ is an upper semicontinuous function in Zariski topology, there is a finite set $B^1 \subset \mathbb{C} \backslash B$ such that the function μ takes its minimal value on the Zariski open set $\mathbb{C} \backslash (B \cup B^1)$. Using the condition $\mu\text{-det}(C, p) < d$, it follows that for all t close enough to 1, the value $\mu(\bar{C}_t, p)$ is equal to $\mu(C, p)$. In other words, $1 \in \mathbb{C} \backslash (B \cup B^1)$. Choose now a path $\gamma \colon [0, 1] \to \mathbb{C}$ such that $\gamma(0) = 0$, $\gamma(1) = 1$, and $\gamma((0, 1]) \subset \mathbb{C} \backslash (B \cup B^1)$. If we define $C_t = \bar{C}_{\gamma(t)}$ for $t \in [0, 1]$, then the family of plane curves C_t has all the properties required to apply (3.2), with $\varepsilon = 1$. To end the proof, we use (3.8). ☐

(3.13) **Remark.** It is not true that *any* irreducible curve C having just one singularity of multiplicity $m < d - 1$ can be deformed into an irreducible curve C_0 having a singularity of multiplicity $d - 1$. Consider, for instance, the quartic curve

$$C \colon x^2 z^2 + 2xy^2 z + y^4 - x^3 y = 0.$$

Then C has just one singular point, namely $p = (0 : 0 : 1)$, which is a singularity of type A_6.

A list of all the isolated singularities of the form

$$C_0 \colon f_3(x, y) + f_4(x, y) = 0,$$

and such that the curve C_0 is irreducible, gives the possibilities D_4, D_5, and E_6, see [BG], p. 273. But the singularity A_6 cannot be deformed into any of these three singularities, for obvious reasons.

(3.14) **Remark.** An example of an irreducible curve $C \subset \mathbb{P}^2$, having just one singular point and such that the group $\pi_1(\mathbb{P}^2 \backslash C)$ is not abelian, is given below in (4.21).

Consider again the fibration φ from (3.5), but this time with the point p *not* situated on the curve C. The fundamental group $\pi_1(\mathbb{P}^1 \backslash \{a_1, \ldots, a_s\})$ is free with $s - 1$ generators $\gamma_1, \ldots, \gamma_{s-1}$, where γ_i is a loop based at a base point $a \in \mathbb{P}^1 \backslash \{a_1, \ldots, a_s\}$ and goes once around the point a_i for $i = 1, \ldots, s - 1$ as in (2.16). Each loop γ_i induces a monodromy automorphism

$$T_i \colon \pi_1(F) \to \pi_1(F) \qquad \text{where} \quad F = L_a \backslash C.$$

Since $L_a \cap C$ consists exactly of $d = \deg(C)$ points, it follows that $\pi_1(F)$ is a free group on $d - 1$ generators $g_1, \ldots, g_{d-1}$, which may be chosen as in (2.16).

It follows that the automorphism T_i is completely determined by giving the elements

$$T_i(g_j) \qquad \text{for} \quad j = 1, \ldots, d - 1.$$

If $[\beta]$ denotes the image in $G = \pi_1(\mathbb{P}^2 \backslash C)$ of an element $\beta \in \pi_1(F)$ under the morphism induced by the inclusion, it follows easily that we have the *monodromy relations*

$$[T_i(g_j)] = [g_j] \qquad \text{for all} \quad j = 1, \ldots, d - 1 \quad \text{and} \quad i = 1, \ldots, s - 1.$$

The proof of (2.21) suggests that these monodromy relations are the only ones (i.e., any other relations can be deduced from them). This is exactly the content of the following famous result, conjectured by O. Zariski [Z1] and proved by van Kampen [vK]. For a modern proof we refer to Cheniot [Ch1] and [Ch2].

(3.15) **Theorem** (van Kampen–Zariski). *The group $G = \pi_1(\mathbb{P}^2 \backslash C)$ admits the following presentation:*

$$G = \langle g_j, j = 1, \ldots, d - 1 \colon T_i(g_j) = g_j, i = 1, \ldots, s - 1, j = 1, \ldots, d - 1 \rangle.$$

(3.16) **Remark.** Assume that the exceptional line L_{a_i} intersects the curve C *nontransversally* only at a point q and let $m = (C, L_{a_i})_q > 1$. As a generic line L_a through p moves toward L_{a_i}, m point in the intersection $L_a \cap C$ will collapse to the point q.

It is clear that among the monodromy relations $T_i(g_j) = g_j$, only those associated with loops g_j around one of these m points may be nontrivial. Indeed, outside a small ball centered at q, the fibration φ is trivial over D, where D is a small disc centered at a_i in $\mathbb{P}^1$. In other words, the interesting relations coming from the monodromy around the exceptional line L_{a_i} can be *localized* at the point q, recall (2.23). It follows that in general the number of nontrivial relations is much less than $(s - 1)(d - 1)$.

Lastly, we present two consequences of Theorem (3.15), both of them already discussed in Zariski's classical paper [Z1].

(3.17) **Corollary.** *Let C be a curve which has an inflectional tangent of order $d = \deg(C)$. Then the fundamental group $\pi_1(\mathbb{P}^2 \backslash C)$ is abelian. In particular, if C is smooth curve, then*

$$\pi_1(\mathbb{P}^2 \backslash C) = \mathbb{Z}/d\mathbb{Z}.$$

Proof. The assumption means that there is a line L_0 in $\mathbb{P}^2$ such that $L_0 \cap C = \{a\}$, and the point a is a smooth point on the curve C. Let p be a point in $L_0 \setminus C$ and apply (3.15). As remarked in (3.16), the monodromy relations obtained by going around the special line L_0 are the same as in (2.22(iii)). Hence all the generators of $\pi_1(\mathbb{P}^2 \setminus C)$ are equal, which implies that this group is abelian.

To prove the statement about smooth curves, first we recall that for any two smooth curves C and C^1—both of degree d—their complements $\mathbb{P}^2 \setminus C$ and $\mathbb{P}^2 \setminus C^1$ are topologically equivalent, see (1.3.4). Then note that the curve

$$C^1 : x^d + xy^{d-1} + z^d = 0$$

is smooth and the line $L_1 : x = 0$ is an inflectional tangent of order d for it.   □

(3.18) **Corollary** (Compare to (1.13)). *Let $C \subset \mathbb{P}^2$ be a nodal curve (i.e., all the singularities of C are nodes A_1). Then the fundamental group $\pi_1(\mathbb{P}^2 \setminus C)$ is abelian.*

Proof (Along the lines proposed by Zariski [Z1]). Assume first that C consists of d lines, namely, $L_1, \ldots, L_d$, meeting only two at a time. Then the generator g_j of $G = \pi_1(\mathbb{P}^2 \setminus C)$ corresponds to a loop in L_a around the intersection point $L_a \cap L_j$ for $j = 1, \ldots, d - 1$. Let L_{ij} be the exceptional line joining the point p with the intersection point $L_i \cap L_j$ for $1 \leq i < j \leq d - 1$. Then the monodromy relations obtained by going around the line L_{ij} imply $g_i g_j = g_j g_i$, use (3.16) and (2.22(i)). Hence all the generators commute with each other and G is an abelian group.

To treat the general case of a nodal curve C, we use the Severi remark (for a complete proof, see Harris [Hs]) that any nodal curve can be deformed into a union of lines, meeting only two at a time. Then the result follows from the first part of this proof via (3.2).   □

(3.19) **Remark.** One may try to obtain a "degenerate van Kampen–Zariski theorem" similar to (3.15) in the situation where the center of projection p is a point *on* the curve C. If $m = \text{mult}_p(C)$, then we get a system of generators for the group G having $d - m - 1$ elements from (3.6) (we have to take p not on a linear component of the curve C, if any exists). Note also that the *monodromy relations* obtained by going round the exceptional lines through p are still relations among these $(d - m - 1)$ generators.

The major drawback of this approach is that the monodromy relations do *not* form a complete set of relations necessary for a presentation of the group G, see (3.20) for a simple example. However, in many cases, we can use this approach to derive valuable information on the group G, see (3.21) for an example.

(3.20) **Example.** Consider the cuspidal cubic $C : xy^2 - z^3 = 0$ and let $p = (1 : 0 : 0)$. Then there is just one exceptional line. It follows that the fibration φ from (3.5) is trivial in this case (the base space $\mathbb{P}^1 \setminus \{a_1\}$ is contractible). Hence the "degenerate van Kampen–Zariski theorem" gives us in this case one generator g and no relation. On the other hand, we have

$$G = \langle g : g^3 = 1 \rangle$$

from (3.8).

(3.21) **Exercise.** Let $C \subset \mathbb{P}^2$ be a curve of degree d and let $p \in C$ be a point with $\operatorname{mult}_p(C) = d - 2$. Assume that one of the following two conditions is fulfilled:

(i) The curve C has a node q, $q \neq p$, and no line through p is an irreducible component for C;
(ii) The curve C is irreducible, $C \setminus \{p\}$ is smooth, and the tangent cone to the singularity (C, p) consists of exactly one line.

Then the fundamental group $G = \pi_1(\mathbb{P}^2 \setminus C)$ is abelian. *Hint.* Apply the "degenerate van Kampen–Zariski theorem" to the curve C and the center of projection p. From (2.6) we get that G can be generated by two elements g_1 and g_2. In case (i), use (3.16) and (2.22(i)) to get $g_1 g_2 = g_2 g_1$. In case (ii), consider the restriction

$$\psi : C \setminus \{p\} \to \mathbb{P}^1$$

of the projection with center p. Since $\operatorname{im} \psi = \mathbb{C}$ is simply-connected, ψ cannot be a covering map. This implies the existence of a tangent to the curve C at a point q, passing through p. Use (3.16) and (2.22(iii)) to deduce $g_1 = g_2$. In both cases (i) and (ii), the monodromy relations obtained are sufficient to imply that G is abelian.

§4. Two Classical Examples

In this section we discuss two examples due to Zariski [Z1]. They present the simplest curves $C \subset \mathbb{P}^2$ whose fundamental groups $\pi_1(\mathbb{P}^2 \setminus C)$ are *not* abelian.

(4.1) **Exercise.** (i) List all the plane curves of degree 1, 2, and 3 up to linear equivalence. For the degree 3 case, we can have a look at the list in [D4], p. 51.

  (ii) Show that all the fundamental groups $G = \pi_1(\mathbb{P}^2 \setminus C)$ are abelian for $\deg(C) \leq 3$, except in the case where C consists of three concurrent lines.

  (iii) Compute all the groups G for $\deg(C) \leq 3$.

Passing to the plane curves of degree 4, let us recall that there are irreducible quartic curves having as singularities three cusps A_2, e.g., the following "normal form":

$$(4.2) \qquad C : f = x^2 y^2 + y^2 z^2 + z^2 x^2 - 2xyz(x + y + z) = 0,$$

see [BG] for details and (4.7).

  The following result says that all the other irreducible quartics are not interesting for our discussion.

(4.3) **Proposition.** *If $C' \subset \mathbb{P}^2$ is an irreducible quartic curve, not a three cuspidal quartic, then the fundamental group $\pi_1(\mathbb{P}^2 \setminus C')$ is abelian.*

Proof. When C' is smooth, we can use (3.17). When C' has a point with multiplicity 3, then use (3.8). Assume now that all the singular points of the curve C' have multiplicity 2, i.e., they are A_k-singularities, for some integers $k \geq 1$. If there is an A_1-singularity, the result follows from (3.11) (if this is the only singularity of C') or from (3.21(i)) (if there are some other singularities). If C' has just one singular point, we can use (3.21(ii)). Hence we can assume that C' has at least two singularities and each of them is of type A_k for some $k \geq 2$.

Let C be an irreducible curve of degree d having m singular points $a_1, \ldots, a_m$. The following simple formula for the Euler characteristic of the curve C is a special case of formula (5.4.4), which is proved in Chapter 5.

$$(4.4) \qquad \chi(C) = 2 - (d-1)(d-2) + \sum_{i=1,m} \mu(C, a_i).$$

Since C is irreducible, it follows that $b_0(C) = b_2(C) = 1$, and hence $\chi(C) \leq 2$. This implies the following *upper bound* for the sum of the Milnor numbers of the singularities which may occur on an irreducible curve of degree d in $\mathbb{P}^2$

$$(4.5) \qquad \sum_{i=1,m} \mu(C, a_i) \leq (d-1)(d-2).$$

In our case, we have $(d-1)(d-2) = 6$ since $d = 4$. From our assumption on C', it follows that either C' is a three cuspidal quartic or C' has exactly two singularities, say a_1 and a_2.

In the second case, we may assume

$$(4.6) \qquad \mu(C', a_1) \leq \mu(C', a_2).$$

We have to show that in this latter case the group $\pi_1(\mathbb{P}^2 \setminus C')$ is abelian. Choose coordinates $(x : y : z)$ on $\mathbb{P}^2$ such that $a_1 = (0 : 0 : 1)$ and C' has an equation

$$Az^2 + 2Bz + C = 0,$$

where A, B, and C are binary forms in x and y of degrees 2, 3, and 4, respectively, and $A = x^2$. As in [BG], we consider the discriminant

$$\Delta = B^2 - AC.$$

The map $\psi: C^1 \setminus \{a_1\} \to \mathbb{P}^1, (x : y : z) \mapsto (y : z)$ is a ramified covering. Its ramification locus

$$R = \{(x : y) \in \mathbb{P}^1; \, \#\psi^{-1}(x, y) < 2\}$$

is given by $x = 0$ or $\Delta(x, y) = 0$. It is easy to see that a root $(\alpha : \beta)$ with $\alpha \neq 0$ of the equation $\Delta = 0$ corresponds to:

(i) a singularity of C' of type A_k at the point

$$q = (\alpha : \beta : -B(\alpha, \beta)/2\alpha^2),$$

when the multiplicity of the root $(\alpha : \beta)$ is $k + 1 \geq 2$; or

(ii) a bitangent line to C' passing through a_1, the second tangent point q having the coordinates given above, when the multiplicity of the root $(\alpha : \beta)$ is one.

In case (ii) we can argue as in the hint to (3.21(ii)) and deduce that the group $\pi_1(\mathbb{P}^2 \setminus C')$ is abelian.

Up to a change of coordinates, we are left with the case

$$\Delta(x, y) = x^{\alpha} y^{\beta}$$

for some positive odd integers α, β with $\beta > 1$, $\alpha + \beta = 6$ (when α, β are even, the curve C^1 is no longer irreducible). The case $\alpha = 1$ is in contradiction with $A = x^2$. When $\alpha = \beta = 3$, it is easy to see that the singularity (C', a_2) is of type A_2 (use the above discussion), while the singularity (C', a_1) is of type A_4, use [BG] or a direct computation. This contradicts our assumption (4.6) and ends the proof of (4.3).   □

(4.7) **Exercise.** (i) Show that any two three-cuspidal quartics Q_1 and Q_2 in $\mathbb{P}^2$ are linearly equivalent, i.e., there is a linear automorphism $h: \mathbb{P}^2 \to \mathbb{P}^2$ such that $h(Q_1) = Q_2$.

(ii) Show that a three-cuspidal quartic Q in $\mathbb{P}^2$ is a rational curve, i.e., Q is birationally equivalent to $\mathbb{P}^1$, see [Hn], p. 30.

Hint. Let Q be a three-cuspidal quartic in $\mathbb{P}^2$ and let $\hat{Q}$ be its dual curve. Show that $\hat{Q}$ is a cuspidal cubic using the Plücker formula (1.2.19). Since properties (i) and (ii) are true for cuspidal cubics, they are also true for their duals, the three-cuspidal quartics.

(4.8) **Proposition.** *Let C be a three-cuspidal quartic curve in $\mathbb{P}^2$. Then we have*

$$\pi_1(\mathbb{P}^2 \setminus C) \simeq \tilde{D}_3 \simeq B_3(\mathbb{P}^1).$$

Here $\tilde{D}_3$ is the binary 3-dihedral group as defined in (2.1(iii)) and $B_3(\mathbb{P}^1)$ is the full braid group on three strings of the projective line $\mathbb{P}^1$ as in (2.11), (2.12).

In particular, the group $G = \pi_1(\mathbb{P}^2 \setminus C)$ is finite of order $|G| = 12$ and

$$G/[G, G] = \mathbb{Z}/4\mathbb{Z},$$

as in (2.1(iii)) and (1.3). This group G is sometimes called the metacyclic group of order 12.

Proof. Since it is easy to see that $\tilde{D}_3 \simeq B_3(\mathbb{P}^1)$, it is enough to show that the group G is isomorphic to one of these two groups. As a result, it is natural to give *two proofs*, as follows.

First Proof. This proof is based on the relation (2.11), the Zariski theorem of Lefschetz type (1.17), and on the following result for which we refer to [DgLi], p. 6.

(4.9) For a generic 2-plane $E \subset \mathbb{P}^n$, its intersection $E \cap \tilde{\Delta}$ with the projective discriminant variety $\tilde{\Delta}$ is an irreducible curve of degree $d = 2n - 2$

with $\kappa = 3(n - 2)$ cusps, $\delta = 2(n - 2)(n - 3)$ nodes, and no other singularities.

We need this result only in the case $n = 3$, and then the reader can obtain it directly using the equation for the discriminant hypersurface $\tilde{\Delta} \subset \mathbb{P}^3$ and the discussion of its geometry given in [D4], p. 58. Use (4.7(i)) to be sure that all the three cuspidal quartics Q in $\mathbb{P}^2$ have the same fundamental group $\pi_1(\mathbb{P}^2 \setminus Q)$.

Second Proof. This proof is more self-contained than the first one, but requires some tedious computations. We apply the van Kampen–Zariski theorem (3.15) to the three cuspidal quartic C from (4.2) and the center of projection $p = (1 : -1 : 0)$. Take the line L_∞: $x + y = 0$ as the line at infinity and work in the affine chart $\mathbb{C}^2 = \mathbb{P}^2 \setminus L_\infty$. Replacing y with $1 - x$, we get the following equation for C in $\mathbb{C}^2$:

$$g(x, z) = x^4 - 2x^3 + (4z^2 + 2z + 1)x^2 - (4z^2 + 2z)x + z^2 = 0.$$

The pencil of lines (L_a), $a \in \mathbb{P}^1$, through p in $\mathbb{P}^2$ consists of the line at infinity L_∞ and the family of lines L_t: $z = t$ in our affine coordinates.

A simple computation gives the following list of exceptional lines through p with respect to the curve C:

(i) L_∞, which passes through the cusp $a_3 = (0 : 0 : 1)$;
(ii) $L_{1/8}$, which is tangent to C at a simple point;
(iii) L_0, which passes through the cusps $a_1 = (1 : 0 : 0)$ and $a_2 = (0 : 1 : 0)$;
(iv) L_{-1}, which is bitangent to C at two simple points.

Let a be a small positive number and let $L_a \cap C = \{b_1, b_2, b_3, b_4\}$, where the points b_1 and b_2 are near the cusp a_1 and the points b_3 and b_4 are near the cusp a_2.

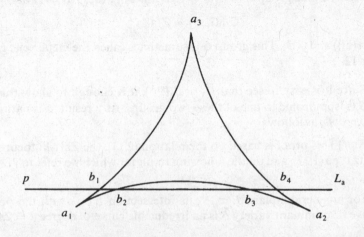

Let g_i be elements in $\pi_1(L_a \backslash C, p)$ corresponding to loops around the points b_i, chosen so that they generate this group and satisfy the relation

(4.10) $$g_4 g_3 g_2 g_1 = 1$$

as in (2.16(ii)).

When the line L_a moves around the exceptional line L_0 we get the following monodromy relations as in (2.22(ii)):

(4.11) $g_1 g_2 g_1 = g_2 g_1 g_2$ and $g_3 g_4 g_3 = g_4 g_3 g_4$.

When the line L_a approaches the line L_1, the points b_1 and b_4 merge together and we get the monodromy relation

(4.12) $$g_1 = g_4$$

as in (2.22(iii)).

When the line L_a approaches the line L_{-1}, the points b_1 and b_4 merge together again, and so do the points b_2 and b_3. Hence we get as above

(4.13) $g_1 = g_4$ and $g_2 = g_3$.

Thus we have by (3.15)

$$G = \langle g_1, g_2 : g_1 g_2 g_1 = g_2 g_1 g_2, g_1 g_2^2 g_1 = 1 \rangle,$$

i.e., $G \simeq B_3(\mathbb{P}^1)$. If we set $\alpha = g_1 g_2$ and $\beta = g_1 g_2 g_1$, we get the following new presentation for G:

$$G = \langle \alpha, \beta : \alpha^3 = \beta^2 = (\beta \alpha)^2 \rangle,$$

i.e., $G \simeq \tilde{D}_3$. We may find still another proof of (4.8) in [O6].   □

(4.14) **Corollary.** *Let F be the affine Milnor fiber associated to a three cuspidal quartic curve in* $\mathbb{P}^2$. *Then*

$$\pi_1(F) = [G, G] = \mathbb{Z}/3\mathbb{Z}.$$

Proof. Use (4.8) and (1.10).   □

(4.15) **Exercise.**

(i) Consider the projective discriminant hypersurface $\tilde{\Delta} \subset \mathbb{P}^4$ and let E be the 2-plane spanned by the binary forms $\tilde{p}_1 = x^4, \tilde{p}_2 = y^4, \tilde{p}_3 = (x + y)^4$.

Show that the intersection curve $E \cap \tilde{\Delta} = C$ has, in suitable coordinates $(a : b : c)$ on $E = \mathbb{P}^2$, the following equation:

$$(ab + bc + ca)^3 - 27a^2 b^2 c^2 = 0.$$

This curve C has one node A_1, three singularities of type E_6, and no other singularities, i.e., the 2-plane E is not generic with respect to $\tilde{\Delta}$, compare with (4.9). Using (3.2), deduce the existence of an epimorphism

$$\pi_1(\mathbb{P}^2 \backslash C) \to B_4(\mathbb{P}^1).$$

(ii) Show that any irreducible section of the projective discriminant $\tilde{\Delta} \subset \mathbb{P}^3$ is a three cuspidal quartic, in contrast to the case (i) above. *Hint.* Use (4.3).

(iii) Show that there is no 2-plane E in $\mathbb{P}^4$ such that the intersection $C = E \cap \tilde{\Delta}$ is irreducible and has exactly one singular point.

Hint. If such a section C exists, the unique singular point p should correspond to the binary form x^4 (up to a change of coordinates). Write the binary forms in the neighborhood of x^4 in the form

$$\tilde{p} = x^4 + a_1 x^3 y + a_2 x^2 y^2 + a_3 x y^3 + a_4 y^4.$$

In these coordinates $(a_1, \ldots, a_4)$, the equation for $\tilde{\Delta}$ is given by a weighted homogeneous polynomial

$$P(a_1, a_2, a_3, a_4) = 0$$

of degree 7 with respect to the weights $\text{wt}(a_i) = i$ for $i = 1. \ldots, 4$, and such that

$$P(0, 0, a_3, a_4) = 27a_3^4 - 256a_4^3$$

defines an E_6-singularity. The sum S of all the Milnor numbers of a generic section of $\tilde{\Delta}$ is, according to (4.9), given by $S = 16$. Use the inequality $16 > 6 = \mu(E_6)$ to reach a contradiction.

The second example we want to discuss in this section is contained in the following result.

(4.16) Proposition. *Let p, q be two integers such that $p \geq 2, q \geq 2$, and $(p, q) = 1$. Consider the irreducible curve*

$$C_{p,q} : (x^p + y^p)^q + (y^q + z^q)^p = 0.$$

Then

$$\pi_1(\mathbb{P}^2 \setminus C_{p,q}) = (\mathbb{Z}/p\mathbb{Z}) * (\mathbb{Z}/q\mathbb{Z}).$$

For $p = 2, q = 3$, we obtain a sextic curve having six cusps A_2 situated on a conic, an example already treated by Zariski [Z1]. The general case of (4.16) was proved by Oka [O4], using some tedious computations. The proof we give below is due to Nemethi [N2]; our contribution is only to make the group-theoretic part more transparent using (2.5). We have chosen this proof not only for its simplicity, but also because it illustrates again the possibility of applying the Zariski theorem of Lefschetz type (1.17) "in the opposite sense, using a convenient hypersurface in order to solve the problem in the plane," see Zariski [Z3], Introduction.

Proof. Consider the following surface in $\mathbb{P}^3$

$$V : (x^p + t^p)^q + (y^q + z^q)^p = 0.$$

It is easy to see that the singular locus $S(V)$ of V consists of pq lines L_{ij} $(i = 1, \ldots, p, j = 1, \ldots, q)$ defined as follows. The line L_{ij} joins the point $A_i =$

$(1 : 0 : 0 : \alpha_i)$ to the point $B_j = (0 : 1 : \beta_j : 0)$ where α_i (resp. β_j) are the roots of the equation $1 + \alpha^p = 0$ (resp. $1 + \beta^q = 0$).

A Whitney stratification for V can be obtained by taking as strata $V \setminus S(V)$, $S(V) \setminus S_0$, and S_0 where S_0 is the set consisting of all the points A_i, B_j. The 2-plane $E : t = y$ is transversal to all these strata since it contains none of the points A_i, B_j and it is not tangent to the smooth part $V \setminus S(V)$. Hence by the Zariski theorem (1.17) we get an isomorphism

$$\pi_1(\mathbb{P}^2 \setminus C_{p,q}) \xrightarrow{\sim} \pi_1(\mathbb{P}^3 \setminus V),$$

where we have identified E with $\mathbb{P}^2$ and the intersection $E \cap V$ with the curve $C_{p,q}$.

Let F be the Milnor fiber associated to the surface V and let $h : F \to F$ be the corresponding monodromy homeomorphism. Using (2.4) we obtain

$$\pi_1(\mathbb{P}^3 \setminus V) = \tilde{G}(F, h).$$

To compute this group, consider the polynomials

$$f^1 = (x^p + t^p)^q \qquad \text{and} \qquad f^2 = (y^q + z^q)^p.$$

Let $F^i : f^i - 1 = 0$ and let $h^i : F^i \to F^i$ be the corresponding Milnor fibers and monodromy homeomorphisms for $i = 1, 2$.

We obtain, from the Thom–Sebastiani theorem (3.3.19), a homotopy equivalence

$$j : F^1 * F^2 \to F$$

such that $h \circ j = j \circ (h^1 * h^2)$. Moreover, using (3.3.19'), it follows that j can be taken to be injective. In other words, we can identify $F^1 * F^2$ with a subspace in F, invariant under h. To obtain a group isomorphism

$$\tilde{G}(F, h) = \tilde{G}(F^1 * F^2, h^1 * h^2)$$

we need the following general result.

(4.17) **Lemma.** *Let S be a connected, locally contractible topological space and let $N : S \to S$ be a homeomorphism of finite order d. Let $T \subset S$ be a subspace such that:*

(i) *T is connected and locally contractible;*
(ii) *$N(T) \subset T$ and the restriction $N \mid T : T \to T$ is a homeomorphism of order d;*
(iii) *the inclusion $j : T \to S$ induces an isomorphism*

$$j_\# : \pi_1(T) \xrightarrow{\sim} \pi_1(S).$$

Then $\tilde{G}(S, N) = \tilde{G}(T, N \mid T)$.

Proof. If $p : \tilde{S} \to S$ is the universal covering space of S, then (iii) implies that $p \mid \tilde{T} : \tilde{T} \to T$ is the universal covering space of the subspace T where $\tilde{T} = p^{-1}(T)$. If $\tilde{N} : \tilde{S} \to \tilde{S}$ is a lifting of the homeomorphism N, it is clear that $\tilde{N}(\tilde{T}) \subset \tilde{T}$, i.e., $\tilde{N} \mid \tilde{T}$ is a lifting of the homeomorphism $N \mid T$. It follows that restriction to $\tilde{T}$ induces a group homomorphism

$$\rho\colon \tilde{G}(S, N) \to \tilde{G}(T, N \mid T).$$

Using (ii), (iii), and basic facts about universal covering spaces, see, for instance, [Sp], pp. 85–87, it follows that ρ is an isomorphism.

It remains now to compute the group $\tilde{G}(F^1 * F^2, h^1 * h^2)$. To do this, recall the notations F_a, h_a from Example (2.5). Note that the Milnor fiber F^1 has q connected components and we can "index" them using the following injective map

$$j^1\colon F_q \to F^1, \qquad j^1\left(\exp\frac{2\pi i k}{q}\right) = \left(\exp\frac{2\pi i k}{pq}, 0\right)$$

for $k = 0, \ldots, q - 1$. We clearly have $h^1 \circ j^1 = j^1 \circ h_q$. We can define, in a similar way, an injective map $j^2\colon F_p \to F^2$ and get the following commutative diagram:

$$
\begin{array}{ccc}
F_q * F_p & \xrightarrow{\ j^1*j^2\ } & F^1 * F^2 \\
{\scriptstyle h_q*h_p}\downarrow & & \downarrow{\scriptstyle h^1*h^2} \\
F_q * F_p & \xrightarrow{\ j^1*j^2\ } & F^1 * F^2
\end{array}
$$

Since $j^1 * j^2$ is again injective, we can identify $F_q * F_p$ with a subspace in $F^1 * F^2$. Recall that by (2.5) we have

$$\tilde{G}(F_q * F_p, h_q * h_p) = (\mathbb{Z}/q\mathbb{Z}) * (\mathbb{Z}/p\mathbb{Z}) = (\mathbb{Z}/p\mathbb{Z}) * (\mathbb{Z}/q\mathbb{Z}).$$

Hence, we can complete the proof by using (4.17) and the following general result.

(4.18) **Lemma.** *Let A, B be two locally path-connected topological spaces and let $A = A_1 \cup \cdots \cup A_a$, $B = B_1 \cup \cdots \cup B_b$ be their decompositions into path-connected components. Choose $\alpha_i \in A_i$, $i = 1, \ldots, a$, and $\beta_j \in B_j$, $j = 1, \ldots, b$, and set $A^0 = \{\alpha_1, \ldots, \alpha_a\}$, $B^0 = \{\beta_1, \ldots, \beta_b\}$. Then the inclusion*

$$j\colon A^0 * B^0 \to A * B$$

induces an isomorphism

$$j_\#\colon \pi_1(A^0 * B^0) \to \pi_1(A * B).$$

Proof. Let $r_A\colon A \to A^0$ be the map defined by

$$r_A(x) = \alpha_i \qquad \text{for} \quad x \in A_i.$$

Let r_B be the analogous map for the space B and note that

$$(r_A * r_B) \circ j = 1.$$

It follows that $j_\#$ is a monomorphism. To make the proof for $j_\#$ surjective more transparent, we assume that the components A_i, B_j have the homotopy type of CW-complexes whose 0-skeletons consist of the points α_i and β_j, respectively.

We shall apply this lemma in the case $A = F^1$, $B = F^2$, and the connected components of these Milnor fibers are homotopy equivalent to bouquets of circles by (3.2.7), i.e., in the case at hand the above assumption is fulfilled. We can even assume that A_i and B_j are CW-complexes as above, since the group $\pi_1(A * B)$ is clearly a homotopy invariant. Then the join $A * B$ has an induced structure of a CW-complex such that its 1-skeleton $(A * B)^1$ is the union of $A^0 * B^0$ with the 1-skeletons A_i^1, B_j^1 for $i = 1, \ldots, a, j = 1, \ldots, b$.

Here and in the sequel we identify the factors A, B with the corresponding subsets in $A * B$. The morphism

$$\pi_1(A * B)^1) \to \pi_1(A * B)$$

is surjective by [Sp], p. 403. Hence it is enough to show that any path $\gamma: [0, 1] \to (A * B)^1$, $\gamma(0) = \gamma(1) = \alpha_1$, can be deformed in $A * B$ to a path γ' lying in the subset $A^0 * B^0$. This path γ can be taken to be "piecewise linear" in an obvious sense. Hence γ can be written as

$$\gamma = \gamma_1 \cdots \gamma_m,$$

where each path γ_i is either:

(i) a closed path in some A_j^1 with $\gamma_i(0) = \gamma_j(1) = \alpha_j$ (resp. in some B_j^1 with $\gamma_i(0) = \gamma_i(1) = \beta_j$); or
(ii) a path lying in $A^0 * B^0$.

Assume, to fix the notations, that γ_i is a closed path in B_j^1. Then we can replace γ_i with the path

$$\gamma_i^1 = [\beta_j, \alpha_1][\alpha_1, \beta_j]\gamma_i[\beta_j, \alpha_1][\alpha_1, \beta_j],$$

where $[a, b]$ denotes the path along the segment $\overline{ab}$ going from a to b.

Now the loop

$$[\alpha_1, \beta_j]\gamma_i[\beta_j\alpha_1]$$

is a loop in the contractible cone $\alpha_1 * B_j$ and hence it is trivial in $A * B$. Therefore γ is equivalent in $A * B$ to the path

$$\gamma_1 \cdots \gamma_{i-1}[\beta_j, \alpha_1][\alpha_1, \beta_j]\gamma_{i+1} \cdots \gamma_m = \gamma_1 \cdots \gamma_{i-1}\gamma_{i+1} \cdots \gamma_m.$$

In this way all the loops of type (i) in the above decomposition of γ can be canceled, without changing the equivalence class of the loop γ in $A * B$.

The resulting path is a path lying in $A^0 * B^0$ and this ends the proof of (4.18).   □

(4.19) **Exercise.** (i) Using (4.9), show that the intersection of the projective discriminant hypersurface $\tilde{\Delta} \subset \mathbb{P}^4$ with a generic 2-plane E is an irreducible curve $C = E \cap \tilde{\Delta}$ of degree 6, having six cusps and four nodes as its singularities.

(ii) Show that
$$\pi_1(E \setminus C) = B_4(\mathbb{P}^1).$$

(iii) Show that the group $B_4(\mathbb{P}^1)$ is not isomorphic to the group $(\mathbb{Z}/2\mathbb{Z}) *$ $(\mathbb{Z}/3\mathbb{Z})$. *Hint.* Use the fact that $\mathrm{Center}(B_4(\mathbb{P}^1)) = \mathbb{Z}/2\mathbb{Z}$, see [DgLi], and the identification
$$(\mathbb{Z}/2\mathbb{Z}) * (\mathbb{Z}/3\mathbb{Z}) = \mathrm{PSL}_2(\mathbb{Z})$$
to show that $\mathrm{Center}((\mathbb{Z}/2\mathbb{Z}) * (\mathbb{Z}/3\mathbb{Z})) = 1$.

(4.20) **Remark.** We should compare the characterization of fibered knots, given by Stallings' theorem (2.1.25), with (4.14) in view of the relation
$$\pi_1(F) = [\pi_1(S_\varepsilon^5 \setminus K), \pi_1(S_\varepsilon^5 \setminus K)].$$

(This is a consequence of (1.5).) On the other hand, the Milnor fiber F corresponding to the plane curve $C_{p,q}$ from (4.16) has a fundamental group $\pi_1(F)$ which is free and finitely generated as in (2.1.25).

(4.21) **Example.** Consider the family of plane sextic curves $C^t: (xy + tz^2)^3 + (xyz + x^3 + y^3)^2 = 0$. The following statements are easy to check:

(i) The curve C^0 is irreducible and has $(0:0:1)$ as its unique singular point.
(ii) The curve C^t for $0 < t < \varepsilon$ has six cusps situated on the conic $xy + tz^2 = 0$. An obvious equisingularity argument and (4.16) show that

$$\pi_1(\mathbb{P}^2 \setminus C^t) = (\mathbb{Z}/2\mathbb{Z}) * (\mathbb{Z}/3\mathbb{Z}).$$

(iii) Using (3.2) we get an epimorphism

$$e: \pi_1(\mathbb{P}^2 \setminus C^0) \to \pi_1(\mathbb{P}^2 \setminus C^t).$$

It follows that $\pi_1(\mathbb{P}^2 \setminus C^0)$ is not an abelian group. In fact, it can be shown that e is an isomorphism, but the details for this will be given elsewhere.

•

CHAPTER 5

Projective Complete Intersections

§1. Topology of the Projective Space $\mathbb{P}^n$

Although the topology of the complex projective space $\mathbb{P}^n$ is well known, we recall in this section some basic facts on it. The reason for doing this is:

(i) to fix some notation useful in the sequel; and
(ii) the topology of the projective complete intersections shares a lot of properties with the topology of $\mathbb{P}^n$.

The n-dimensional complex projective space $\mathbb{P}^n$ is a compact complex manifold and has a natural orientation, see [GH], p. 18. This orientation leads to an identification

$$(1.1) \qquad H^{2n}(\mathbb{P}^n) = \mathbb{Z}; \qquad \omega \mapsto \langle \omega, [\mathbb{P}^n] \rangle,$$

where $\omega \in H^{2n}(\mathbb{P}^n)$, $[\mathbb{P}^n]$ is the fundamental class of the manifold $\mathbb{P}^n$ and generates $H_{2n}(\mathbb{P}^n)$, and $\langle \ , \ \rangle$ denotes the evaluation pairing

$$H^{2n}(\mathbb{P}^n) \times H_{2n}(\mathbb{P}^n) \to \mathbb{Z}.$$

Indeed, using the universal-coefficient theorem for cohomology, see [Sp], p. 243, we get a natural identification

$$H^{2n}(\mathbb{P}^n) = \operatorname{Hom}(H_{2n}(\mathbb{P}^n), \mathbb{Z}).$$

The projective space $\mathbb{P}^n$ is a CW-complex of dimension $2n$, having exactly one cell e^{2i} in each of the even dimensions, i.e.,

$$(1.2) \qquad \mathbb{P}^n = e^0 \cup e^2 \cup \cdots \cup e^{2n}.$$

This decomposition can be obtained as follows.

(1.3) **Exercise.**
(i) Consider the Hopf map

$$H: S^{2n-1} \to \mathbb{P}^{n-1}, \qquad h(x_1, \ldots, x_n) = (x_1 : \ldots : x_n),$$

where S^{2n-1} is the unit sphere in $\mathbb{C}^n$.

Show that the n-dimensional projective space $\mathbb{P}^n$ can be obtained from the $(n-1)$-dimensional projective space $\mathbb{P}^{n-1}$ by attaching a $2n$-cell via the Hopf map H. *Hint.* If necessary, have a look at [Gn], p. 86.

(ii) Consider the function $f\colon \mathbb{P}^n \to \mathbb{R}$ defined by

$$f(x) = \frac{c_0|x_0|^2 + c_1|x_1|^2 + \cdots + c_n|x_n|^2}{|x_0|^2 + |x_1|^2 + \cdots + |x_n|^2},$$

where $c_0, \ldots, c_n$ are distinct real numbers.

Show that f is a Morse function with one critical point of index $2i$ for each $i = 0, 1, \ldots, n$. *Hint.* If necessary, have a look at [M3], p. 26.

(1.4) Corollary.

$$\pi_1(\mathbb{P}^n) = 0.$$

This follows from (1.2) and also from the remark that the Hopf map is in fact an S^1-bundle map

(1.5) $S^1 \to S^{2n+1} \xrightarrow{H} \mathbb{P}^n.$

The Gysin sequence in cohomology associated to the map (1.5) gives the following result.

(1.6) Proposition. *The integral cohomology algebra $H_\cdot(\mathbb{P}^n)$ for $n \geq 1$ is a truncated polynomial algebra*

$$H_\cdot(\mathbb{P}^n) = \mathbb{Z}[\alpha]/(\alpha^{n+1})$$

generated by an element α of degree 2.

More precisely, let $j\colon \mathbb{P}^1 \to \mathbb{P}^n$ be the embedding $(x_0 : x_1) \mapsto$ $(x_0 : x_1 : 0 : \cdots : 0)$. Then using (1.3(i)) inductively, it follows that

$$j^*\colon H^2(\mathbb{P}^n) \to H^2(\mathbb{P}^1)$$

is an isomorphism. The generator α in (1.6) is chosen such that $j^*(\alpha) = 1$ under the identification (1.1).

Consider now the exact sequence of sheaves on $\mathbb{P}^n$, the so-called *exponential sequence*

$$0 \to \mathbb{Z} \to \mathcal{O} \xrightarrow{\exp} \mathcal{O}^* \to 0.$$

This induces an isomorphism

$$\mathrm{Pic}(\mathbb{P}^n) = H^1(\mathbb{P}^n, \mathcal{O}^*) \xrightarrow[\sim]{c_1} H^2(\mathbb{P}^n)$$

showing that every line bundle $L \in \mathrm{Pic}(\mathbb{P}^n)$ is determined by its first Chern class $c_1(L) \in H^2(\mathbb{P}^n)$.

In particular, the *hyperplane bundle* $\mathcal{O}(1) = [H]$ satisfies

(1.7) $c_1(\mathcal{O}(1)) = \alpha,$

see, for instance, [Hz1], pp. 58–59. The inverse (or dual) line bundle

$$\mathcal{O}(-1) = [-H] = [H]'$$

is just the universal (canonical, tautological) line bundle on $\mathbb{P}^n$ with first Chern given by

(1.8) $$c_1(\mathcal{O}(-1)) = -\alpha.$$

We have the following exact sequence of vector bundles on $\mathbb{P}^n$, involving the tangent bundle $T\mathbb{P}^n$ of the projective space $\mathbb{P}^n$, see [GH], p. 409, or [Hn], p. 182,

$$0 \to \mathcal{O} \to \mathcal{O}(1)^{n+1} \to T\mathbb{P}^n \to 0.$$

It follows that the total Chern class of the projective space $\mathbb{P}^n$ is given by

(1.9) $$c(\mathbb{P}^n) = c(T\mathbb{P}^n) = (1 + \alpha)^{n+1}.$$

Let $V \subset \mathbb{P}^{n+1}$ be a reduced hypersurface of degree d and let Z be its singular locus. Then $V \setminus Z$ is a smooth complex submanifold in $\mathbb{P}^{n+1} \setminus Z$ and hence has a normal bundle $\nu_{V \setminus Z}$. This line bundle on $V \setminus Z$ is determined by the following *adjunction formula*, see [GH], p. 146 (they consider only the case V smooth, but the proof carries over to our more general setting)

(1.10) $$\nu_{V \setminus Z} = \mathcal{O}(d)|_{V \setminus Z},$$

i.e., $\nu_{V \setminus Z}$ is the restriction of the line bundle $\mathcal{O}(d)$ to the subvariety $V \setminus Z$.

As a very special case, consider the hyperplane $H: x_0 = 0$ in $\mathbb{P}^{n+1}$. Then $\nu_H = \mathcal{O}(1)|_H$, i.e., the hyperplane bundle $\mathcal{O}(1)$ on $\mathbb{P}^n$ can be regarded as the normal line bundle ν_H under the identification $\mathbb{P}^n = H$. Moreover, let $a = (1 : 0 : \cdots : 0)$ and consider the projection with center a

$$p: \mathbb{P}^{n+1} \setminus \{a\} \to H = \mathbb{P}^n.$$

Then p is exactly the projection in the hyperplane bundle $\mathcal{O}(1)$, see [Hz1], p. 59.

We end this section with some general remarks on line bundles and their associated S^1-bundles. Let L be a line bundle on a complex analytic space M. The multiplication by scalars in $\mathbb{C}$ induces an action of the circle

$$S^1 = \{\lambda \in \mathbb{C}; |\lambda| = 1\}$$

on the total space $T(L)$ of the line bundle L. Put a Hermitian metric $(\ ,\ )$ on L and consider the principal S^1-bundle $S(L)$ consisting of all the vectors $v \in T(L)$ with $(v, v) = 1$. Since any two Hermitian metric $(\ ,\ )_1$ and $(\ ,\ )_2$ on L can be put in a family

$$t(\ ,\ )_1 + (1 - t)(\ ,\ )_2, \quad t \in [0, 1],$$

of Hermitian metrics, it follows that the homeomorphism class of the principal S^1-bundle $S(L)$ does not depend on the choice of the metric on L (use (1.3.5)).

(1.11) **Exercise.** (i) Show that there is a homeomorphism

$$S(L) \to S(L'),$$

where L' denotes the dual (inverse) bundle of L. *Hint.* Use a Hermitian metric on L to get an $\mathbb{R}$-linear isomorphism $L \to L'$.

(ii) Show that there is a homeomorphism

$$S(L)/G(p) \to S(L^{\otimes d})$$

for any positive integer d, where $G(p)$ is the finite cyclic subgroup in S^1 consisting of all the elements of order d. *Hint.* A homeomorphism is induced by the map

$$v \to v \otimes \cdots \otimes v.$$

§2. Topology of Complete Intersections

From now on in this chapter we consider a (set-theoretic global) *complete intersection*

$$(2.1) \qquad\qquad V: f_1 = \cdots = f_c = 0 \quad \text{in } \mathbb{P}^{n+c}$$

with $n \geq 1$. This means that the codimension of the variety V is equal to the number of equations f_i used to define it. In our notation, we have

$$\dim V = n, \qquad \text{codim } V = c.$$

When $c = 1$, we set $f_1 = f$ and V is called a hypersurface. If V is a complete intersection in $\mathbb{P}^{n+c}$, then the germ (V, x) is a complete intersection singularity in the sense of (1.6.1) for any point $x \in V$. The converse of this remark is false, see (2.5) below.

The *affine cone* CV over the variety V is the affine variety defined by

$$(2.2) \qquad\qquad CV: f_1 = \cdots = f_c = 0 \quad \text{in } \mathbb{C}^{n+c+1}.$$

Let S denote the unit sphere in $\mathbb{C}^{n+c+1}$ and let $K_V = S \cap CV$ denote the associated link. The restriction of the Hopf bundle (1.5) to V is called the *Hopf bundle of the variety V.*

$$(2.3) \qquad\qquad S^1 \to K_V \xrightarrow{H_V} V.$$

This is exactly the S^1-bundle associated to the line bundle $L = \mathcal{O}(-1)|_V$ on V. Using the exact homotopy sequence of a fibration and (3.2.12) we get:

(2.4) **Corollary.** *The complete intersection V is simply-connected for $n = \dim V \geq 2$ and connected for $n = 1$.*

(2.5) **Exercise.** (i) Show that the disjoint union of two lines in $\mathbb{P}^3$ is not a complete intersection (in spite of the fact that it is a locally complete intersection).

(ii) Let C_1 and C_2 be two smooth irreducible projective curves. Show that $C_1 \times C_2$ is isomorphic to a complete intersection in some projective space only when $C_1 = C_2 = \mathbb{P}^1$. *Hint.* Recall that $\pi_1(C) = 0 \Leftrightarrow C = \mathbb{P}^1$. Next we use the map $\mathbb{P}^1 \times \mathbb{P}^1 \to \mathbb{P}^3$

$$(a_0 : a_1), (b_0 : b_1)) \mapsto (a_0 b_0, a_0 b_1, a_1 b_0, a_1 b_1)$$

to show that $\mathbb{P}^1 \times \mathbb{P}^1$ is isomorphic to the quadratic hypersurface

$$x_0 x_3 - x_1 x_2 = 0.$$

Another strong restriction on the topology of a complete intersection is given by the following result.

(2.6) Lefschetz Theorem. *Let $j: V \to \mathbb{P}^{n+c}$ be the inclusion. Then*

$$j^k : H^k(\mathbb{P}^{n+c}) \to H^k(V)$$

is an isomorphism for $k < n$ and a primitive monomorphism for $k = n$.

Proof. Let $\mathbb{P} = \mathbb{P}^{n+c}$, $U = \mathbb{P} \setminus V$. The exact sequence in cohomology of the pair $(\mathbb{P}, V)$ and the Alexander duality isomorphism

$$H^s(\mathbb{P}, V) = H_{2n+2c-s}(U)$$

show that it is enough to prove:

(i) $H_m(U) = 0$ for $m \geq n + 2c$;
(ii) $H_{n+2c-1}(U)$ is torsion free.

We can write $U = U_1 \cup \cdots \cup U_c$ where

$$U_i = \mathbb{P} \setminus \{x \in \mathbb{P};\ f_i(x) = 0\}.$$

Each of these open sets U_i is affine by (1.6.7(ii)). Hence $H_m(U_i) = 0$ for $m > n + c$ and $H_{n+c}(U)$ is torsion free by (1.6.8). We end the proof using induction on c and the Mayer–Vietoris sequence for the open cover $U_1 \cup \cdots \cup U_c$. $\square$

(2.7) Exercise. (i) Give a different proof for (2.6) using the Gysin sequence associated with the Hopf bundle H_V and (3.2.12).

(ii) Show that the inclusion $j: V \to \mathbb{P}^{n+c}$ is an n-homotopy equivalence. *Hint.* There are at least two ways to proceed in (ii):

(a) use the general result in (1.6.5);
(b) use (2.4), (2.6), and the Whitehead theorem, see [Sp], pp. 399 and 405.

The structure of the cohomology groups $H^k(V)$ in dimensions $k \geq n$ can be very different from the cohomology groups $H^k(\mathbb{P}^n)$. Assume from now on that V is reduced and let

$$V = V_1 \cup \cdots \cup V_p$$

be the decomposition of V into irreducible components. Then the degrees

$d = \deg V$ and $e_i = \deg(V_i)$, $i = 1, \ldots, p$, are related by the following formula

(2.8) $$d = e_1 + \cdots + e_p.$$

Moreover, using the Lefschetz Duality Theorem, see [Sp], p. 297, it follows that

(2.9) $$H^{2n}(V) = H^{2n}(V_1) \oplus \cdots \oplus H^{2n}(V_p) = \mathbb{Z}^p$$

$$\omega \mapsto (\langle \omega, [V_1] \rangle, \ldots, \langle \omega, [V_p] \rangle)$$

as in (1.1) essentially.

With this identification, the morphism

$$j^{2n} \colon H^{2n}(\mathbb{P}^{n+c}) \to H^{2n}(V)$$

is given by the following formula

(2.10) $$\mathbb{Z} \ni a \mapsto (e_1 a, \ldots, e_p a) \in \mathbb{Z}^p.$$

This can be seen as follows. Let E_m denote a generic linear space in $\mathbb{P}^{n+c}$ of dimension m. Then

$$e_i = \#(V_i \cap E_c) = ([V_i], [E_c])$$

the intersection number of the cycles $[V_i]$ and $[E_c]$ in $H_\cdot(\mathbb{P}^{n+c})$. It follows that $[V_i] = e_i[E_n]$ and hence

$$\langle j^{2n}(\alpha^n), [V_i] \rangle = \langle \alpha^n, [V_i] \rangle = \langle \alpha^n, e_i[E_n] \rangle = e_i.$$

Let $s = \dim V_{\text{sing}}$ be the dimension of the singular locus of V (as usual, $\dim \varnothing = -1$). Note that we have $d = \deg(V) = d_1 \cdots d_c$ where $d_i = \deg f_i$.

(2.11) **Theorem.**

$$H^m(V) = H^m(\mathbb{P}^{n+c}) \qquad for \quad n + s + 2 \le m \le 2n,$$

and the morphism $j^m \colon H^m(\mathbb{P}^{n+c}) \to H^m(V)$ *is multiplication by* $d = \deg V$.

Proof. The general case is quite technical and we refer to Kato [Ka] for a proof in an even more general setting. Here we give a proof for the *hypersurface case*, based on the Kato–Matsumoto result (3.2.2). For simplicity of notation, put $S = S^{2n+3}$, $\mathbb{P} = \mathbb{P}^{n+1}$. Consider the Milnor fibration

$$F \to S \backslash K_V \to S^1$$

associated to the hypersurface singularity $(CV, 0)$. Then, by (3.2.2) we deduce that the Milnor fiber F is $(n - s - 1)$-connected. The corresponding Wang sequence (3.1.18) shows that

$$H_k(S \backslash K_V) = 0 \qquad for \quad 2 \le k \le n - s - 1.$$

Thus we get

$$H^{2n-k+2}(K_V) = H^{2n-k+3}(S, K_V) = 0$$

from the Alexander duality.

Consider the commutative diagram of Hopf fibrations:

$$
\begin{array}{ccc}
S & \longrightarrow & \mathbb{P} \\
\uparrow & & \uparrow \\
\big| & & \big| \\
K_V & \longrightarrow & V
\end{array}
$$

The associated Gysin sequences in cohomology lead to the following diagram:

$$
\begin{array}{ccccccccc}
\longrightarrow & H^{2p+1}(S) & \longrightarrow & H^{2p}(\mathbb{P}) & \xrightarrow[\sim]{\psi} & H^{2p+2}(\mathbb{P}) & \longrightarrow & H^{2p+2}(S) & \longrightarrow \\
 & \big\downarrow & & \big\downarrow{\scriptstyle j^{2p}} & & \big\downarrow{\scriptstyle j^{2p+2}} & & \big\downarrow & \\
\longrightarrow & H^{2p+1}(K_V) & \longrightarrow & H^{2p}(V) & \xrightarrow[\sim]{\psi_V} & H^{2p+2}(V) & \longrightarrow & H^{2p+2}(K_V) & \longrightarrow
\end{array}
$$

Here ψ (resp. ψ_V) is the cup product with α (resp. with $\alpha_V = j^2(\alpha)$), see [Sp], p. 261, For $2p \geq n + s + 2$, $2p \leq 2n - 2$, it follows that both ψ and ψ_V are isomorphisms. The result follows from (2.10) by decreasing induction on p, as soon as we prove $H^{2n-1}(V) = 0$ (assume $n \geq s + 3$). But this follows from the above Gysin sequence since we have

$$0 = H^{2n}(K_V) \to H^{2n-1}(V) \xrightarrow{\psi_V} 0. \qquad \square$$

(2.12) **Corollary.** *Let V be an n-dimensional reduced complete intersection in some projective space $\mathbb{P}^N$ which has the same $\mathbb{Z}$-cohomology algebra as $\mathbb{P}^n$. If $n \geq 2$, then V is isomorphic as an algebraic variety to $\mathbb{P}^n$.*

Proof. The ring isomorphism $H^{\cdot}(V) \simeq H^{\cdot}(\mathbb{P}^n)$ implies:

(i) V is irreducible by (2.9);

(ii) $H^2(V) = \mathbb{Z}$ is generated by $\alpha_V = j^2(\alpha)$ where $j: V \to \mathbb{P}^N$ is the inclusion, by (2.6);

(iii) α_V^n is a generator for $H^{2n}(V)$.

On the other hand, it follows from (2.10) that $\alpha_V^n = (\deg V) \cdot g$ for some generator g of $H^{2n}(V) = \mathbb{Z}$. This is possible only when $\deg(V) = 1$, i.e., when all the equations f_i are linear. In this case, it is clear that V is a linear subspace in $\mathbb{P}^N$ and hence $V \simeq \mathbb{P}^n$ as algebraic varieties. $\qquad \square$

(2.13) **Exercise.** Consider the cuspidal curve $C: x^{d-1}y - z^d = 0$ in $\mathbb{P}^2$. Use projection from the singular point $(0:1:0)$ to show that C is homeomorphic to $\mathbb{P}^1$. In particular, C and $\mathbb{P}^1$ have the same cohomology algebra. This shows that the condition $n \geq 2$ in (2.12) is indeed necessary.

(2.14) **Remark.** In the hypersurface case ($c = 1$), consider the complement $U = \mathbb{P}^{n+1} \setminus V$. Assume that $s \leq n - 2$ so that the Milnor fiber F is simply-connected. Then the principal $\mathbb{Z}/d\mathbb{Z}$-bundle $F \to U$ (recall (4.18)) is just the universal covering space for the complement U. It follows that

$$\pi_1(U) = \mathbb{Z}/d\mathbb{Z} \quad \text{and} \quad \pi_i(U) = 0 \quad \text{for} \quad i = 0 \text{ or } 2 \leq i \leq n - s - 1.$$

In view of these formulas, it is natural to try to compare the complement U to the Eilenberg–MacLane space $K(\mathbb{Z}/d\mathbb{Z}, 1)$. Using a general result about $K(\pi, n)$ spaces, see [Sp], p. 428, we get a map

$$\varphi: U \to K(\mathbb{Z}/d\mathbb{Z}, 1)$$

which is a $(n - s - 1)$-equivalence. Moreover, recall that the homology of the space $K(\mathbb{Z}/d\mathbb{Z}, 1)$ is given by the following formula, see [Sp], p. 503:

$$(2.15) \qquad H_m(K(\mathbb{Z}/d\mathbb{Z}, 1)) = \begin{cases} \mathbb{Z}, & m = 0, \\ 0, & m \text{ even}, m > 0, \\ \mathbb{Z}/d\mathbb{Z}, & m \text{ odd}. \end{cases}$$

(2.16) **Exercise.** (i) In which way is (2.15) compatible with (2.11)?

(ii) Consider the line bundle $L = \mathcal{O}(1)/U$. Show that $\beta = c_1(L)$ is a generator for $H^2(U) = \mathbb{Z}/d\mathbb{Z}$ when $s \leq n - 2$. Deduce that U is not a parallelizable manifold when $n + 2$ is not a multiple of d. (This is interesting, since $U = F/\langle h \rangle$ and the Milnor fiber F is parallelizable by (3.1.23).)

(iii) Show that $\pi_1(\mathbb{P}^{n+c} \setminus V) = 0$ for $c > 1$.

The following result shows that (co)homology with coefficients in a field (we take $\mathbb{C}$ for simplicity, but $\mathbb{Q}$ or $\mathbb{R}$ are equally good) is much easier to study than the integer (co)homology considered until now.

(2.17) **Lemma.**

(i) $j^k: H^k(\mathbb{P}^{n+c}; \mathbb{C}) \to H^k(V; \mathbb{C})$ is a monomorphism for all k with $0 \leq k \leq 2n$.

(ii) $j_k: H_k(V; \mathbb{C}) \to H_k(\mathbb{P}^{n+c}; \mathbb{C})$ is an epimorphism for all k with $0 \leq k \leq 2n$.

Proof. Since (i) and (ii) are dual statements, it is enough to prove (i). But (i) is equivalent to

$$\alpha_V^k \neq 0 \quad \text{for} \quad 0 \leq k \leq n.$$

And this clearly holds, since $\alpha_V^n \in H^{2n}(V)$ corresponds to the element $e = (e_1, \ldots, e_p)$ in the notations from (2.10). This element e remains nonzero in

$$H^{2n}(V; \mathbb{C}) = H^{2n}(V) \otimes \mathbb{C}. \qquad \square$$

(2.18) **Definition.**

(i) $H_0^k(V; \mathbb{C}) = \text{coker } j^k$ is called the *primitive kth cohomology* group of the projective variety V;

(ii) $H_k^0(V; \mathbb{C}) = \ker j_k$ is call the *primitive* kth *homology* group of the projective variety V.

Alexander duality implies that the primitive cohomology of the complete intersection V is related to the homology of the complement $U = \mathbb{P}^{n+c} \setminus V$ by the equality

(2.19) $$H_0^k(V; \mathbb{C}) = H_{2n+2c-k-1}(U; \mathbb{C}).$$

In the hypersurface case, $U = F/\langle h \rangle$ and hence

(2.20) $$H_*(U; \mathbb{C}) = H_*(F; \mathbb{C})^{h_*},$$

the fixed part under the homology monodromy operator, see, for instance, [Bd], p. 120.

(2.21) **Exercise.** Consider the hypersurface $V: f(x) = x_0 x_1 \cdots x_{n+1} = 0$:

(i) Show that the corresponding Milnor fiber $F: f - 1 = 0$ has the homotopy type of a $(n + 1)$-dimensional torus

$$F \sim T^{n+1} = (S^1)^{n+1}.$$

(ii) Show that the corresponding monodromy operator h_* is the identity.

(iii) Deduce that

$$\dim H_0^{n+k}(V; \mathbb{C}) = \dim H^{n-k+1}(T^{n+1}) = \binom{n+1}{k}.$$

(iv) Reprove (iii), using the isomorphism

$$\mathbb{P}^{n+1} \setminus V = (\mathbb{C}^*)^{n+1}.$$

(2.22) **Corollary.** *A hypersurface* $V \subset \mathbb{P}^{n+1}$ *has the same* $\mathbb{C}$-*homology groups as* $\mathbb{P}^n$ *if and only if its corresponding monodromy operator*

$$\tilde{h}_*: \tilde{H}_*(F; \mathbb{C}) \to \tilde{H}_*(F; \mathbb{C}),$$

acting on the reduced homology of the associated Milnor fiber, has no eigenvalue equal to 1.

(2.23) **Exercise.** Show that the hypersurface

$$V_n: x_0 x_1 \cdots x_n + x_{n+1}^{n+1} = 0$$

has the same $\mathbb{C}$-homology groups as $\mathbb{P}^n$. *Hint.* Use (2.21(ii)), (3.4.12), and the Thom–Sebastiani formula (3.3.21). Note that the singular locus of V_n has dimension $n - 2$. For $n = 2$, V_2 is a cubic surface in $\mathbb{P}^3$ having three cusps as singularities. The $\mathbb{Z}$-homology of this surface is computed below in (4.8). In particular, $H_.(V) \neq H_.(\mathbb{P}^2)$.

A more general example similar to (2.23), but leading to hypersurfaces of any degree and with a finite number of singularities, is the following, see also [BD].

(2.24) **Proposition.** *For any dimension $n \geq 2$, degree $d \geq 3$, and integer a with $1 \leq a < d - 1$ consider the hypersurface*

$$V_n^{d,a} : x_0^a x_1^{d-a} + x_1 x_2^{d-1} + \cdots + x_{n-1} x_n^{d-1} + x_{n+1}^d = 0.$$

Then:

(i) *the hypersurface $V_n^{d,a}$ has one singular point for $a = 1$ and two singular points for $a > 1$;*

(ii) $H.(V_n^{d,a}; \mathbb{C}) = H.(\mathbb{P}^n; \mathbb{C})$ *for* $(a, d) = 1$.

Proof. (i) A direct computation shows that $p_1 = (1 : 0 : \cdots : 0)$ is the only singular point when $a = 1$, and that p_1 and $p_2 = (0 : 1 : 0 : \cdots : 0)$ are the only singular points for $a > 1$.

(ii) Consider the polynomial

$$g = x_0^a x_1^{d-a} + x_1 x_2^{d-1} + \cdots + x_{n-1} x_n^{d-1}.$$

Let G be the associated Milnor fiber and let $h_g : G \to G$, $h_g(x) = \lambda \cdot x$ with $\lambda = \exp(2\pi i/d)$, be the corresponding monodromy homeomorphism. We show that h_g is isotopic to the identity of G for $(a, d) = 1$. To do this, we look for a $\mathbb{C}^*$-action on $\mathbb{C}^{n+1}$ such that G is invariant with respect to this action, and the map h_g can be written as "multiplication" by some number $\mu \in \mathbb{C}^*$. Since $\mathbb{C}^*$ is connected we can choose a path μ_t from 1 to μ. The multiplication by μ_t provides us with the requested isotopy. Hence we look for an action

$$t \cdot x = (t^{w_0} x_0, \ldots, t^{w_n} x_n)$$

with *integer* weights w_i such that $\deg(g) = 0$ with respect to these weights. Explicitly, this means

$$aw_0 + (d - a)w_1 = w_1 + (d - 1)w_2 = \cdots = w_{n-1} + (d - 1)w_n = 0.$$

A solution to these equations is given by $w_0 = (1 - d)^{n-1}(a - d)$ and $w_j = (1 - d)^{n-j}a$ for $j = 1, \ldots, n$. As the integers a and d are coprime by assumption, there is an integer b such that $ab \equiv 1 \pmod{d}$. Since all the weights satisfy $w_j \equiv a \pmod{d}$, it follows that $\mu = \lambda^b$ has all the required properties.

Then it follows as in (2.23) that the monodromy operator of the function

$$g + x_{n+1}^d$$

has no eigenvalue equal to 1. This ends the proof in view of (2.22). □

We end this section with a discussion about some nice "tubular" neighborhoods for our complete intersection V in $\mathbb{P} = \mathbb{P}^{n+c}$. Details as well as proofs can be found in Durfee's paper [Df6].

Consider the real rational function

$$\varphi: \mathbb{P} \to \mathbb{R}_+, \qquad \varphi(x) = \frac{|f_1(x)|^{2d/d_1} + \cdots + |f_c(x)|^{2d/d_c}}{(|x_0|^2 + |x_1|^2 + \cdots + |x_{n+c}|^2)^d},$$

where $d_i = \deg f_i$ for $i = 1, \dots, c$ and $d = d_1, \dots, d_c$. Note that this function φ is well defined and

$$\varphi^{-1}(0) = V.$$

There is a positive real number $\delta_0 < 0$ such that φ has no critical value in $(0, \delta_0)$. For $\delta \in (0, \delta_0)$ we set

(2.25) $T_V = \varphi^{-1}([0, \delta])$

and refer to T_V as a closed "tubular" neighborhood for the subvariety V in $\mathbb{P}$. T_V is a closed manifold with boundary

$$\partial T_V = \varphi^{-1}(\delta).$$

The inclusions

$$V \to T_V \qquad \text{and} \qquad \mathbb{P} \setminus T_V \to \mathbb{P} \setminus V$$

are homotopy equivalences.

Consider the map

$$f = (f_1, \dots, f_c): \mathbb{C}^{n+c+1} \to \mathbb{C}^c$$

and let ψ be the restriction $f|S$ where S is the unit sphere in $\mathbb{C}^{n+c+1}$. Let $\tilde{S}$ be the "sphere" in $\mathbb{C}^c$ given in the equation

$$\tilde{S}: \rho(y) = |y_1|^{2d/d_1} + \cdots + |y_c|^{2d/d_c} = \delta$$

with $\delta \in (0, \delta_0)$ as above and δ a regular value for map $\rho \circ \psi$. It follows that

(2.26) $E = \psi^{-1}(\delta)$

is a smooth manifold in S.

The restriction of the Hopf map H fibers E over ∂T_V. In particular,

(2.27) $\partial T_V = E/S^1$.

In the hypersurface case, the map

$$E \xrightarrow{\psi} \tilde{S} \simeq S^1$$

is a fibration with fiber $\partial \overline{F}$, the boundary of the Milnor fiber associated to the hypersurface singularity $(CV, 0)$ in $(\mathbb{C}^{n+2}, 0)$. Using this remark, (2.27) takes the following more precise form:

(2.28) $\partial T_V = \partial \overline{F}/\langle h \rangle$,

where $\langle h \rangle$ denotes the finite cyclic group of order d generated by the restriction of the monodromy homeomorphism h to $\partial \overline{F}$.

(2.29) Examples. (i) Let V be a smooth hypersurface in $\mathbb{P}^{n+1}$, with deg $V = d$. Then T_V is a tubular neighborhood in the proper sense, i.e., it is associated to the normal bundle ν_V of V in $\mathbb{P}^{n+1}$. Hence ∂T_V is the S^1-bundle associated to the line bundle $\mathcal{O}(d)|_V$ by (1.10). On the other hand, $\partial \bar{F}$ can be identified in this case with the link K_V. Hence $\partial \bar{F}$ is the S^1-bundle associated to the line bundle $\mathcal{O}(-1)|_V$. The homeomorphism (2.28) can then be obtained from our general considerations in (1.11). In particular, when V is a hyperplane, ∂T_V is a $(2n + 1)$-sphere, the sphere at infinity in the affine piece $\mathbb{P}^{n+1} \setminus V = \mathbb{C}^{n+1}$.

(ii) Let V be defined by $x_0, \ldots, x_{n+1} = 0$ in $\mathbb{P}^{n+1}$. It follows from (3.1.15) that

$$\partial \bar{F} = S^n \times (S^1)^{n+1}.$$

The monodromy homeomorphism $h: \partial \bar{F} \to \partial \bar{F}$ acts as the identity on S^n and as a translation of order d on the torus $(S^1)^{n+1}$. It follows that

$$\partial T_V = \partial \bar{F}/\langle h \rangle = S^n \times (S^1)^{n+1}.$$

In particular for $n = 1$, the curve V consists of the three coordinate lines in $\mathbb{P}^2$, and ∂T_V is the three-dimensional torus $(S^1)^3$.

(2.30) Exercise. (i) Let V be a smooth conic in $\mathbb{P}^2$. Show that $\partial T_V = L(4, 1)$, the lens space of type $(4, 1)$. *Hint.* Note that the corresponding link K_V can be written as $S^3/(\mathbb{Z}/2\mathbb{Z})$ and then use (2.29(i)).

(ii) Consider the hypersurface $V: x_0 x_1 = 0$ in $\mathbb{P}^{n+1}$, i.e., V consists of two hyperplanes. Show that there is a homeomorphism

$$\partial T_V = S^{2n} \times S^1.$$

Hint. Regard S^{2n} as being obtained from two cones CS^{2n-1} over S^{2n-1}, by glueing their bases as in the following illustration:

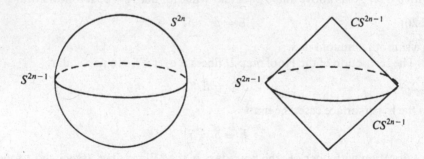

The boundary ∂T_V is contained in the affine chart $x_0 = 1$ and we can use arg x_1 as a coordinate for S^1.

(2.31) Proposition. *Let V be a hypersurface in $\mathbb{P}^{n+1}$. Then the inclusion*

$$\partial T_V \to \mathbb{P}^{n+1} \setminus V$$

is an n-equivalence.

Proof. This follows from (3.2.4) and the homotopy exact sequences of the following two $(\mathbb{Z}/d\mathbb{Z})$-fibrations:

$$
\begin{array}{ccc}
\partial\bar{F} & \longrightarrow & \bar{F} \simeq F \\
\downarrow & & \downarrow \\
\partial\bar{F}/\langle h\rangle = \partial T_V & \longrightarrow & U = F/\langle h\rangle
\end{array}
$$

Here $U = \mathbb{P}^{n+1}\backslash V$. $\square$

§3. Smooth Complete Intersections

In this section we assume that the complete intersection V in $\mathbb{P} = \mathbb{P}^{n+c}$ from (2.1) is smooth. Then the diffeomorphism type of the smooth manifold V, as well as that of the embedding $V \subset \mathbb{P}$, are completely determined by the *multidegree* $\mathbf{d} = (d_1, \ldots, d_c)$ of V where $d_i = \deg f_i$ for $i = 1, \ldots, c$. (This follows from (1.3.4) in the case $c = 1$ and can be proved similarly for $c > 1$.) Hence it is natural to try to compute the algebraic topology invariants of the variety V in terms of the multidegree $\mathbf{d}$. And this is exactly what we are going to do in this section.

(3.1) Lemma. *The integral homology and the integral cohomology of a smooth complete intersection V are torsion free.*

Proof. It is enough to prove this statement for the cohomology groups $H^{\cdot}(V)$ because of Poincaré duality. It follows from (2.6) and (2.11) that

$$(3.2) \qquad H^k(V) = H^k(\mathbb{P}^n) \qquad \text{for} \quad k \neq n.$$

For the middle dimension we have $\text{Tors } H^n(V) = \text{Tors } H_n(V) = \text{Tors } H^{n+1}(V) = 0$. Here the first equality comes from Poincaré duality, while the second is a general relation between the torsion parts in homology and cohomology, see [Gn], p. 136. Hence the integral homology $H_{\cdot}(V)$ is completely determined if we are able to compute the rank of the free $\mathbb{Z}$-module $H_n(V)$, i.e., the nth *Betti number* $b_n(V)$ of V. To do this, we compute the Euler characteristic of V

$$\chi(V) = \sum_{k=0,2n} (-1)^k b_k(V)$$

in two ways. First, using (3.2) we get

$$(3.3) \qquad \chi(V) = n + 1 + (-1)^n b_n^0(V),$$

where $b_n^0(V) = \dim H_0^n(V)$ as in (2.18). Second, we know that

$$(3.4) \qquad \chi(V) = \langle c_n(V), [V]\rangle,$$

where $c_n(V)$ is the top Chern class of the tangent bundle TV of V, see [Hz1], p. 70.

To calculate this Chern class, we remark that the normal bundle v_V of the complex submanifold V in $\mathbb{P}$ is the direct sum of the normal bundles v_i of the smooth parts of the hypersurfaces $f_i = 0$ restricted to V. In other words, we have

$$(3.5) \qquad\qquad T\mathbb{P}^{n+c}|_V = TV \oplus v_1 \oplus \cdots \oplus v_c.$$

Taking the total Chern class and using (1.10) we get

$$(3.6) \qquad (1 + \alpha_V)^{n+c+1} = c(V)(1 + d_1\alpha_V)\cdots(1 + d_c\alpha_V),$$

where $\alpha_V = j^2(\alpha) \in H^2(V)$, $j: V \hookrightarrow \mathbb{P}^{n+c}$. Hence $c_n(V)$ is the homogeneous part of degree $2n$ in the product

$$(1 + \alpha_V)^{n+c+1}(1 - d_1\alpha_V + d_1^2\alpha_V^2 - \cdots)\cdots(1 - d_c\alpha_V + d_c^2\alpha_V^2 - \cdots).$$

To use the formula (3.4), recall that

$$\alpha_V^n = d_1 \ldots d_c g \qquad \text{for } g \text{ a generator of } H^{2n}(V),$$

as in (2.10), (2.11). $\qquad\qquad\qquad\qquad\qquad\qquad\qquad\qquad\qquad\qquad\qquad$ □

(3.7) **Exercise.** (i) Let V be a hypersurface of degree d in $\mathbb{P}^{n+1}$ (i.e., $c = 1$, $d_1 = d$). Show that

$$c_n(V) = \left[\binom{n+2}{2} - d\binom{n+2}{3} + \cdots + (-1)^n d^n \binom{n+2}{n+2}\right]\alpha_V^n$$

and

$$\chi(V) = \frac{1}{d}[(1 - d)^{n+2} - 1] + n + 2.$$

(ii) Let V be as in (i) above and let $U = \mathbb{P}^{n+1} \setminus V$. Use the fact that

$$U = F/\langle h \rangle \qquad \text{and} \qquad \chi(F) = 1 + (-1)^{n+1}(d - 1)^{n+2}$$

to find $\chi(V)$ in a new way.

(iii) Let V be a smooth curve ($n = 1$). Show that

$$\chi(V) = (c + 2 - d_1 - \cdots - d_c)d_1 \ldots d_c.$$

Use the formula $\chi(V) = 2 - 2g(V)$ to show that a curve V with genus $g(V) = 2$ cannot be a complete intersection.

(iv) Let V be a smooth surface ($n = 2$). Then show that

$$c_2(V) = \left[\sum d_i^2 + \sum_{i \neq j} d_i d_j - (c + 3)\sum d_i + \binom{c+3}{2}\right]\alpha_V^2,$$

$$c_1(V) = (c + 3 - d_1 - \cdots - d_c)\alpha_V,$$

$$\chi(V) = d_1 \ldots d_c \left[\sum d_i^2 + \sum_{i \neq j} d_i d_j - (c + 3)\sum d_i + \binom{c+3}{2}\right].$$

Recall that for a surface V, the line bundle $\mathscr{K}_V = \Omega_V^2$ of 2-forms on V is called the *canonical bundle* of V. We have an adjunction formula for this bundle, namely,

$$(3.8) \qquad \mathscr{K}_V = \mathscr{K}_{\mathbb{P}^{c+2}}|_V \otimes v_1 \otimes \cdots \otimes v_c,$$

see [GH], p. 147. The first Chern class of $\mathscr{K}_V$ is called the *canonical class*. The formula (3.8) together with the formula

$$(3.9) \qquad c_1(\mathscr{K}_{\mathbb{P}^n}) = \mathcal{O}(-n-1),$$

see [GH], p. 146, imply that

$$(3.10) \qquad c_1(\mathscr{K}_V) = (\textstyle\sum d_i - c - 3)\alpha_V = -c_1(V).$$

Consider now the integral cohomology algebra $H^{\cdot}(V)$ of our smooth complete intersection V. This provides us with a new invariant, namely, the lattice

$$(\, , \,) : H^n(V) \times H^n(V) \to \mathbb{Z},$$

$$(\alpha, \beta) \mapsto \langle \alpha \cup \beta, [V] \rangle.$$

When $n = \dim V$ is odd, this is a unimodular skew-symmetric lattice. With respect to a suitable basis in $H^n(V)$, the bilinear form $(\, , \,)$ is given by the matrix

$$I_{2m} = \left(\begin{array}{c|c} 0 & I \\ \hline -I & 0 \end{array}\right),$$

where $2m = b_n(V)$, see (A7).

When $n = \dim V$ is even, we have a symmetric unimodular lattice and there are two interesting invariants associated to it. The first one of them is the *index* $\tau(V)$ *of the manifold* V, which is by definition the index of the lattice $H^n(V)$. The Hirzebruch Index Theorem [Hz1], p. 86, says that this index can be computed in terms of the *Pontrjagin classes* of V. More precisely, if $n = 2k$, we have

$$(3.11) \qquad \tau(V) = \langle L_k(p_1, \ldots, p_k), [V] \rangle,$$

where L_k are the famous L-polynomials and $p_i = p_i(V)$ are the Pontrjagin classes of V. A list of the first L-polynomials can be found in Hirzebruch [Hz1], p. 12. In particular,

$$(3.12) \qquad L_1(p_1) = \tfrac{1}{3}p_1, \qquad L_2(p_1, p_2) = \tfrac{1}{45}(7p_2 - p_1^2).$$

The Pontrjagin classes $p_i \in H^{4i}(V)$ can easily be computed using the formula

$$(3.13) \quad p_0 - p_1 + p_2 - \cdots = (c_0 + c_1 + c_2 + \cdots)(c_0 - c_1 + c_2 - \cdots),$$

where $c_i = c_i(V)$ are the corresponding Chern classes, see [Hz1], p. 67.

(3.14) **Exercise.** (i) Let V be a two-dimensional smooth complete intersection. Show that

$$\tau(V) = \tfrac{1}{3}(c + 3 - \textstyle\sum d_i^2)d_1 \ldots d_c.$$

(ii) Use this formula for $\tau(V)$ to show that

$$\tau(V) < 0$$

except in the following two cases:

(a) $d_1 = \cdots = d_c = 1$, $V = \mathbb{P}^2$, $\tau(V) = 1$;
(b) $d_1 = 2, d_2 = \cdots = d_c = 1$ (i.e., V is a quadratic surface in $\mathbb{P}^3$), $\tau(V) = 0$.

When $n = \dim V = 2k$, a similar inequality holds, namely,

(3.15) $(-1)^k \tau(V) > 0$

with the exceptions:

(a') $V = \mathbb{P}^{4l+2}, l \geq 0$;
(b') V is a degree 2 hypersurface of dimension $4l + 2, l \geq 0$.

For details, see Libgober [Li1] and Wood [Wd2].

The second important invariant which comes from the lattice $H^n(V)$, when $n = 2k$, is its parity, i.e., whether it is an even lattice as defined in (A2).

(3.16) **Proposition.** *The lattice $H^{2k}(V)$ of the 2k-dimensional smooth complete intersection V of multidegree* $\mathbf{d} = (d_1, \ldots, d_c)$ *is even if and only if the integer*

$$\binom{k + s}{k}$$

is even, where $s = \#\{d_j; d_j \text{ is even}\}$.

(3.17) **Corollary.** *In the surface case, the lattice $H^2(V)$ is even if and only if the integer*

$$d_1 + \cdots + d_c - c - 3$$

is even. (This integer is the degree of the canonical bundle $\mathscr{K}_V$, see (3.10).)

To prove (3.16), note first that we can consider the homology lattice

$$(\, , \,) \colon H_{2k}(V) \times H_{2k}(V) \to \mathbb{Z},$$

where the pairing is given by the intersection number of cycles. But the intersection of cycles in homology is dual to the cup product of the corresponding cocyles in cohomology, see [GH], p. 59. It follows that the Poincaré isomorphism is in fact a lattice isomorphism. Hence we can prove (3.16) for the homology lattice $H_{2k}(V)$.

Consider the distinguished element

(3.18) $h = [V \cap E] \in H_{2k}(V)$,

where E is a generic linear space in $\mathbb{P}^{n+c}$ with codim $E = k$. The cycle h is dual to the primitive element α_V^k in $H^{2k}(V)$. Hence

(3.19) $(h, h) = (\alpha_V^k, \alpha_V^k) = d_1 \ldots d_c.$

We have the following result, see [LiW1].

(3.20) **Theorem.** *Let $h^\perp$ be the orthogonal complement of h in $H_{2k}(V)$. Then:*

(i) *the sublattice $h^\perp$ is even;*
(ii) *there is an element $y \in H_{2k}(V)$ such that $(h, y) = 1$ and*

$$(y, y) \equiv \binom{k + s}{k} \bmod 2,$$

where $s = \#\{d_j; d_j \text{ is even}\}$.

Proof of (3.16). Assuming (3.20), the lattice $H_{2k}(V)$ has the following direct sum decomposition:

$$H_{2k}(V) = h^\perp + \mathbb{Z}y$$

(nonorthogonal sum). Indeed, every element $v \in H_{2k}(V)$ can be written as

$$v = (v - a_v y) + a_v y,$$

where $a_v = (v, h)$. The result obviously follows from this decomposition.

Proof of (3.20). We first prove (ii). Recall that the inclusion $j: V \to \mathbb{P}$ is a $2k$-equivalence (2.7(ii)). Hence the induced map

$$j_\#: [\mathbb{P}^k, V] \to [\mathbb{P}^k, \mathbb{P}]$$

between these sets of homotopy classes of maps in onto, see [Sp], p. 405. In particular, there is a map

$$l: \mathbb{P}^k \to V$$

such that $j \circ l$ is homotopic to a linear embedding of $\mathbb{P}^k$ into $\mathbb{P}$. Let $y = l_*([\mathbb{P}^k]) \in H_{2k}(V)$. Then

$$(y, h) = (j \circ l(\mathbb{P}^k), E) = 1,$$

where the first intersection pairing is in $H_.(V)$ and the second one is in $H_.(\mathbb{P})$.

Since dim $V = 2$ dim $\mathbb{P}^k$, it follows that any map $\mathbb{P}^k \to V$ can be approximated, within the same homotopy class, by a smooth immersion, see [Hs], p. 53. Hence we can take our map l above to be a smooth immersion. For any smooth immersion $g: X \to Y$, let $v(g)$ be the *normal bundle* of g, i.e., $v(g)$ is the vector bundle on X such that the following sequence

$$0 \to TX \xrightarrow{dg} g^*(TY) \to v(g) \to 0$$

is exact. With this notation, the composition of immersions

$$\mathbb{P}^k \xrightarrow{l} V \xrightarrow{j} \mathbb{P}^{2k+c}$$

leads to the following relation:

$$(3.21) \qquad T\mathbb{P}^k \oplus v(l) \oplus l^*v(j) = (j \circ l)^*(T\mathbb{P}^{2k+c})$$

of *real* vector bundles on $\mathbb{P}^k$.

Recall that the Stiefel–Whitney class

$$W(A) \in H^{\cdot}(X; \mathbb{Z}/2\mathbb{Z})$$

of complex vector bundle A over a space X is just the mod 2 reduction of its Chern class $c(A)$, i.e., $W(A)$ is the image of $c(A)$ under the natural morphism

$$H^{\cdot}(X; \mathbb{Z}) \to H^{\cdot}(X; \mathbb{Z}/2\mathbb{Z}),$$

see [Hz1], p. 73. Using this remark, (3.21), and (3.5) ($v_V = v(j)$) it follows that the (total) Stiefel–Whitney class of the normal bundle $v(l)$ is given by

$$(3.22) \qquad W(v(l)) = (1 + \beta)^{k+s},$$

where β is the generator of $H^2(\mathbb{P}^k; \mathbb{Z}_2)$. In particular,

$$W_{2k}(v(l)) = \binom{k + s}{k} \beta^k.$$

Finally, we have

$$(y, y) = \langle e(v(l)), [\mathbb{P}^k] \rangle \equiv \langle W_{2k}(v(l)), [\mathbb{P}^k] \rangle \equiv \binom{k + s}{k} \mod 2.$$

Here $e(v(l))$ denotes the *Euler class* of the normal bundle $v(l)$, and the first equality comes from the definition of the *Euler number* as the self-intersection of the zero section, see [Hs], p. 133. The second equality comes from the relation

$$e(A) = c_m(A)$$

valid for a complex vector bundle A of rank m, see [Hz1], p. 72, and the relation between $c(A)$ and $W(A)$ discussed above. (All this is applied to the vector bundles in (3.21) different from $v(l)$, since they are complex vector bundles.) This ends the proof of (3.20(ii)).

Now we prove (3.20(i)) in a different way from [LiW1]. This new approach sets the background for our considerations in the next section too. Let $H \subset \mathbb{P}^{n+c}$ be a hyperplane such that the intersection $H \cap V = W$ is smooth with $n = 2k$. Consider the ICIS

$$(3.23) \qquad (X, 0) = (CW, 0)$$

at the origin of $\mathbb{C}^{n+c}$. Let F be the corresponding Milnor fiber and let $L = H_n(F)$ (resp. $\bar{L} = L/\mathrm{Rad}\, L$) be the associated Milnor lattice (resp. the reduced Milnor lattice). It is clear from (3.3.26) that the reduced Milnor lattice $\bar{L}$ is even. Hence to prove (3.20(i)) it is enough to prove the following.

(3.24) Proposition.

(i) $U = V \setminus W$ is homeomorphic to the Milnor fiber F;

(ii) There is a lattice isomorphism

$$\bar{L} \simeq h^{\perp}$$

induced by the morphism

$$L = H_n(V \setminus W) \xrightarrow{i_*} H_n(V),$$

where $i: V \setminus W \to V$ is the inclusion.

Proof. We can assume that $x_0 = 0$ is an equation for the transversal hyperplane H. Then the variety U is given in the affine space $\mathbb{C}^m$ ($m = 2k + c$) by the equations

$$g_j(\bar{x}) = f_j(1, \bar{x}) = P_j(\bar{x}) + Q_j(\bar{x}) = 0$$

for $j = 1, \ldots, c$, $\bar{x} = (x_1, \ldots, x_m)$, and P_j, Q_j polynomials with $\deg P_j < \deg Q_j = d_j$. Note that

$$Q_1 = \cdots = Q_c = 0$$

are exactly the equations for the conic singularity $(X, 0)$.

For a real number $r > 0$ we can set $y = r^{-1}\bar{x}$ and define

$$U_r = \{y \in \mathbb{C}^m; \, g_j(ry) = 0 \text{ for all } j = 1, \ldots, c\}.$$

It follows that multiplication by r induces an homeomorphism

$$U_r \cap B_\varepsilon \xrightarrow{\sim} U \cap B_{\varepsilon r},$$

where $B_a = \{y \in \mathbb{C}^m; \, |y| < a\}$ is the open ball of radius a centered at the origin. On the other hand, we may write

$$U_r \cap B_\varepsilon = \{y \in B_\varepsilon; \, Q_j^r(y) = 0 \text{ for } j = 1, \ldots, c\},$$

where $Q_j^r(y) = Q_j(y) + P_j(ry)r^{-d_j}$.

For r large enough, the equation $Q_j^r = 0$ is a small deformation of the equation $Q_j = 0$. It follows that $U_r \cap B_\varepsilon$ is the Milnor fiber of the ICIS $(X, 0)$ as introduced in (3.3.26). Using the cylindric structure of affine varieties (1.6.9), it follows that for r large enough

$$U \cap B_{\varepsilon r} \quad \text{is homeomorphic to } U.$$

This ends the proof of (3.24(i)). To prove (ii), we compute the integral homology of V using the Mayer–Vietoris sequence corresponding to the cover $V = U \cup T$, where T is a small tubular neighborhood of the hyperplane section W in V. This Mayer–Vietoris sequence contains the morphism

$$j = (j_1, j_2): H_n(U \cap T) \to H_n(U) + H_n(T).$$

Since the inclusion $W \to T$ is a homotopy equivalence, we can replace $H_{\cdot}(T)$ by $H_{\cdot}(W)$.

Similarly, $U \cap T$ is homotopy equivalent to the link K_W corresponding to the smooth complete intersection W in $H = \mathbb{P}^{m-1}$. In fact, $U \cap T$ is homotopy equivalent to the total space of the S^1-bundle corresponding to the line bundle $\mathcal{O}(1)|_W$ by (1.10), K_W is the total space of the S^1-bundle associated with the line bundle $\mathcal{O}(-1)|_W$, and we can use (1.11(i)). See also (2.29(i)). Using these identifications and the first part (i), the above-mentioned morphism becomes

$$j = (j_1, j_2): H_n(K_W) \to H_n(\bar{F}) + H_n(W),$$

where $j_1: H_n(K_W) \to H_n(\bar{F})$ is induced by the inclusion $K_W = \partial \bar{F} \to \bar{F}$ and $j_2 = p_*$, and where p is the projection map of the Hopf bundle $K_W \to W$. It follows from (2.3.6') that j_1 is precisely the inclusion $\text{Rad } L \to L$.

The Gysin homology sequence associated with the Hopf bundle K_W implies $j_2 = p_* = 0$. We thus get an exact sequence

$$(3.25) \qquad 0 \to \bar{L} \oplus H_n(W) \xrightarrow{s} H_n(V) \to H_{n-1}(K_W) \xrightarrow{p_*} H_{n-1}(W).$$

Since $n = 2k$ is even, it follows that

$$H_n(W) = \mathbb{Z} \qquad \text{and} \qquad \ker p_* = \mathbb{Z}/d\mathbb{Z},$$

where $d = d_1 \ldots d_c = \deg(V)$.

The inclusion $W \to V$ gives an identification

$$s(H_n(W)) = \mathbb{Z}h,$$

and the inclusion $U \to V$ shows that

$$s(\bar{L}) \subset h^\perp.$$

Hence we get the inclusions

$$\text{im}(s) \subset h^\perp \oplus \mathbb{Z}h \subset H_n(V).$$

Both subgroups have index d in $H_n(V)$ (use (A34) and (A5)) and hence they are equal

$$\text{im}(s) = h^\perp \oplus \mathbb{Z}h.$$

But this is possible only if $s(\bar{L}) = h^\perp$ and we are done.   □

(3.26) **Remarks.** (i) The same proof as above shows that for a smooth n-dimensional complete intersection V, with n odd, the associated reduced Milnor lattice $\bar{L}$ is isomorphic to the unimodular skew-symmetric lattice $H_n(V)$. In particular, in this case, the lattice $\bar{L}$ is unimodular too.

(ii) Let F by any smooth complete intersection in the affine space $\mathbb{C}^{n+c}$ such that its projective closure $\bar{F}$ and the part at infinity F_∞ of this closure are both smooth (projective) complete intersection of multidegree

$$\mathbf{d} = (d_1, \ldots, d_c).$$

Then from Alexander duality we get

$$\chi(F) = \chi(\bar{F}) - \chi(F_\infty).$$

Consider now the special case when F is the affine Milnor fiber associated with the conical singularity $(X, 0)$ above, namely,

$$F = \{x \in \mathbb{C}^{n+c}; f_1(x) = a_1, \ldots, f_c(x) = a_c\}$$

for a generic point $(a_1, \ldots, a_c) \in \mathbb{C}^c$. If $\mu(X, 0) = b_n(F)$ denotes the corresponding Milnor number, then we get in this way the following formula:

$$(3.27) \qquad \mu(X, 0) = b_n(V_n^{\mathbf{d}}) + b_{n-1}(V_{n-1}^{\mathbf{d}}) - 1,$$

where $V_m^{\mathbf{d}}$ denotes a smooth m-dimensional projective complete intersection of multidegree $\mathbf{d}$.

(3.28) **Exercise.** Use formula (3.27) to compute $\mu(X, 0)$ for $n = 1$, 2 using (3.7(iii)) and (3.7(iv)). *Hint.* By Bezout's theorem we have

$$b_0(V_0^{\mathbf{d}}) = d_1 \ldots d_c = \deg(V_m^{\mathbf{d}}).$$

(3.29) **Open Problem.** For the Milnor fiber F of a $2k$-dimensional ICIS $(X, 0)$ define the index $\tau(F)$ as the difference $\mu_+ - \mu_-$, where (μ_-, μ_0, μ_+) is the signature of the Milnor lattice $H_{2k}(F)$. It is an old conjecture by Durfee [Df3] that

$$(-1)^k \tau(F) \geq 0.$$

When $(X, 0)$ is a homogeneous singularity, this is true by (3.24) and (3.15). When $(X, 0)$ is a two-dimensional weighted homogeneous singularity, the conjecture is again true, see Xu–Yau [XY].

The recapitulate, in the even-dimensional case $n = 2k$ we know how to compute the parity of the lattice $H^n(V)$ (even or odd) and its signature. With few exceptions this lattice is indefinite, and so these invariants determine completely its isomorphism class by Milnor [M2]. For more details and a study of the pair $(H_n(V), h)$, we refer to Libgober–Wood [LiW1].

Let us now compare the topology of two smooth complete intersections

$$V = V(d_1, \ldots, d_c) \qquad \text{and} \qquad V' = V(d_1', \ldots, d_{c'}')$$

of the same dimension n, but with different codimensions and multidegrees.

(3.30) **Exercise** ($n = 1$). (i) Show that the curves $V(1)$ and $V(2)$ in $\mathbb{P}^2$ are both diffeomorphic to the sphere S^2.

(ii) Show that the cubic curve $V(3)$ in $\mathbb{P}^2$ and the curve $V(2, 2)$ in $\mathbb{P}^3$ are both diffeomorphic to the torus $T^2 = (S^1)^2$. *Hint.* Two oriented 2-manifolds are diffeomorphic if and only if they have the same genus, see [Hs], p. 191.

The surface case ($n = 2$) is much more subtle and we present it following Ebeling [E4], to which we refer for proofs and complete references.

(3.31) **Proposition.** *Two smooth complete intersection surfaces V and V' are homeomorphic if and only if*

$$\chi(V) = \chi(V'), \qquad \tau(V) = \tau(V') \qquad and \qquad n_V \equiv n_{V'} \, (mod \; 2),$$

where $n_V = d_1 + \cdots + d_c - c - 3$, $n_{V'} = d'_1 + \cdots + d'_{c'} - c' - 3$,   *are the degrees of the corresponding canonical bundles, see* (3.10).

(3.32) **Proposition.** *Assume that the smooth complete intersection surfaces V and V' satisfy the conditions:*

(i) $n_V \neq 0$ *and* $n_V \neq n_{V'}$;
(ii) $\tau(V) + \chi(V) - 4 = 8a$ *for some integer* $a > 0$.

Then there is no diffeomorphism $V \to V'$ preserving the natural orientations.

(3.33) **Example** (K3 Surfaces). Consider the following complete intersection surfaces: $S_1 = V(4)$ in $\mathbb{P}^3$, $S_2 = V(3, 2)$ in $\mathbb{P}^4$, and $S_3 = V(2, 2, 2)$ in $\mathbb{P}^5$. Then we have

$$\chi(S_1) = \chi(S_2) = \chi(S_3) = 24 \quad \text{by (3.7(iv))},$$

$$\tau(S_1) = \tau(S_2) = \tau(S_3) = -16 \quad \text{by (3.14(i))},$$

$$n_{S_1} = n_{S_2} = n_{S_3} = 0.$$

It follows from (3.31) that all the surfaces S_1, S_2, and S_3 are homeomorphic to each other. In fact, they are even diffeomorphic, since any two K3 surfaces are diffeomorphic, see [BPV], p. 257.

(3.34) **Example** (Ebeling [E4], Libgober–Wood [LiW3]). The surfaces $X = V(10, 7, 7, 6, 3, 3)$ in $\mathbb{P}^8$ and $Y = V(9, 5, 3, 3, 3, 3, 3, 2, 2)$ in $\mathbb{P}^{11}$ are homeomorphic but not diffeomorphic. In this case,

$$\chi(X) = \chi(Y) = 12\,859\,560,$$

$$\tau(X) = \tau(Y) = -2\,143\,260,$$

$$n_X = 27, \qquad n_Y = 21.$$

Note that $\deg(X) \neq \deg(Y)$ and compare with the following:

(3.35) **Exercise.** Let V, V' be two n-dimensional smooth complete intersections which are homotopy equivalent. If $n \geq 3$, then $\deg(V) = \deg(V')$. *Hint.* A homotopy equivalence $h: V \to V'$ must satisfy $h^*(\alpha_{V'}) = \pm \alpha_V$.

For a discussion on the diffeomorphism types of smooth complete intersections in higher dimensions we refer to Libgober–Wood [LiW2].

§4. Complete Intersections with Isolated Singularities

In this section we consider an n-dimensional complete intersection V in $\mathbb{P} = \mathbb{P}^{n+c}$ as in (2.1) and we assume that V has only isolated singularities ($n \geq 1$).

Let $a_1, \ldots, a_p$ be the singular points of V and let $L_1, \ldots, L_p$ denote the associated Milnor lattices. The main theme of this section is that some information (and sometimes complete information) on the integer homology $H.(V)$ of V can be obtained using these Milnor lattices L_i.

To state the result, we repeat the construction from (3.23), (3.24). Namely, we take a hyperplane H in $\mathbb{P}$ such that hyperplane section $W = V \cap H$ is smooth. Let $(X, 0) = (CW, 0)$ be the associated cone singularity which is an ICIS. Let L (resp. $\bar{L}$) be its Milnor lattice (resp. reduced Milnor lattice). The complement $U = V \setminus W$ can be identified as in (3.24) with a (singular) fiber in the miniversal deformation of the singularity $(X, 0)$. The singularities of U are precisely the points $a_1, \ldots, a_p$. We deduce from (3.3.25) and (3.3.26) the existence of a primitive embedding of lattices

$$(4.1) \qquad \psi_V : L_1 \oplus \cdots \oplus L_p \to L,$$

such that coker $\psi_V = H_n(U)$. Let us consider the lattice morphism

$$(4.2) \qquad \varphi_V : L_1 \oplus \cdots \oplus L_p \to \bar{L}$$

obtained from ψ_V by composing with the projection $L \to \bar{L} = L/\mathrm{Rad}\, L$.

The integral homology $H.(V)$ is determined by this morphism φ_V as follows:

(4.3) Theorem.

(i) $H_j(V) = H_j(\mathbb{P}^n)$ for $j \neq n, n+1$;
(ii) $H_n(V) = H_n(\mathbb{P}^n) + \mathrm{coker}(\varphi_V)$;
(iii) $H_{n+1}(V) = H_{n+1}(\mathbb{P}^n) + \ker(\varphi_V)$.

Proof. The first claim follows from (2.6) and (2.11). To investigate what happens in the middle dimensions, we use the Mayer–Vietoris sequence associated with the cover $V = U \cup T$, where T is a small tubular neighborhood of the complex submanifold W in V. We use the same identifications and notations as in the proof of (3.24). In particular, we can determine the morphism

$$j_2 = p_* : H_n(K_W) \to H_n(W)$$

from the Gysin sequence of the Hopf bundle $p : K_W \to W$ associated with the *smooth* complete intersection W.

The other component j_1 corresponds to the composition

$$H_n(K_W) = \mathrm{Rad}\, L \to L \to H_n(U).$$

Using (4.1), it follows that j_1 has the same kernel and cokernel as φ_V. The advantage of replacing the morphism j_1 by φ_V comes from the fact that the latter is a lattice morphism.

We give the details of the proof only for the case $n = 2k$ even, which is slightly more delicate than the odd dimensional case. In this case, $H_{n+1}(W) = 0$ and the above-mentioned Mayer–Vietoris sequence shows that

$$H_{n+1}(V) = \ker j = \ker j_1 = \ker \varphi_V$$

(use $j_2 = 0$). Moreover, we get the following exact sequence

$$0 \to \text{coker } j \xrightarrow{s} H_n(V) \to H_{n-1}(K_W) \xrightarrow{p_*} H_{n-1}(W).$$

Here $\text{coker } j = \text{coker } \varphi_V + H_n(W)$ and

$$\ker p_* = \mathbb{Z}/d\mathbb{Z} \quad \text{where} \quad d = d_1 \ldots d_c = \deg(V)$$

as in the proof of (3.24). Again, as in the proof of (3.24), we have

$$s(H_n(W)) = \mathbb{Z}h',$$

where $h' \in H_n(V)$ is the cycle corresponding to the intersection $V \cap E$, for E a linear subspace in $\mathbb{P}$ with codim $E = k$.

Consider now the cap product by α_V^k

$$\alpha_V^k \cap : H_n(V) \to H_0(V) = \mathbb{Z},$$

and recall that the cap product by α_V corresponds to the intersection with a hyperplane, see [L2]. There is an element $y \in H_n(V)$ defined exactly as in the proof of (3.20(ii)) such that

$$\alpha_V^k \cap y = 1.$$

It follows that

$$H_n(V) = A + \mathbb{Z}y,$$

where $A = \ker(\alpha_V^k \cap \cdot)$.

Using the above geometric interpretation of the cap product by α_V it follows that

$$s(\text{coker } \varphi_V) \subset A.$$

Hence we get the inclusions

$$\text{im}(s) \subset A + \mathbb{Z}h' \subset H_n(V).$$

Since both subgroups have the same index d in $H_n(V)$ it follows that $s(\text{coker } \varphi_V) = A$. (Use the relation $\alpha_V^k \cap h' = d$.) This ends the proof of the theorem. $\qquad\square$

(4.4) Corollary.

(i) $H_{n+1}(V)$ is torsion free;
(ii) Let V_0 be a smooth complete intersection in $\mathbb{P}$ having the same dimension and multidegree as V. Then

$$\chi(V) = \chi(V_0) + (-1)^{n+1} \sum_{i=1,p} \mu(V, a_i),$$

where $\mu(V, a_i) = \text{rk } L_i$ is the Milnor number of the singularity (V, a_i).

Proof. The first claim (i) follows from (4.3(iii)), but also from (2.11) since

$$\text{Tors } H_{n+1}(V) = \text{Tors } H^{n+2}(V) = 0.$$

The second claim (ii) follows from (3.24) and (4.3). But it can also be proved directly, as follows. Let V_0 be a small deformation of the complete intersection V, such that V_0 is smooth

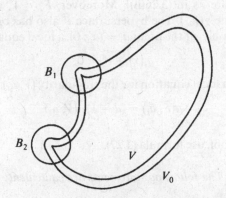

Let B_i be a small open ball in $\mathbb{P}$ centered at the point a_i such that:

(α) $\bar{B}_i \cap V$ is contractible by the local conic structure of algebraic sets (1.5.1);

(β) $\bar{B}_i \cap V_0$ can be identified with the (closed) Milnor fiber of the singularity (V, a_i).

The pieces $V' = V \setminus (\bigcup B_i)$ and $V'_0 = V_0 \setminus (\bigcup B_i)$ are homeomorphic by an argument based on Ehresmann's Fibration Theorem as in the proof of (1.3.2) (just replace the ambient projective space $\mathbb{P}$ with the compact manifold with boundary $\mathbb{P} \setminus (\bigcup B_i)$).

The Mayer–Vietoris sequences for the closed covering $V = V' \cup \bigcup_i (\bar{B}_i \cap V)$ shows that

$$\chi(V) = \chi(V') + p.$$

Indeed, $\chi(\bar{B}_i \cap V) = 1$ since this set is contractible, and $\chi(\partial \bar{B}_i \cap V) = 0$ since the Euler characterisic of any odd-dimensional compact oriented manifold is zero.

On the other hand, the Mayer–Vietoris sequence for the closed covering $V_0 = V'_0 \cup \bigcup (\bar{B}_i \cap V_0)$ shows that

$$\chi(V_0) = \chi(V'_0) + p + (-1)^n \sum \mu(V, a_i).$$

Since $\chi(V') = \chi(V'_0)$ by our remark above, we are done. $\square$

(4.5) **Corollary.** *Let $V: f = 0$ be a hypersurface in $\mathbb{P}^{n+1}$ of degree d, having only isolated singularities, say at the points $a_1, \dots, a_p$. Let $F: f - 1 = 0$ be the corresponding Milnor fiber. Then*

$$\chi(F) = 1 + (-1)^{n+1} \left[(d-1)^{n+2} - d \sum_{i=1,p} \mu(V, a_i) \right].$$

Proof. Using the Lefschetz Duality Theorem, see [Sp], p. 297, it follows that

$$\chi(F) = \chi(\overline{F}) - \chi(F_\infty),$$

where $\overline{F}$ and F_∞ are as in (3.26(ii)). Moreover, $F_\infty = V$, and hence we can use (4.5) to compute $\chi(F_\infty)$. The hypersurface $\overline{F}$ also has only isolated singularities. More precisely, at the point $\bar{a}_i = (a_i : 0)$, a local equation for $\overline{F}$ is just

$$g_i(x) + t^d = 0,$$

where $g_i(x) = 0$ is a local equation for the singularity (V, a_i). It follows that

$$\mu(\overline{F}, \bar{a}_i) = (d - 1)\mu(V, a_i)$$

for all $i = 1, \ldots, p$.

To finish the proof, use formula (3.27).   □

(4.6) Corollary. (A) *The following statements are equivalent*:

 (i) *the* (*singular*) *complete intersection V is a $\mathbb{Q}$-homology manifold*;
 (ii) *all the Milnor lattices L_i are nondegenerate*.

When these equivalent statements hold, then

$$H_{n+1}(V) = 0.$$

 (B) *The following statements are equivalent*:

 (i) *the* (*singular*) *complete intersection V is a $\mathbb{Z}$-homology manifold*;
 (ii) *all the Milnor lattices L_i are unimodular*.

When these equivalent statements hold, then

$$H_{n+1}(V) = \text{Tors } H_n(V) = 0.$$

Proof. The equivalence (i) ⇔ (ii) in both cases is just the same as in (3.4.7). The last claim in both (A) and (B) follows from (4.3) (and (A19) for (B)).   □

We now give some concrete examples to show that in many cases Theorem (4.3) completely determines the integral homology $H_.(V)$, even if we are not able to determine the lattice morphism φ_V. For an even dimensional complete intersection V, which has only isolated singularities with nondegenerate Milnor lattices, we can use (A23), (A26), (A27), and (A28) to get several situations where Tors $H_n(V) = 0$. Note that these situations are not covered by (4.6(B)).

(4.7) Examples. Assume that V is as above and that either:

 (i) V has at most three singular points, all of which are of type A_1; or
 (ii) V has a single singularity, which is of type A_k for $k \le 7$, or D_m for $m \ne 0$ (mod 8), or E_6, or E_7.

Then Tors $H_n(V) = 0$.

To consider a more geometric example, let us look at the cubic surfaces in $\mathbb{P}^3$ with isolated singularities. The combinations of singularities which may occur on such a cubic surface V are listed in Bruce–Wall [BW] and are the following:

$$A_1, 2A_1, 3A_1, \underline{4A_1}; A_2, 2A_2, \underline{3A_2};$$

$$A_1A_2, A_1A_3, A_1A_4, \underline{A_1A_5}; 2A_1A_2, \underline{2A_1A_3};$$

$$A_3, A_4, A_5, D_4, D_5, E_6;$$

All the combinations of singularities in this list, except the four underlined ones, can produce no torsion in $H_2(V)$ by a similar argument to that in (4.7). The following result shows that these four cases indeed produce nontrivial torsion.

(4.8) Proposition. *The integral homology of a cubic surface $V \subset \mathbb{P}^3$ with isolated singularities has no torsion, except in the following cases:*

Singularities on V	Tors $H_2(V)$
$4A_1$	$\mathbb{Z}/2\mathbb{Z}$
$3A_2$	$\mathbb{Z}/3\mathbb{Z}$
A_1A_5	$\mathbb{Z}/2\mathbb{Z}$
$2A_1A_3$	$\mathbb{Z}/2\mathbb{Z}$

Proof. We give details only in the case $4A_1$, the other cases being similar. The reduced Milnor lattice $\bar{L}$ is in our case the lattice E_6, see (A12). Choose a basis $e_1, \ldots, e_6$ for this lattice such that the products (e_i, e_j) are given by the usual Dynkin diagram of type E_6, namely,

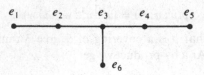

The lattice $4A_1$ has a basis $f_1, \ldots, f_4$ with $(f_i, f_i) = -2$ and $(f_i, f_j) = 0$ for $i \neq j$.

Since the Weyl group $\mathrm{Aut}(E_6)$ acts transitively on the set of vectors v with $(v, v) = -2$, it follows that we can take $\varphi_V(f_1) = e_6$. It follows that $f_1^\perp = \langle f_2, f_3, f_4 \rangle$ is embedded via φ_v into $e_6^\perp$, the orthogonal complement of e_6. A basis for $e_6^\perp$ is given by the vectors $e_2, e_1, \bar{e}_3, e_5, e_4$ where $\bar{e}_3 = 2e_3 + e_2 + e_4 + e_6$. The Dynkin diagram corresponding to this basis is exactly the Dynkin diagram of type A_5. Using the explicit description of the lattice (or root system) A_5, it follows that the embedding

$$\varphi_V: 3A_1 = f_1^\perp \to e_6^\perp = A_5$$

is equivalent up to an automorphism $u \in \text{Aut}(A_5)$ to the obvious embedding $f_2 \to e_2, f_3 \mapsto \bar{e}_3, f_4 \mapsto e_4$. Moreover, since u is a composition of reflections, u extends to an automorphism of the whole lattice E_6 such that $u(e_6) = e_6$. It follows that, up to an automorphism of the lattice E_6, we have

$$\text{im } \varphi_V = \langle e_2, \bar{e}_3, e_4, e_6 \rangle = \langle e_2, e_4, e_6, 2e_3 \rangle.$$

Hence $\text{Tors } H_2(V) = \text{Tors}(E_6/\text{im } \varphi_V) = \mathbb{Z}/2\mathbb{Z}$. □

(4.9) **Remark.** Let V be a complete intersection surface, such that $S = V_{\text{sing}}$ is finite and $b_3(V) = 0$. Then

$$H_1(V \backslash S) = H^3(V, S) = \text{Tors } H_2(V).$$

Hence there is a unique covering

$$p: \tilde{V} \to V$$

from a normal surface $\tilde{V}$ to our surface V of degree $|\text{Tors } H_2(V)|$, and ramified exactly at the singular points S when

$$\text{Tors } H_2(V) \neq 0.$$

More precisely, p is the covering associated in the usual way with the subgroup

$$\ker(\pi_1(V \backslash S) \to H_1(V \backslash S)),$$

see, for example, [Sp], p. 82.

To illustrate this, we describe the case of the cubic surface V in $\mathbb{P}^3$ with three cusps A_2. An equation for V is

$$V: xyz - t^3 = 0.$$

Consider the map $p: \mathbb{P}^2 \to X$

(4.10) $$p(u : v : w) = (u^3 : v^3 : w^3 : uvw).$$

It is easy to check that p is a covering of degree 3 ramified exactly at the singular points of V. As a by-product we get

(4.11) $$\pi_1(V \backslash S) = \mathbb{Z}/3\mathbb{Z}.$$

Note that V can be regarded as a quotient variety $\mathbb{P}^2/(\mathbb{Z}/3\mathbb{Z})$ where $\mathbb{Z}/3\mathbb{Z}$ acts on $\mathbb{P}^2$ via the formula

$$\hat{1} \cdot (u : v : w) = (\lambda u : \lambda^2 v : w)$$

with $\lambda = \exp(2\pi i/3)$.

On next example is related to the hypersurfaces

$$V_n^{d,a}: x_0^a x_1^{d-a} + x_1 x_2^{d-1} + \cdots + x_{n-1} x_n^{d-1} + x_{n+1}^d = 0$$

in $\mathbb{P}^{n+1}$, already considered in (2.24).

(4.12) **Proposition.** *Assume that* $n \geq 2$, $d \geq 3$, *and* $(d, a) = (d - 1, a) = 1$. *Then the hypersurface* $V_n^{d,a}$ *is a* $\mathbb{Z}$-*homology projective space, i.e.,*

$$H_.(V_n^{d,a}) = H_.(\mathbb{P}^n).$$

Proof. Let p_1 and p_2 be the two singular points of the hypersurface $V_n^{d,a}$ for $a > 1$ as in the proof of (2.24) (for $a = 1$ the only singular point is p_1). Let L_1 and L_2 be the corresponding Milnor lattices. We prove that the associated morphism

$$\varphi: L_1 \oplus L_2 \to \bar{L}$$

is an isomorphism. There are two distinct cases to consider.

Case 1 (n odd). Then the reduced Milnor lattice $\bar{L}$ is a unimodular skew-symmetric lattice by (3.26(i)). A detailed computation based on the paper by Orlik and Randell [OR] shows that in this case the Milnor lattices L_1 and L_2 are unimodular too (these details can be found in [BD]). Since rk L_1 + rk L_2 = rk $\bar{L}$ (this follows from (2.24)), it follows that φ is necessarily an isomorphism. Note that in this case $V_n^{d,a}$ is a $\mathbb{Z}$-homology manifold.

Case 2 (n even). In this case we have

$$\det \bar{L} = \det L_1 = d, \qquad \det L_2 = 1,$$

(the details for this are again in [BD]). Since by (2.24) φ is a monomorphism, it follows from (A5) that φ is also an epimorphism. Note that in this case $V_n^{d,a}$ is not a $\mathbb{Z}$-homology manifold. ☐

The following result shows that the above condition $(d - 1, a) = 1$ is necessary. It also provides us with examples of surfaces V having large torsion in $H_2(V)$.

(4.13) **Proposition.** *Consider the surfaces*

$$V^{d,a}: x^a y^{d-a} + yz^{d-1} + t^d = 0,$$

where $(d, a) = 1$, $d - 1 > a \geq 1$. *Then*

$$\text{Tors } H_2(V^{d,a}) = (\mathbb{Z}/d\mathbb{Z})^{k-1},$$

where $k = (d - 1, a)$ *is the greatest common divisor of* $d - 1$ *and* a.

Proof. Let $U = \mathbb{P}^3 \setminus V^{d,a}$, and note that

$$\text{Tors } H_2(U) = \text{Tors } H^4(\mathbb{P}^3, V) = \text{Tors } H^3(V) = \text{Tors } H_2(V),$$

where $V = V^{d,a}$. To compute Tors $H_2(U)$ we use the Gysin homology sequence of the Hopf fibration over U

$$S^1 \to S^7 \setminus K_V \to U.$$

It gives the following exact sequence

(4.14) $$0 \to \mathbb{Z}/d\mathbb{Z} \to H_2(S^7 \setminus K_V) \to H_2(U) \to 0.$$

Let F be the Milnor fiber corresponding to V and let $h: F \to F$ be the associated monodromy homeomorphism. The corresponding Wang sequence (3.1.18) implies

$$H_2(S^7 \setminus K_V) = \operatorname{coker}(h_* - I)$$

(use (3.2.2) to deduce $H_1(F) = 0$). To determine h_* we use the Thom–Sebastiani construction (3.3.21).

Indeed, we have $F = F_1 * F_2$ where $F_1: f_1 - 1 = 0$ with $f_1(x, y, z) = x^a y^{d-a} + yz^{d-1}$, and $F_2: f_2 - 1 = 0$ with $f_2(t) = t^d$. It follows from the proof of (2.24) that $h_{1*} = 1$. Moreover, we have

$$b_1(F_1) = \dim H_1(F_1; \mathbb{C}) = \dim H_1(F_1; \mathbb{C})^{h*} = \dim H_1(\tilde{U}; \mathbb{C}),$$

where $\tilde{U} = \mathbb{P}^2 \setminus C$, $C: f_1 = 0$. Since C has exactly $k + 1$ irreducible components (one of them being the line $y = 0$) it follows from (4.13) that

$$b_1(F_1) = k, \qquad H_1(\tilde{U}) = \mathbb{Z}^k.$$

Consider the fundamental groups

$$H = \pi_1(F_1), \qquad G = \pi_1(\tilde{U}).$$

Since $h_{1*} = 1$, it follows that the commutator subgroups coincide, namely,

$$[H, H] = [G, G]$$

under the inclusion $H \to G$ coming from the covering map $F_1 \to \tilde{U}$, see [N2].

It follows that

$$H_1(F_1) = \frac{H}{[H, H]} \hookrightarrow \frac{G}{[G, G]} = H_1(\tilde{U}) = \mathbb{Z}^k$$

and hence the group $H_1(F_1)$ is torsion free. As a result

$$H_1(F_1) = \mathbb{Z}^k.$$

Note that $\tilde{H}_0(F_2) = \mathbb{Z}^{d-1}$, and the corresponding monodromy operator

$$h_{2*}: \tilde{H}_0(F_2) \to \tilde{H}_0(F_0)$$

is such that $\operatorname{coker}(h_{2*} - 1) = \mathbb{Z}/d\mathbb{Z}$ (use (3.4.12) or (A26)).

Then note that h_* acting on $H_2(F)$ is the tensor product $h_{1*} \otimes h_{2*}$, and that this tensor product in turn can be identified with the direct sum

$$h_{2*} \oplus \cdots \oplus h_{2*}$$

of k copies of the homomorphism h_{2*}. It follows that

$$H_2(S^7 \setminus K_V) = \operatorname{coker}(h_* - I) = (\mathbb{Z}/d\mathbb{Z})^k.$$

In view of (4.14), we are done by applying the following easy fact.   □

(4.15) **Exercise.** If an abelian group G fits into an exact sequence

$$0 \to \mathbb{Z}/d\mathbb{Z} \xrightarrow{i} (\mathbb{Z}/d\mathbb{Z})^k \to G \to 0,$$

then $G \simeq (\mathbb{Z}/d\mathbb{Z})^{k-1}$. *Hint.* Lift the inclusion i to a homomorphism $\tilde{i}: \mathbb{Z} \to \mathbb{Z}^k$ and show that the element $\tilde{i}(1)$ can be written as λu, where u is a primitive element in $\mathbb{Z}^k$ and λ is an integer relatively prime to d.

(4.16) **Remark.** We can prove a slightly weaker version of (4.13), namely,

$$|\text{Tors } H_2(V^{d,a})| = d^{k-1}$$

and

$$\text{Tors } H_2(V^{d,a}) \subset (\mathbb{Z}/d\mathbb{Z})^{2k-1}$$

as follows. Using the results in [BaK], pp. 285–286, it is easy to see that

$$D(L_1) = (\mathbb{Z}/d\mathbb{Z})^k, \qquad D(L_2) = (\mathbb{Z}/d\mathbb{Z})^{k-1}.$$

Since $\det \bar{L} = d$, the first claim follows from (A5), while the second is a consequence of (A23).

As shown by this remark, there are cases when methods different from (4.3) can yield more precise results.

Before giving another example of this sort, we discuss a little of the topology of *projective cones*. To simplify this discussion, assume that V is a hypersurface in $\mathbb{P}^n$ given by an equation $f(x_0, \dots, x_n) = 0$ and that V has only isolated singularities, say at the points $a_1, \dots, a_p$. Consider the associated projective cone over V, namely, the hypersurface $\tilde{V}$ given in $\mathbb{P}^{n+1}$ by exactly the same equation

$$(4.17) \qquad\qquad \tilde{V}: f(x_0, \dots, x_n) = 0$$

as above (the new variable x_{n+1} does not occur at all).

Let L_i be the line joining the point $\bar{a}_i = (a_i : 0)$ to the point $\bar{a}_\infty = (0 : 0 : \cdots : 0 : 1)$ in $\mathbb{P}^{n+1}$. Let $L = L_1 \cup \cdots \cup L_p$ be the union of these lines. Since $\tilde{V} \setminus L$ is smooth, we can apply the Lefschetz Duality Theorem and get

$$H^k(\tilde{V}, L) = H_{2n-k}(\tilde{V} \setminus L).$$

Note that there is a retraction $\tilde{V} \setminus L \to V^0$ where $V^0 = V \setminus \{a_1, \dots, a_k\}$ is the smooth part of V. Indeed, since $\tilde{V}$ can be regarded as the union of the line L_x joining a point $x \in V$ to the point $\bar{a}_\infty$, this claim is obvious. By Lefschetz duality again

$$H_{2n-k}(V^0) = H^{k-2}(V, A),$$

where $A = \{a_1, \dots, a_p\}$. Putting all this together (with some special care in the cases $k = 2, 3$) yields

$$(4.18) \qquad\qquad H^k(\tilde{V}) = H^{k-2}(V) \qquad \text{for all } k \geq 2.$$

(4.19). **Exercise.** Assume that V is smooth. Then $\tilde{V}$ has exactly one singular point, namely, $\bar{a}_\infty$. Use (4.3) to prove (4.18).

Consider now the hypersurface

(4.20) $W_n^d: x_0 x_1^{d-1} + x_1 x_2^{d-1} + \cdots + x_n x_{n+1}^{d-1} = 0,$ $d \geq 3,$

in $\mathbb{P}^{n+1}$. Then W_n^d has a unique singular point, namely,

$$a_1 = (1:0:\cdots:0).$$

(4.21) **Proposition.**

(i) *For n odd, the integral cohomology of the hypersurface W_n^d is given by*

$$H^k(W_n^d) = \begin{cases} H^k(\mathbb{P}^n) & \text{for } k \neq n+1, \\ \mathbb{Z}^2 & k = n+1. \end{cases}$$

(ii) *For n even, the integral cohomology of the hypersurface W_n^d is given by*

$$H^k(W_n^d) = \begin{cases} H^k(\mathbb{P}^n) & \text{for } k \neq n, \\ \mathbb{Z}^2 & k = n. \end{cases}$$

Proof. Let H be the hyperplane in $\mathbb{P}^{n+1}$ given by the equation $x_1 = 0$. The intersection $W_n^d \cap H$ is exactly the projective cone $\tilde{W}_{n-2}^d$ over the hypersurface W_{n-2}^d. The complement $W_n^d \backslash H$ is isomorphism to the affine space $\mathbb{C}^n$ (just make $x_1 = 1$ in the equation for W_n^d). By Lefschetz duality it follows that

$$H^k(W_n^d, \tilde{W}_{n-2}^d) = H_{2n-k}(\mathbb{C}^n) = 0$$

for $0 \leq k \leq 2n$. Hence

$$H^k(W_n^d) \simeq H^k(\tilde{W}_{n-2}^d) = H^{k-2}(W_{n-2}^d)$$

from (4.18). The result can then be proved by induction on n, using as the starting point the homeomorphisms

$$W_0^d = \{2 \text{ point}\}, \qquad W_1^d = S^2 \vee S^2,$$

(the latter is obtained as in (2.13)). $\square$

(4.22) **Exercise.** Consider the three-dimensional hypersurface W_3^3 of degree 3 as above.

(i) Show that (W_3, a_1) is a singularity of type S_{11} in Arnold's classification and hence its associated Milnor lattice is

$$L_1 = I_{10} \oplus (\mathbb{Z}, (\ )_0).$$

(ii) Show that the reduced Milnor lattice $\bar{L}$ in this case is I_{10}.
(iii) Compute the homology $H.(W_3^3)$ using these facts and (4.3).

Hint. See (A16).

This exercise is meant to show that Theorem (4.3) can be useful even when some of the Milnor lattices L_i are degenerate.

We end this section with some remarks on the cup products in the integral cohomology algebra $H^{\cdot}(V)$ of a complete intersection V having only isolated singularities. Note that the group $H^n(V)$ is torsion free ($n = \dim V$), and hence the cup product followed by evaluation on the fundamental class $[V]$ gives rise to a pairing

(4.23) $\alpha \colon H^n(V) \times H^n(V) \to \mathbb{Z}.$

This lattice $(H^n(V), \alpha)$ can be described more precisely as follows. Let V_0 be a nearby smooth complete intersection, having the same multidegree as V (as in the proof of (4.4)). Using (3.24) and (3.26(i)), we can identify the reduced Milnor lattice $\bar{L}$ corresponding to V with a sublattice in $H_n(V_0)$. Let $I = \operatorname{im} \varphi_V$ be the smaller sublattice in $H_n(V_0)$ obtained in this way.

The statement in (4.3(ii)) can be restated as:

(4.24) **Proposition.** *There is a natural isomorphism*

$$H_n(V) = H_n(V_0)/I.$$

Recall that the Poincaré isomorphism

$$P \colon H_n(V_0) \to H^n(V_0) = \operatorname{Hom}(H_n(V_0), \mathbb{Z})$$

is given by $P(u) = (u, \ )$, where $(\ ,\ )$ is the intersection from on $H_n(V_0)$.

(4.25) **Corollary.** *There is a natural isomorphism of lattices*

$$H^n(V) = I^{\perp} = \{u \in H_n(V_0); (u, v) = 0 \text{ for all } v \in I\}.$$

Proof. Use (4.24) and the isomorphism

$$H^n(V) = \operatorname{Hom}(H_n(V), \mathbb{Z})$$

coming from the universal coefficient theorem for cohomology, see [Sp], p. 243.   □

Using (A34) and (4.25) we get the following:

(4.26) **Corollary.** *Assume that all the lattices L_i are nondegenerate. Then the lattice $H^n(V)$ is itself nondegenerate and*

$$\det H^n(V) = \det L_1 \cdots \det L_p \cdot |\operatorname{Tors} H_n(V)|^{-2}.$$

(4.27) **Example.** Consider the complete intersections $V_1, V_2,$ and V_3 of the same even dimension n and the same multidegree, having the following combinations of singular points: $3A_1, A_1 A_2,$ and A_3, respectively. Then the integral homology groups $H_{\cdot}(V_i)$, $i = 1, 2, 3$, are all the same, since we can show as in

(4.7) that Tors $H_n(V_i) = 0$. But the varieties V_i have *distinct homotopy types*, since using (4.26) we get the following table:

i	1	2	3
det $H^n(V_i)$	8	6	4

(4.28) **Remark.** Let X be a smooth affine hypersurface in $\mathbb{C}^{n+1}$. Let $\overline{X}$ be the closure of X in the projective space $\mathbb{P}^{n+1}$ and let X_∞ be the part at infinity of X, i.e., $X_\infty = \overline{X} \cap H_\infty$ where $H_\infty = \mathbb{P}^{n+1} \backslash \mathbb{C}^{n+1}$ is the hyperplane at infinity. Some topological invariants of the *affine* hypersurface X can be computed from the corresponding invariants of the *projective* hypersurfaces $\overline{X}$ and X_∞ in a simple direct way, e.g., from the Lefschetz duality we get

$$\chi(X) = \chi(\overline{X}) - \chi(X_\infty).$$

The (co)homology groups of X depend not only on the (co)homology groups of $\overline{X}$ and X_∞, but also on the homomorphisms between these groups induced by the inclusion $j: X_\infty \to \overline{X}$. To determine explicitly these homomorphisms can be a hard problem, as the following example shows.

(4.29) **Example.** Consider the surface

$$W = W_2^d : x_0 x_1^{d-1} + x_1 x_2^{d-1} + x_2 x_3^{d-1} = 0, \qquad d \geq 3,$$

in $\mathbb{P}^3$, a special case of (4.20). Consider the hyperplanes $H_i : x_i = 0$ for $i = 1, 2$. Note that both hyperplanes H_i contain the unique singular point $a_1 = (1:0:0:0)$ of the surface W. Hence the complements

$$X_1 = W \backslash H_1 \qquad \text{and} \qquad X_2 = W \backslash H_2,$$

are both smooth hypersurfaces in $\mathbb{C}^3 = \mathbb{P}^3 \backslash H_i$. Note that $\overline{X}_i = W$ for $i = 1, 2$ and that $X_{1,\infty} = W \cap H_1$ coincides with $X_{2,\infty} = W \cap H_2$. Indeed,

$$W \cap H_1 = W \cap H_2 = \mathbb{P}^1 \vee \mathbb{P}^1,$$

the union of two lines meeting at a point. Let $j_i : X_{i,\infty} \to W$ be the inclusion for $i = 1, 2$ and consider the induced morphisms

$$j_i^2 : H^2(W) \to H^2(X_{i,\infty}).$$

The exact sequence of the pair $(W, X_{i,\infty})$ in cohomology and Lefschetz duality imply

$$H_1(X_i) = \text{coker } j_i^2, \qquad H_2(X_i) = \ker j_i^2,$$

for $i = 1, 2$. For $i = 1$, the equation of X_1 shows that $X_1 = \mathbb{C}^2$, a fact already used in the proof of (4.21). Now we show that

(4.30) $H_k(X_2) = \mathbb{Z}$ for $k = 0, 1, 2$.

To do this, recall the following general result from the theory of $\mathbb{Z}/p\mathbb{Z}$ actions (Smith Theory), see [Bd], pp. 376–377.

(4.31) **Proposition.** *Let $G = \mathbb{Z}/p\mathbb{Z}$, with p a prime number, act on a space X such that the induced action on the integral cohomology $H^{\cdot}(X)$ is trivial. Let X^G denote the set of fixed points under this G-action. Then, for any integer $k \geq 0$, we have*

$$\sum_{j \geq 0} \dim H^{k+2j}(X^G; G) \leq \sum_{j \geq 0} \dim H^{k+2j}(X; G),$$

where $\dim$ *means dimension of a vector space over the field G.*

We take $X = X_2: x_0 x_1^{d-1} + x_1 + x_3^{d-1} = 0$ in $\mathbb{C}^2$. Note that $\mathbb{C}^*$ acts on X via the multiplication rule

$$t * (x_0, x_1, x_3) = (t^{-(d-1)(d-2)} x_0, t^{d-1} x_1, t x_3).$$

Let p be a prime number which divides $d - 1 \geq 2$. Consider the $G = \mathbb{Z}/p\mathbb{Z}$ action on X given by

$$\hat{a} \cdot (x_0, x_1, x_3) = \exp\left(\frac{2\pi i a}{p}\right) * (x_0, x_1, x_3).$$

Since this action of G is embedded in a $\mathbb{C}^*$-action, it follows that the induced G-action on $H^{\cdot}(X)$ is trivial. Moreover, we have

$$X^G = X \cap \{x_3 = 0\} = \mathbb{C} \cup \mathbb{C}^*,$$

where $\mathbb{C}$ is the line $x_1 = 0$, while $\mathbb{C}^*$ is the curve $x_0 x_1^{d-2} + 1 = 0$ (both in the plane given by $x_3 = 0$). Using (4.31) with $k = 1$, it follows that

$$\dim H^1(X; G) \geq 1.$$

Using the Universal Coefficient Theorem for cohomology and the equality $\chi(X) = 1$, it follows that

(4.32) $\qquad\qquad \mathrm{rk}\, H_2(X) = \mathrm{rk}\, H_1(X) = \mathrm{rk}\, H^1(X) \neq 0.$

Consider now the diagram of inclusions

$$\mathbb{P}^1 \vee \mathbb{P}^1 \xrightarrow{\;j_2\;} W$$

$$j' \searrow \qquad \swarrow j''$$

$$\mathbb{P}^3$$

Since both $\mathbb{P}^1$ are embedded linearly in $\mathbb{P}^3$, it follows that

$$(j')^2(\alpha) = (1, 1),$$

where $\alpha \in H^2(\mathbb{P}^3)$ is the standard generator and

$$H^2(\mathbb{P}^1 \vee \mathbb{P}^1) = H^2(\mathbb{P}^1) \oplus H^2(\mathbb{P}^1) = \mathbb{Z}^2$$

(use (2.10)). The above diagram shows that

$$(1, 1) \in \text{im}(j_2^2).$$

On the other hand, (4.32) implies that this subgroup $\text{im}(j_2^2)$ has rank at most 1. It follows that the element $(1, 1)$ generates this subgroup and hence

$$\ker j_2^2 \cong \text{coker } j_2^2 \cong \mathbb{Z},$$

as claimed in (4.30).

(4.33) **Exercise.** Consider the smooth surface in $\mathbb{C}^3$

$$X = X^{d,a} \colon x + x^{d-1}y + y^{d-a}z^a = 0$$

for $d \geq 3, d > a \geq 1$, and $(a, d - 1) = 1$.

(i) Show that for $a = 1$ the surface X is contractible. *Hint.* Show that the polynomial mapping

$$f \colon \mathbb{C}^3 \to \mathbb{C}, \qquad f(x, y, z) = x + x^{d-1}y + y^{d-1}z,$$

is a smooth fibration with fiber X (recall the proof of (1.4.1)).

In fact, using a result due to Russell [Ru] and Sathaye [Sy], we can show that for $a = 1$ there is an algebraic automorphism of $\mathbb{C}^3$ which carries the pair $(\mathbb{C}^3, X)$ onto the pair $(\mathbb{C}^3, H)$ where H is the plane given by the equation $x = 0$. The special case $d = 3$ is already considered by Nagata [Ng], p. 16.

(ii) Show that the projective closure $\overline{X}$ has two singularities for $a > 1$, both of which have nondegenerate Milnor lattices (use the results in [BaK], pp. 285–286). Deduce that

$$H^2(\overline{X}; \mathbb{C}) = \mathbb{C}^2, \qquad H^3(\overline{X}; \mathbb{C}) = 0.$$

(iii) Show that for $a > 1$ we have

$$H_k(X^{d,a}; \mathbb{C}) = \mathbb{C} \qquad \text{for} \quad k = 0, 1, 2.$$

Hint. Use the same method as in proving (4.30).

(4.34) **Exercise.** (i) Let G be a group acting on a space X and let $H \subset G$ be a normal subgroup. Prove the following equality among fixed points sets

$$X^G = (X^H)^{G/H},$$

where the action of the quotient group G/H on the subspace X^H is induced by the action of G on the space X. Use this to prove that the result (4.31) is true for any group $G = \mathbb{Z}/p^\alpha \mathbb{Z}$, α a positive integer.

(ii) Consider the Milnor fiber

$$F \colon x_0^a x_1^{d-a} + x_1 x_2^{d-1} + \cdots + x_{n-1} x_n^{d-1} = 1$$

with $d \geq 3, d - 1 \geq a \geq 1$, $(d, a) = 1$ from the proof of (2.24). Show that F is

$(n-2)$-connected and that

$$\mathrm{rk}\, H_{n-1}(F) \geq k \qquad \text{for } n \text{ odd}$$

and

$$\mathrm{rk}\, H_{n-1}(F) \geq k - 1 \qquad \text{for } n \text{ even},$$

where $k = (d - 1, a)$. *Hint.* Since $d - 1 \geq a$, there is a prime number p such that

$$p^\alpha | d - 1,\ p^{\alpha+1} \nmid d - 1,\ p^{\alpha+1} \nmid a, \qquad \alpha > 0.$$

Let β be the largest integer ≥ 0 such that $p^\beta | a$. Use the $\mathbb{C}^*$ action in (2.24) to induce a G-action on F, where $G = \mathbb{Z}/p^{\alpha(n-2)+\beta}\mathbb{Z}$ such that

$$F^G = F \cap \{x \in \mathbb{C}^{n+1};\ x_3 = \cdots = x_n = 0\}.$$

Then use the equality $F^G = F_1$ where F_1 is the Milnor fiber considered in the proof of (4.13).

(4.35) **Exercise.** For an affine variety X with dim $X > 0$ recall from (1.6.12) the notation $X_R = X \cap S_R$ for its link at infinity, $R \gg 0$.
  (i) Let X and Y be two connected affine varieties such that

$$\dim X > 0, \qquad \dim Y > 0, \qquad \pi_1(X) = \pi_1(Y) = 0.$$

Show that

$$\pi_1((X \times Y)_R) = 0.$$

Hint. Use (1.6.9) and van Kampen's theorem (4.2.17).
  (ii) Show that a smooth affine variety X is diffeomorphic to $\mathbb{C}^n$ where $n = \dim X \geq 2$ if and only if the following two conditions hold:

(α) X is contractible;
(β) $\pi_1(X_R) = 0$.

Hint. In the case $n = 2$, use Ramanujam's paper [Rm]. For $n \geq 3$ the claim follows from Theorem 5.1 in Smale's paper [Sm].
  (iii) Let a and d be positive integers such that $1 < a < d$ and $(a, d) = 1$. Consider the polynomial

$$f_{d,a}(x, y, z) = ((xz + 1)^d - (yz + 1)^a)/z,$$

and the associated surface in $\mathbb{C}^3$

$$V_{d,a}:\ f_{d,a}(x, y, z) - 1 = 0.$$

Then tom Dieck and Petrie [tDP] have shown that $V_{d,a}$ is a smooth contractible surface with logarithmic Kodaira dimension

$$\bar{\kappa}(V_{d,a}) = 1$$

(for the definition and basic properties of this subtle invariant refer, for instance, to Iitaka [Ii]).

Use (i) and (ii) to deduce that any product

$$Y = V_{d,a} \times V_{d',a'}$$

of two such surfaces is diffeomorphic to $\mathbb{C}^4$. Since

$$\bar{\kappa}(Y) = \bar{\kappa}(V_{d,a}) + \bar{\kappa}(V_{d',a'}) = 2 \neq -\infty = \bar{\kappa}(\mathbb{C}^4)$$

it follows that Y is not isomorphic to $\mathbb{C}^4$ as an algebraic variety.

For different examples of such "exotic algebraic structures" on $\mathbb{C}^n$ ($n \geq 3$), see Zaidenberg [Za].

(4.36) **Open Problem.** Consider the hypersurface $Y_{d,a}$ in $\mathbb{C}^4$ given by

$$Y_{d,a}: x + x^{d-1}y + y^{d-a}z^a + t^d = 0$$

for positive integers a, d such that $1 < a < d$, $(d, a) = (d - 1, a) = 1$. Use (1.6.13), (4.12), and (4.35(ii)) to deduce that $Y_{d,a}$ is diffeomorphic to $\mathbb{C}^3$. Does there exist a pair (d, a) as above such that $Y_{d,a}$ is isomorphic to $\mathbb{C}^3$ as an algebraic variety? Compare this problem with (4.33).

CHAPTER 6

de Rham Cohomology of Hypersurface Complements

§1. Differential Forms on Hypersurface Complements

In this chapter we work with *regular* differential forms in the sense of Algebraic Geometry or, depending on the context, in the sense of Analytic Geometry. Fix a positive integer $n \geq 0$ and consider the affine space $\mathbb{C}^{n+1}$. Then a (regular algebraic) differential p-form on $\mathbb{C}^{n+1}$ is just a differential form

$$(1.1) \qquad \omega = \sum_I c_I \, dx_{i_1} \wedge \cdots \wedge dx_{i_p},$$

where $I = (i_1, \ldots, i_p)$ and $c_I \in \mathbb{C}[x_0, \ldots, x_n]$, i.e., we allow only polynomial coefficients. Let Ω^p denote the $\mathbb{C}$-vector space of all these p-forms on $\mathbb{C}^{n+1}$. Let S denote the polynomial ring $\mathbb{C}[x_0, \ldots, x_n]$. Note that Ω^p is in fact an S-module. We also fix a set of strictly positive integer weights

$$(1.2) \qquad \operatorname{wt}(x_i) = w_i \qquad \text{for} \quad i = 0, \ldots, n.$$

The ring S becomes a *graded ring* by setting

$$(1.3) \qquad \deg(x_0^{a_0}, \ldots, x_n^{a_n}) = a_0 w_0 + \cdots + a_n w_n.$$

The S-module Ω^p becomes a *graded S-module* by setting

$$(1.4)$$
$$\deg(x_0^{a_0}, \ldots, x_n^{a_n} \, dx_{i_1} \wedge \cdots \wedge dx_{i_p}) = a_0 w_0 + \cdots + a_n w_n + w_{i_1} + \cdots + w_{i_p}.$$

For any positive integer m, we let S_m, Ω_m^p denote the corresponding homogeneous components of degree m. We also use the equivalent notation

$$\omega \in \Omega_m^p \quad \Leftrightarrow \quad |\omega| = m.$$

The differential form ω in (1.1) can be regarded as an element in $H^0(\mathbb{C}^{n+1}, \Omega_{\mathbb{C}^{n+1}}^p)$, i.e., as a *global section* of the sheaf on $\mathbb{C}^{n+1}$ associated to the S-module Ω^p, see [Hn], p. 110.

Consider a polynomial $f \in S$ and the associated *principal open subset*

$$(1.5) \qquad D(f) = \{x \in \mathbb{C}^{n+1}; f(x) \neq 0\}.$$

Using [Hn], loc. cit., it follows that we have the following simple description of the sections of the sheaf $\Omega^p_{\mathbb{C}^{n+1}}$ over the open set $D(f)$

$$(1.6) \qquad\qquad H^0(D(f), \Omega^p_{\mathbb{C}^{n+1}}) = \Omega^p_f,$$

where Ω^p_f is the *localization* of the S-module Ω^p with respect to the multiplicative system $\{f^s; s \geq 0\}$.

(1.7) **Exercise.** Let $j: X \to \mathbb{C}^{n+1}$ denote the inclusion of a smooth closed subvariety X in $\mathbb{C}^{n+1}$. Let Ω^p_X be the sheaf of regular differential p-forms on X. Show that the restriction morphism

$$j^*: H^0(\mathbb{C}^{n+1}, \Omega^p) \to H^0(X, \Omega^p_X)$$

is surjective. *Hint.* Consider the exact sequence of coherent sheaves on $\mathbb{C}^{n+1}$ and, respectively, on X:

$$0 \to I_X \otimes \Omega^p_{\mathbb{C}^{n+1}} \to \Omega^p_{\mathbb{C}^{n+1}} \xrightarrow{\alpha} \mathcal{O}_X \otimes \Omega^p_{\mathbb{C}^{n+1}} \to 0,$$

$$0 \to \ker\beta \to \mathcal{O}_X \otimes \Omega^p_{\mathbb{C}^{n+1}} \xrightarrow{\beta} \Omega^p_X \to 0,$$

where I_X is the defining ideal sheaf for X and the morphisms α and β are defined in the obvious way.

The restriction morphism j^* is just the composition

$$H^0(\Omega^p_{\mathbb{C}^{n+1}}) \xrightarrow{\bar{\alpha}} H^0(\mathcal{O}_X \otimes \Omega^p_{\mathbb{C}^{n+1}}) \xrightarrow{\bar{\beta}} H^0(\Omega^p_X).$$

Prove that both $\bar{\alpha}$ and $\bar{\beta}$ are surjective homomorphisms, using the following standard vanishing result

$$H^1(I_X \otimes \Omega^p_{\mathbb{C}^{n+1}}) = H^1(\ker\beta) = 0,$$

see [Hn], p. 215.

Let us now turn our attention to differential forms on the projective space $\mathbb{P}^n$ (see, for instance, [G]). Recall first the definition of regular functions on a principal open set in $\mathbb{P}^n$. Let $f \in S_N$ (with respect to the trivial weights $w_0 = \cdots = w_n = 1$) and consider the associated *principal open set*

$$(1.8) \qquad\qquad U(f) = \{x \in \mathbb{P}^n; f(x) \neq 0\}.$$

Then a regular function $\varphi \in H^0(U(f), \mathcal{O}_{\mathbb{P}^n})$ is just a quotient

$$(1.9) \qquad\qquad \varphi = \frac{g}{f^s}, \qquad g \in S_{sN},$$

for some integer $s \geq 0$.

Consider the natural projection

$$\pi: D(f) \to U(f), \qquad (x_0, \ldots, x_n) \mapsto (x_0 : \cdots : x_n).$$

Let $\omega \in H^0(U(f), \Omega^p_{\mathbb{P}^n})$ and set $\alpha = \pi^*\omega$. Then $\alpha \in H^0(D(f), \Omega^p_{\mathbb{C}^{n+1}})$ and satisfies, in addition, the following two conditions:

(1.10)

   (i) α is invariant with respect to the $\mathbb{C}^*$-action on $D(f)$;

   (ii) $\Delta\alpha = 0$ where Δ is the *contraction* with the Euler vector field;

(1.11)
$$E = \sum_{j=0,n} x_j \frac{\partial}{\partial xj}.$$

More explicitly, this means the following. Consider the $\mathbb{C}^*$-action on $\mathbb{C}^{n+1}$ associated with the usual projective space $\mathbb{P}^n$, i.e.,

$$\lambda \cdot x = (\lambda x_0, \dots, \lambda x_n).$$

Define $\phi_\lambda(x) = \lambda \cdot x$ for $\lambda \in \mathbb{C}^*$, $x \in \mathbb{C}^{n+1}$. Note that the open set $D(f)$ is $\mathbb{C}^*$-invariant. Condition (1.10(i)) means formally that

(1.12)
$$\phi_\lambda^*(\alpha) = \alpha \qquad \text{for all} \quad \lambda \in \mathbb{C}^*.$$

This relation is a direct consequence of the obvious formula: $\pi \circ \phi_\lambda = \pi$.

    Concerning (1.10(ii)), recall that we can define *geometrically* the contraction $\Delta_V \alpha$ of a differential p-form α with respect to any vector field V as being the $(p-1)$-form given by the following punctual formula

(1.13)
$$\Delta_V(\alpha)(x)(v_1, \dots, v_{p-1}) = \alpha(V(x), v_1, \dots, v_{p-1}).$$

The Euler vector field E is clearly tangent to the fibers of the projection map π and hence

$$d\pi_x(E(x)) = 0 \qquad \text{for all} \quad x \in D(f),$$

$$(\Delta\alpha)(x)(v_1, \dots, v_{p-1}) = \pi^*(\omega)(x)(E(x), v_1, \dots, v_{p-1})$$

$$= \omega(\pi(x))(d\pi_x(E(x)), d\pi_x(v_1), \dots) = 0.$$

Now assume conversely that we have a section $\alpha \in H^0(D(f), \Omega^p_{\mathbb{C}^{n+1}})$ satisfying (1.10(i)) and (1.10(ii)). Then we can define the p-form $\bar\omega$ on the open set $U(f)$ by the formula

$$\bar\omega(\bar{x})(\bar{v}_1, \dots, \bar{v}_p) = \alpha(x)(v_1, \dots, v_p),$$

where $\bar{x} \in U(f)$, $x \in D(f)$, with $\pi(x) = \bar{x}$ and $\bar{v}_j \in T_{\bar{x}} U(f)$, $v_j \in T_x D(f)$, with

$$d_x \pi(v_j) = \bar{v}_j \qquad \text{for} \quad j = 1, \dots, p.$$

We can easily check that this p-form $\bar\omega$ is well defined and that $\pi^*\bar\omega = \alpha$. It follows that there is a bijection between the sections $\omega \in H^0(U(f), \Omega^p_{\mathbb{P}^n})$ and the sections $\alpha \in H^0(D(f), \Omega^p_{\mathbb{C}^{n+1}})$ satisfying (1.10(i)) and (1.10(ii)). It is usual (and we will do this in the sequel) to identify the p-form ω with the associated p-form $\alpha = \pi^*\omega$. Using (1.6) and (1.10(i)) it follows that

(1.14)
$$\alpha = \frac{\beta}{f^s} \qquad \text{for some} \quad \beta \in \Omega^p_{sN}.$$

The condition $\Delta\alpha = 0$ is clearly equivalent to the condition $\Delta\beta = 0$.

    The contraction Δ has the following algebraic properties.

(1.15) **Lemma.** (A) *There is a unique S-linear operator* $\Delta: \Omega_m^p \to \Omega_m^{p-1}$ *satisfying the properties:*

(i) $\Delta(\omega \wedge \omega') = \Delta(\omega) \wedge \omega' + (-1)^p \omega \wedge \Delta(\omega')$ *for* $\omega \in \Omega^p$, $\omega' \in \Omega^q$;
(ii) $\Delta(df) = Nf$ *for any* $f \in S_N$.

  (B) *For a homogeneous differential form* $\omega \in \Omega_m^p$ *we have*

$$\Delta d(\omega) + d\Delta(\omega) = |\omega|\omega.$$

  (C) *The sequence*

$$0 \to \Omega^{n+1} \overset{\Delta}{\to} \Omega^n \overset{\Delta}{\to} \cdots \overset{\Delta}{\to} \Omega^1 \overset{\Delta}{\to} \Omega^0 \to 0$$

is exact, except for the last term where

$$\mathrm{im}\{\Delta: \Omega^1 \to \Omega^0\} = (x_0, \ldots, x_n),$$

the maximal ideal in S generated by $x_0, \ldots, x_n$.

For a proof of (1.15) (elementary, but tedious) we refer to [Dg1]. In particular, (C) follows by identifying the above sequence with the *Koszul complex* of the regular sequence $x_0, \ldots, x_n$ in S. Using (1.14) and (1.15(C)) we get the following.

(1.16) **Proposition.** *Any differential p-form* ω *(for* $p > 0$*) on the open set* $U(f)$ *can be written as*

$$\omega = \frac{\Delta(\gamma)}{f^s}$$

for some integer $s > 0$ *and* $\gamma \in \Omega_{sN}^{p+1}$.

(1.17) **Example.** The S-module Ω^{n+1} is free of rank 1 and we can take the $(n+1)$-form

$$\omega_{n+1} = dx_0 \wedge \cdots \wedge dx_n$$

as a generator. Define

$$\Omega = \Delta(\omega_{n+1}) = \sum_{j=0,n} (-1)^j x_j \, dx_0 \wedge \cdots \wedge \hat{dx_j} \wedge \cdots \wedge dx_n,$$

where the hat ($\hat{}$) shows that the factor dx_j is deleted.

Using (1.16) we infer that any n-form on the open set $U(f)$ has the form

$$\omega = \frac{h\Omega}{f^s} \qquad \text{for} \quad h \in S_{sN-n-1}, \quad s > 0.$$

It is interesting to see how we can express the differential of the form ω in (1.16)

$$d\omega = d\left(\frac{\Delta(\gamma)}{f^s}\right)$$

in a similar form

$$dω = \frac{Δ(δ)}{f^{s+1}} \quad \text{for some} \quad δ ∈ Ω^{p+2}_{(s+1)N}.$$

We have

$$dω = \frac{f \, dΔ(γ) - s \, df ∧ Δ(γ)}{f^{s+1}}.$$

Using (1.15) we get

$$dΔ(γ) = sNγ - Δ(dγ)$$

and

$$df ∧ Δ(γ) = Nfγ - Δ(df ∧ γ).$$

These formulas imply

$$dω = -\frac{Δ(f \, dγ - s \, df ∧ γ)}{f^{s+1}}.$$

The numerator here, as well as the considerations of Dolgachev in [Dg1], p. 61, lead us to introduce the $\mathbb{C}$-linear maps

$$d_f : Ω^p → Ω^{p+1} \quad \text{for all} \quad p ≥ 0$$

defined by the formula

$$(1.18) \qquad d_f(ω) = f \, dω - \frac{|ω|}{N} \, df ∧ ω$$

for a homogeneous differential form $ω$. This differentiation d_f satisfies the following commutativity property with the contraction operator $Δ$ (see [Dg1], loc. cit.)

$$(1.19) \qquad d_f Δ(ω) = -Δ(d_f(ω)).$$

In view of all this we may write

$$(1.20) \qquad d\left(\frac{Δ(γ)}{f^s}\right) = \frac{d_f Δ(γ)}{f^{s+1}} = -\frac{Δ(d_f(ω))}{f^{s+1}}.$$

To pass from the usual projective space $\mathbb{P}^n$ to a weighted projective space $\mathbb{P} = \mathbb{P}(\mathbf{w})$, $\mathbf{w} = (w_0, \ldots, w_n)$, we make the following "minor" changes:

(i) Use the gradings on the ring S and on the modules $Ω^p$ associated to the weights $\mathbf{w}$ as in (1.3), (1.4).

(ii) Use the associated weighted Euler vector field

$$E = \sum_{j=0,n} w_j x_j \frac{∂}{∂x_j}.$$

The corresponding contraction $Δ$ still verifies the properties (1.15).

(iii) Take (1.16) as the *definition* of regular differential forms on the principal open set

$$U(f) = \{x ∈ \mathbb{P}; f(x) ≠ 0\}.$$

Note that this affine variety $U(f)$ may have quotient singularities (see Appendix B) and hence some care is needed in considering differential forms on $U(f)$.

This definition of the regular differential forms on $U(f)$ implies all the expected properties for these differential forms, see [Dg1] and [S6]. In particular, the resulting de Rham complex of sheaves $(\Omega^*_{U(f)}, d)$ is a resolution for the constant sheaf $\mathbb{C}$.

In this chapter all the cohomology groups $H^*(X)$ are with $\mathbb{C}$-coefficients. Then we have the following fundamental result.

(1.21) **Theorem** (Grothendieck). *Let X be an affine variety and assume that either X is smooth or $X = U(f)$, a principal open set in $\mathbb{P}(\mathbf{w})$. Let $(A^*, d) = (H^0(X, \Omega^*_X), d)$ be the de Rham complex of the regular global differential forms on X. Then*

$$H^s(X) = H^s(A^*, d) \qquad \text{for any} \quad s \geq 0.$$

For a proof (trivial in the language of hypercohomology), see Grothendieck [Gr], or [GH], p. 453, in the smooth case, and Steenbrink [S6] in the case $X = U(f)$. We have also the following *analytic* Grothendieck theorem, see [Gr].

(1.22) **Theorem.** *Let X be a complex Stein manifold and let $Y \subset X$ be a closed hypersurface. Let (A^*, d) be the de Rham complex of the global sections in the sheaves $\Omega^*_X[Y]$ of analytic differential forms on $U = X \backslash Y$ which have polar singularities along Y. Then*

$$H^s(U) = H^s(A^*, d) \qquad \text{for any} \quad s \geq 0.$$

It is perhaps useful to recall here the definition of the sheaf $\Omega^p_X[Y]$. If X_0 is an open set in X, then a section $\omega \in H^0(X_0, \Omega^p_X[Y])$ is an analytic differential p-form on $X_0 \backslash Y$ satisfying the following property. Any point $y \in X_0 \cap Y$ has a neighborhood U_y in X such that the equality

$$(1.23) \qquad\qquad\qquad \omega = \frac{\beta}{f^s}$$

holds on $U_y \backslash Y$, where β is an analytic p-form on U_y and $f = 0$ is an equation for the hypersurface Y in the open set U_y.

(1.24) **Examples.** (i) Take $U = \mathbb{C}^n \backslash \{0\}$ for $n > 1$. Then U is *not* an affine variety. We clearly have $H^{2n-1}(U) = H^{2n-1}(S^{2n-1}) = \mathbb{C}$. On the other hand, since $2n - 1 > n$, there are no regular (algebraic or analytic) $(2n - 1)$-forms on U.

Hence the Grothendieck theorem (1.21) cannot work for nonaffine varieties. What remains true, in general, is that

$$H^s(X) = \mathbb{H}^s(X, \Omega_X),$$

i.e., the cohomology of any variety X can be computed as the hypercohomology of the de Rham complex $\Omega_X^{\cdot}$ of regular differential forms on X, see [GH], p. 448.

(ii) Take $U = \mathbb{C} \backslash A$ where $A = \{a_1, \ldots, a_p\}$ is a finite set of points. Then U is affine and the associated Gysin sequence (2.2.14) shows that the Poincaré residue map

$$H^1(U) \overset{R}{\to} H^0(A)$$

is an isomorphism. This shows that a $\mathbb{C}$-basis for the vector space $H^1(U)$ is given by the 1-forms

$$\omega_j = \frac{dx}{2\pi i(x - a_j)} \qquad \text{for} \quad j = 1, \ldots, p.$$

Here and in the sequel $i = \sqrt{-1}$.

(iii) Take $U = \mathbb{C} \backslash \mathbb{Z}$. Then U has the homotopy type of a countable bouquet of circles S^1 and hence

$$H_1(U) = \mathbb{C}^{(\mathbb{Z})},$$

i.e., $H_1(U)$ is the *direct sum* of countably many copies of $\mathbb{C}$. It follows that passing to cohomology we get

$$H^1(U) = \operatorname{Hom}(H_1(U), \mathbb{C}) = \mathbb{C}^{\mathbb{Z}},$$

the *direct product* of countable many copies of $\mathbb{C}$.

To find the differential form corresponding to a sequence

$$\lambda = (\ldots, \lambda_{-1}, \lambda_0, \lambda_1, \ldots) \in \mathbb{C}^{\mathbb{Z}} = H^1(U)$$

we can proceed as follows. First use the Mittag–Leffler theorem (see, for instance, [KK], p. 238) to find a meromorphic function g on $\mathbb{C}$ whose principal part at the point k is

$$\frac{\lambda_k}{2\pi i(x - k)}$$

for all $k \in \mathbb{Z}$ and such tht g has no other poles. Consider now the differential form

$$\omega_\lambda = g(x)\, dx \in H^0(\mathbb{C}, \Omega_{\mathbb{C}}^1[\mathbb{Z}]).$$

Then the residue of the form ω_λ at a point k is precisely λ_k, using our convention for residues (recall (2.2.15)). Note that some cohomology classes in $H^1(U)$ cannot be represented by rational differential forms

$$\omega = \frac{P}{Q}\, dx \qquad \text{with} \quad P, Q \in \mathbb{C}[x].$$

(iv) Recall the analytic description of the cohomology $H^*(U)$ given in (2.2.20), where $U = B \backslash C$ with B a small open ball at the origin of $\mathbb{C}^2$ and $(C, 0)$ a plane curve singularity.

Note that in all the examples (ii), (iii), and (iv) above only differential forms with poles of order 1 have been used. To make this more precise, we give the following:

(1.25) **Definition.** (i) (Projective Case). Let $\omega \in H^0(U(f), \Omega_{\mathbb{P}}^p)$ be a rational differential form as in (1.16). The minimal value of the positive integer s for which such a representation of ω exists is called the *order of the pole of the differential form ω along the hypersurface $V: f = 0$*. This order will be denoted in the sequel by $\mathrm{ord}_V(\omega)$.

(ii) (Stein Case). Let $\omega \in H^0(X, \Omega_X^p[Y])$ be a holomorphic differential form on $X \backslash Y$ with polar singularities along the hypersurface Y as in (1.22). The minimal value of the positive integer s, such that the form ω has a local representative of the form (1.23) for all $y \in Y$, is called the *order of the pole of the form ω along the hypersurface Y*. This element in $\mathbb{N} \cup \{\infty\}$ is denoted by $\mathrm{ord}_Y(\omega)$.

(1.26) **Exercise.** Use the Mittag–Leffler theorem as in (1.24(iii)) above to construct an example with $\mathrm{ord}_Y(\omega) = \infty$.

We now want to define a basic concept, namely, the polar filtration on the de Rham complexes of type (A^*, d). Consider first the projective case. Let (A_0^*, d) be the de Rham complex associated to the hypersurface $V: f = 0$ in the weighted projective space $\mathbb{P} = \mathbb{P}(\mathbf{w})$. In other words, setting $U = U(f)$, we have

$$(1.27) \qquad\qquad A_0^m = H^0(U, \Omega_{\mathbb{P}}^m).$$

(1.28) **Definition.** The *polar filtration* on the complex (A_0^*, d) is the decreasing filtration P defined by

$$P^s A_0^m = \{\omega \in A_0^m; \, \mathrm{ord}_V(\omega) \le m - s + 1\}$$

for $m - s + 1 \ge 0$ and $P^s A_0^m = 0$ for $m - s + 1 < 0$.

The filtration P has the following obvious properties:

(1.29) **Lemma.**

(i) *P is a filtration of complexes, i.e.,*

$$d(P^s A_0^m) \subset P^s A_0^{m+1};$$

(ii) *P is a decreasing filtration, bounded above*

$$0 = P^{n+1} A_0^* \subset P^n A_0^* \subset \cdots \subset P^0 A_0^* \subset P^{-1} A_0^* \subset \cdots;$$

(iii) *P is exhaustive*

$$A_0^* = \bigcup_{s \in \mathbb{Z}} P^s A_0^*.$$

The polar filtration P on the complex A_0^* induces the *polar filtration* P on the cohomology groups $H^*(U) = H^*(A_0^*)$ by the formula

(1.30) $$P^s H^m(U) = \text{im}\{H^m(P^s A_0^*) \to H^m(A^*)\}.$$

Here the cohomology morphism is induced by the inclusion of complexes $P^s A^* \to A^*$. Since U is an algebraic variety, the cohomology groups $H^m(U)$ also have a *Hodge filtration* F which is part of the mixed Hodge structure constructed by Deligne [De1]. We refer the reader, if necessary, to the brief elementary introduction to mixed Hodge theory given in Appendix C in this book. Hence on the cohomology groups $H^m(U)$ we have two decreasing filtrations, the polar filtration P and the Hodge filtration F. It is a natural question to see how these two filtrations are related to one another.

(1.31) **Theorem** (Deligne–Dimca). *For any hypersurface complement $U = \mathbb{P}\backslash V$ we have*

$$P^s H^m(U) \supset F^s H^m(U)$$

for any integers s and m.

Proof. The case of hypersurface complements in the usual projective space $\mathbb{P} = \mathbb{P}^n$ is a special case of the main result in [DD]. To treat the case of a weighted projective space $\mathbb{P} = \mathbb{P}(w_0, \ldots, w_n)$ let $p: \mathbb{P}^n \to \mathbb{P}$ be the natural projection, namely,

$$p(x_0 : \cdots : x_n) = (x_0^{w_0} : \cdots : x_n^{w_n}).$$

If $f = 0$ is the equation for V in $\mathbb{P}$, then $\tilde{f} = p^*(f)$ is a homogeneous polynomial. The hypersurface complement $\tilde{U} = U(\tilde{f}) \subset \mathbb{P}^n$ is just the preimage $p^{-1}(U)$.

The finite group $G(\mathbf{w})$ associated with the ramified covering p acts on the cohomology group $H^m(\tilde{U})$ such that the fixed part is exactly im $p^* \simeq H^m(U)$. Since the monomorphism

$$p^*: H^m(U) \to H^m(\tilde{U})$$

preserves both filtrations P and F, the result for U follows from the result for $\tilde{U}$. $\square$

(1.32) **Corollary.** *Any element in the cohomology group $H^m(U)$ can be represented by a rational differential form $\omega \in A_0^m$ such that*

$$\text{ord}_V(\omega) \leq m.$$

Proof. This statement follows from (1.31) as soon as we prove

$$F^1 H^m(U) = H^m(U).$$

As above, it is enough to prove this for open sets U in the usual projective space $\mathbb{P}^n$. In this special case, we can use the duality of mixed Hodge numbers explained in Appendix C to derive

$$h^{0,q}(H^m(U)) = h^{n,n-q}(H_0^{2n-m-1}(V)) = 0$$

for all q (since $n > \dim V = n - 1$). This implies $F^1 H^m(U) = F^0(H^m(U)) = H^m(U)$.   □

(1.33) Remark. The equality

$$P^s H^m(U) = F^s H^m(U)$$

holds when U is the complement of a quasismooth hypersurface. The case $U \subset \mathbb{P}^n$ is treated by Griffiths [G], while the weighted case can be deduced similarly, using Steenbrink's results in [S2] and our result (2.17) below. To show that the above equality does not always hold, consider the following example. Let $V: x(xy(x + y) + z^3) = 0$ be the union of a smooth cubic curve in $\mathbb{P}^2$ with one of its inflectional tangents. It is not difficult to show that

$$2 = \dim P^2 H^2(U) > \dim F^2 H^2(U) = 1,$$

where $U = \mathbb{P}^2 \backslash V$ as usual.

We consider now the polar filtration in the Stein case. More precisely, we restrict our attention to the following special case: X is an open ball centered at the origin of $\mathbb{C}^n$ with radius $r > 0$ and Y is a hypersurface in X with equation $f = 0$. Moreover, we assume that the hypersurface Y has a conic structure in the ball X as in (1.5.1).

Let X_m be the open ball at the origin of $\mathbb{C}^n$ with radius r/m for $m \geq 1$. Consider the sequence of de Rham compexes

$$A_m^* = H^0(X_m, \Omega_{X_m}^*[X_m \cap Y]).$$

The restriction of differential forms induces monomorphisms

$$\rho_{m,m'}: A_m^* \to A_{m'}^* \quad \text{for} \quad m' > m.$$

Due to the conic structure of the hypersurface Y in $X = X_1$ (and hence also in all the smaller balls X_m for $m > 1$), we have the following commutative diagram:

(1.34)

$$
\begin{array}{ccc}
H^*(A_m^*) & \xrightarrow{\bar{\rho}_{m,m'}} & H^*(A_{m'}^*) \\
{\scriptstyle G}\Big\downarrow{\scriptstyle \wr} & & {\scriptstyle G}\Big\downarrow{\scriptstyle \wr} \\
H^*(X_m \backslash Y) & \xrightarrow{i_{m,m'}^*} & H^*(X_{m'} \backslash Y).
\end{array}
$$

Here the vertical isomorphisms come from the Grothendieck theorem (1.22), the top homomorphisms are induced by the restrictions $\rho_{m,m'}$, and the bottom isomorphisms are induced by the inclusions

$$i_{m,m'}: X_m \backslash Y \to X_{m'} \backslash Y.$$

As a result, it follows that $\bar{\rho}_{m,m'}$ are isomorphisms for all $m' > m \geq 1$. The direct limit

$$(1.35) \qquad\qquad A_\infty^* = \varinjlim_m A_m^*$$

of the family of complexes $(A_m^*, \rho_{m,m'})$ can be identified with the de Rham complex of germs at the origin of $\mathbb{C}^n$ of meromorphic differential forms with polar singularities along the hypersurface Y.

In down-to-earth terms, an element in A_∞^* has a representative ω as in (1.23), where f is the germ at the origin of the defining equation $f = 0$ for Y and the coefficients of the differential form β are function germs in $\mathcal{O}_n$.

(1.36) **Exercise.** Show that the inclusions

$$A_m^* \to A_\infty^*$$

induce isomorphisms

$$H^*(A_m^*) \simeq H^*(A_\infty^*)$$

for all $m \geq 1$.

In view of this result it is natural to work with the limit de Rham complex A_∞^* instead of one of the individual complexes A_m^* depending on both r and m. This limit complex A_∞^* has a natural *polar filtration P* defined by

$$(1.37) \qquad\qquad P^s A_\infty^m = \{\omega \in A_\infty^m; \operatorname{ord}_Y(\omega) \leq m - s + 1\}$$

for $m - s + 1 \geq 0$ and $P^s A_\infty^m = 0$ for $m - s + 1 < 0$.

This filtration has all the properties (1.29) of the global projective case. There is an associated *polar filtration P* on the cohomology groups $H^m(X \setminus Y) = H^m(A_\infty^*)$ defined by

$$(1.38) \qquad\qquad P^s H^m(X \setminus Y) = \operatorname{im}\{H^m(P^s A_\infty^*) \to H^m(A_\infty^*)\}.$$

Assume from now on that $(Y, 0)$ is an isolated singularity. Then the Gysin sequence of the pair $(X \setminus \{0\}, Y \setminus \{0\})$ contains the residue morphism

$$H^m(X \setminus Y) \overset{R}{\to} H^{m-1}(Y \setminus \{0\})$$

(only the dimensions $m = n$ and $m = n - 1$ are interesting). Both these cohomology groups have HMS such that the residue morphism R is a MHS morphism of type $(-1, -1)$, see Appendix C for some details. In particular, the cohomology groups $H^m(X \setminus Y)$ have a Hodge filtration F and we may ask whether the result (1.31) extends to this local situation. This local case is only partially similar to the global case, as the following results show.

(1.39) **Proposition** (Dimca [D8], Karpishpan [Kp]).

$$P^s H^n(X \setminus Y) \supset F^s H^n(X \setminus Y)$$

for any integer s.

(1.40) **Proposition** (Karpishpan [Kp]).

$$P^s H^{n-1}(X \setminus Y) \subset F^s H^{n-1}(X \setminus Y)$$

for any integer s.

In fact, a proof for (1.39) is given below, see (4.21).

Let us come back now to our global algebraic varieties. Let $f \in S$ be a weighted homogeneous polynomial of type $(\mathbf{w}; N)$, $U \subset \mathbb{P}(\mathbf{w})$ the associated principal open set, and $F: f - 1 = 0$ the corresponding Milnor fiber in $\mathbb{C}^{n+1}$. The monodromy homeomorphism $h: F \to F$ has finite order N, recall (3.1.19). Consider the finite cyclic group of order N

(1.41) $$G = \{\lambda \in \mathbb{C}; \lambda^N = 1\}$$

and the obvious action

$$G \times F \to F, \qquad (\lambda, x) \mapsto \lambda \cdot x,$$

induced by the $\mathbb{C}^*$-action of $\mathbb{C}^{n+1}$ associated to the weights $\mathbf{w}$. Since $U = F/G$ we get

(1.42) $$H^0(U, \mathcal{O}_U) = H^0(F, \mathcal{O}_F)^G,$$

i.e., the coordinate ring of the affine quotient variety U is the subring of G-invariant functions in $H^0(F, \mathcal{O}_F)$. We can identify the group of characters

$$G^* = \text{Hom}(G, \mathbb{C}^*)$$

of the group G with the quotient group $\mathbb{Z}/N\mathbb{Z}$ as follows:

(1.43) $$\mathbb{Z}/N\mathbb{Z} \ni a \mapsto (\lambda \mapsto \lambda^a) \in G^*.$$

For any such character $a \in \mathbb{Z}/N\mathbb{Z}$ we define the corresponding eigenspace

(1.44) $$H^0(F, \mathcal{O}_F)_a = \left\{ g \in H^0(F, \mathcal{O}_F); h^*(g) = \exp\left(\frac{2\pi i a}{N}\right) \cdot g \right\}.$$

With this notation we can rewrite relation (1.42) as

(1.45) $$H^0(U, \mathcal{O}_U) = H^0(F, \mathcal{O}_F)_0.$$

What about the eigenspaces corresponding to the characters $a \neq 0$?

To answer this question, first note that we can write

$$H^0(U, \mathcal{O}_U) = H^0(U, \mathcal{O}_\mathbb{P}).$$

Then recall that we can define the twisted sheaves $\mathcal{O}_\mathbb{P}(t)$ for any $t \in \mathbb{Z}$ such that

$$H^0(U, \mathcal{O}_\mathbb{P}(t)) = \left\{ \frac{g}{f^s}; g \in S_{sN+t} \text{ for some } s \geq 0 \right\},$$

see [Dg1]. A moment's thought shows that we have

(1.46) $$H^0(U, \mathcal{O}_\mathbb{P}(a)) = H^0(F, \mathcal{O}_F)_a$$

for all $a \in \mathbb{Z}$. Here and in the sequel we use the same notation for an integer a and for its class in $\mathbb{Z}/N\mathbb{Z}$.

Let Ω_F^* be the de Rham complex of sheaves of regular algebraic differential forms on the Milnor fiber F. The group G acts on this complex Ω_F^* via the monodromy homeomorphism h. Note that the Grothendieck theorem is compatible with the G-eigenspace decompositions, namely,

$$(1.47) \qquad H^m(F)_a = H^m(H^0(F, \Omega_F^*)_a)$$

for all $m \geq 0$ and $a \in \mathbb{Z}/N\mathbb{Z}$. (The definition of these G-eigenspaces is exactly as in (1.44).) On the weighted projective space $\mathbb{P}$ we also have twisted sheaves of differential forms $\Omega_{\mathbb{P}}^*(t)$. Using these we have in analogy to (1.46)

$$(1.48) \qquad H^0(U, \Omega_{\mathbb{P}}^*(a)) = H^0(F, \Omega_F^*)_a$$

for any a. Hence it is natural to consider the complexes A_a^* for $a \in \mathbb{Z}/N\mathbb{Z}$ defined by

$$(1.49) \qquad A_a^m = H^0(U, \Omega_{\mathbb{P}}^m(a)) = \left\{ \frac{\Delta(\gamma)}{f^s}; \gamma \in \Omega_{sN+a}^{m+1} \text{ for some } s \geq 0 \right\}.$$

The differential d in the complex A_a^* can be given by the formulas (1.18) and (1.20) above (the complex A_0^* is hence a special case of these complexes A_a^*). The complex A_a^* has a natural polar filtration generalizing (1.28)

$$(1.50) \qquad P^s A_a^m = \left\{ \frac{\Delta(\gamma)}{f^k}; |\gamma| = kN + a \text{ for } k \leq m - s + 1 \right\}$$

for $m - s + 1 \geq 0$ and $P^s A_a^m = 0$ for $m - s + 1 < 0$. Here and in the sequel the elements $\{-N+1, \ldots, -1, 0\}$ are chosen as a set of representatives for the quotient group $\mathbb{Z}/N\mathbb{Z}$, i.e., we consider from now on $-N < a \leq 0$. It is also useful to consider the sum complex

$$(1.51) \qquad A^* = \bigoplus_{a \in \mathbb{Z}/N\mathbb{Z}} A_a^*$$

endowed with the induced (sum) differential d and the induced (sum) polar filtration P. This polar filtration P satisfies all the properties in (1.29).

(1.52) **Exercise.** Let Δ be the contraction with the Euler vector field E as above and consider the n-form on the Milnor fiber F

$$\Omega = j^*(\Delta(dx_0 \wedge \cdots \wedge dx_n)),$$

where $j: F \to \mathbb{C}^{n+1}$ is the inclusion. Show that the n-form Ω has no zeros on F. In particular, any n-form ω on F can be written as

$$\omega = h\Omega$$

for some $h \in H^0(F, \mathcal{O}_F)$. Compare with (1.17).

§2. Spectral Sequences and Koszul Complexes

In the previous section we have considered some de Rham complexes A_a^* as well as their sum $A^* = \bigoplus A_a^*$. Each of these complexes has been endowed with a polar filtration P, satisfying the properties (1.29). To any filtration with such properties on a complex we can associate a *spectral sequence*, see, for instance, [Mc], p. 44. Since the complexes A_0^* and A^* are in fact differential graded *algebras*, it is natural to try to obtain spectral sequences of algebras, see [Mc], p. 59. This becomes possible by shifting our polar filtration P a little, namely, we define new decreasing filtrations $\bar{P}$ on the complexes A_a^* and A^* by setting

$$(2.1) \qquad\qquad \bar{P}^s = P^{s+1}.$$

Then the exterior product of differential forms induces a map

$$\bar{P}^s A^* \otimes \bar{P}^t A^* \to \bar{P}^{s+t} A^*.$$

Using a remark in [Mc], p. 59, we get the following *geometric spectral sequences*.

(2.2) Proposition.

(i) *For any $a \in \mathbb{Z}/N\mathbb{Z}$ there is an E_1-spectral sequence $(E_r(f)_a, d_r)$ with*

$$E_1^{s,t}(f)_a = H^{s+t}(\bar{P}^s A_a^* / \bar{P}^{s+1} A_a^*),$$

which converges to the Milnor fiber cohomology eigenspace $H^*(F)_a$.

(ii) *There is an E_1-spectral sequence $(E_r(f), d_r)$ with*

$$E_1^{s,t}(f) = H^{s+t}(\bar{P}^s A^* / \bar{P}^{s+1} A^*)$$

which converges to the Milnor fiber cohomology $H^*(F)$.

In addition, the spectral sequence $E_r(f)_0$ (resp. $E_r(f)$) is a spectral sequence of algebras converging to $H^*(U) = H^*(F)_0$ (resp. to $H^*(F)$) as an algebra.

We pass now to the construction of some purely *algebraic spectral sequences*. For $a \in \{-N+1,\ldots,-1,0\}$ let (B_a, d', d'') be the double complex defined by

$$(2.3) \qquad\qquad B_a^{s,t} = \Omega_{tN+a}^{s+t+1}, \qquad s,t \in \mathbb{Z},$$

$d' = d$, $d''(\omega) = -|\omega| N^{-1} df \wedge \omega$ for a homogeneous differential form ω. Let (B_a^*, D_f) be the total complex associated to the double complex B_a, namely,

$$(2.4) \qquad\qquad B_a^m = \bigoplus_{s+t=m} B_a^{s,t}, \qquad D_f = d' + d''.$$

Note that the sum complex $B^* = \bigoplus B_a^*$ can be identified with the de Rham complex Ω^* (up to a shift of indices) as a graded group, but its differential D_f is quite different from the exterior differential d.

We call this complex (Ω^*, D_f) the *de Rham–Koszul complex* of the weighted homogeneous polynomial $f \in S_N$. This name comes from the fact that the Koszul complex of the sequence of elements $(\partial f/\partial x_0, \ldots, \partial f/\partial x_n)$ in the polynomial ring S can be identified with the complex

$$(2.5) \qquad K_f^* : 0 \to \Omega^0 \xrightarrow{df} \Omega^1 \xrightarrow{df} \Omega^1 \cdots \xrightarrow{df} \Omega^{n+1} \to 0,$$

see, for instance, Matsumura [Ms], pp. 132–135. In other words, our complex (Ω^*, D_f) has a differential which is a linear combination of the exterior differential d with the differential from the Koszul complex K_f^*.

Consider the decreasing filtration F on the complex B_a^* given by

$$(2.6) \qquad F^s B_a^m = \bigoplus_{k \geq s} B_a^{k, m-k} \qquad \text{for} \quad s \in \mathbb{Z},$$

and the corresponding sum filtration F on the sum complex (Ω^*, D_f). Consider also the following morphisms (up to sign) of complexes

$$(2.7) \qquad \delta : B_a^* \to A_a^*, \qquad \bar{\delta} : B^* \to A^*$$

$$\bar{\delta}(\omega) = \Delta(\omega) \cdot f^{-t} \qquad \text{for} \quad \omega \in B_a^{s,t}.$$

(To check that $\bar{\delta}$ commutes up to sign with the differentials in A_a^* and B_a^* we have to use the formulas (1.18), (1.20).)

Since these morphisms $\bar{\delta}$ are clearly compatible with the F and $\bar{P}$ filtrations, we get the following result, compare with [GH], p. 443.

(2.8) **Proposition.** *For any $a \in \mathbb{Z}/N\mathbb{Z}$ there is an E_1-spectral sequence $('E_r(f)_a, d_r)$ with*

$$'E_1^{s,t}(f)_a = H^{s+t}(F^s B_a^* / F^{s+1} B_a^*),$$

which converges to the cohomology $H^*(B_a^*, D_f)$. The morphism $\bar{\delta}$ induces a morphism of spectral sequences

$$\delta_r : ('E_r(f)_a, d_r) \to (E_r(f)_a, d_r).$$

Summing these objects for $a = -N + 1, \ldots, 0$ (as was done already in (2.2(ii))) we get a spectral sequence $('E_r(f), d_r)$ converging to the cohomology $H^*(\Omega^*, D_f)$ and a morphism of spectral sequences

$$\delta_r : ('E_r(f), d_r) \to (E_r(f)_a, d_r).$$

The relation between the geometric spectral sequences from (2.2) and these algebraic spectral sequences in (2.8) is given by the following basic result. Let $\tilde{E}_r(f)_0$ (resp. $\tilde{E}_r(f)$) denote the *reduced* spectral sequence associated to $E_r(f)_0$ (resp. $E_r(f)$) which is obtained by replacing the terms at the origin $E_r^{0,0}(f)_0 = E_r^{0,0}(f) = \mathbb{C}$ (for all r) by zero. For $a \neq 0$ we set $\tilde{E}_r(f)_a = E_r(f)_a$. In the same way, define the reduced spectral sequences $'\tilde{E}(f)_0$ (resp. $'\tilde{E}(f)$) by killing the one-dimensional vector spaces in $\tilde{E}_r^{-1,0}(f)_0$ and $\tilde{E}_r^{0,0}(f)_0$ spanned by 1 and df, respectively. Then we have natural morphisms

$$\tilde{\delta}_r : '\tilde{E}_r(f)_a \to \tilde{E}_r(f)_a, \qquad \tilde{\delta}_r : '\tilde{E}_r(f) \to \tilde{E}_r(f),$$

induced by the morphisms δ_r.

(2.9) Theorem. *The morphisms* $\tilde{\delta}_r$ *are isomorphisms for all* $r \geq 1$ *and they induce isomorphisms* $H^k(B_a^*, D_f) = H^k(F)_a$ *and* $H^{k+1}(\Omega^*, D_f) = H^k(F)$ *for any* $k \geq 0$.

Proof. Since $F^{n+1}B^* = F^{n+1}B_a^* = \bar{P}^{n+1}A_a^* = \bar{P}^{n+1}A^* = 0$, the filtrations F and $\bar{P}$ are strongly convergent, as in [Mc], p. 50. Hence it is enough to show that the morphism $\tilde{\delta}_1$ is an isomorphism. The vertical columns in $'E_1(f)_a$ correspond to certain homogeneous components in the cohomology groups of the Koszul complex K_f^*. More precisely, we have

$$(2.10) \qquad\qquad 'E_1(f)_a^{s,t} = H^{s+t+1}(K_f^*)_{tN+a}.$$

To describe the vertical columns in the second spectral sequence $\tilde{E}^1(f)$ is more subtle.

Note that fK_f^* is a subcomplex in K_f^* and let $\bar{K}^*$ denote the quotient complex K_f^*/fK_f^*. Then there is a map $\bar{\Delta}^j \colon \bar{K}^j \to \bar{K}^{j-1}$ induced by the contraction Δ and $\tilde{K}^* = \ker \bar{\Delta}^*$ is a subcomplex in $\bar{K}^*$. Let $\tilde{\Delta}$ denote the composition

$$K_f^* \to \bar{K}^* \overset{\bar{\Delta}}{\to} \tilde{K}^{*-1}.$$

Then the vertical columns in $\tilde{E}_1(f)$ correspond to certain homogeneous components in the cohomology groups $H^*(\tilde{K}^*)$, compare with [Dg1], Sect. 4.1. The morphism $\tilde{\delta}_1$ corresponds to (certain homogeneous components of) the morphism

$$\tilde{\Delta}^* \colon H^*(K_f^*) \to H^*(\tilde{K}^{*-1}).$$

A well-defined inverse for this morphism is given by the map

$$\nabla \colon H^*(\tilde{K}^{*-1}) \to H^*(K_f),$$

$$\nabla(\Delta(\omega)) = [df \wedge \Delta(\omega)/Nf].$$

To verify this, use that $df \wedge \omega = 0$ implies $0 = \Delta(df \wedge \omega) = Nf\omega - df \wedge \Delta(\omega)$. Special attention should also be paid to the case of 0-forms and 1-forms, this being the reason to consider our reduced spectral sequences. □

(2.11) Remark. The isomorphism $\delta \colon H^*(\Omega^*, D_f) \to H^*(F)$ can be described explicitly as follows:

$$\delta[\omega] = [j^*\Delta(\omega)],$$

where $j \colon F \to \mathbb{C}^{n+1}$ is the inclusion. A even more "geometric" isomorphism involving the cohomology groups $H^*(\Omega^*, D_f)$ can be obtained as follows. The Gysin sequence of the smooth hypersurface F in $\mathbb{C}^{n+1}$ gives an isomorphism

$$H^{k+1}(\mathbb{C}^{n+1}\backslash F) \overset{R}{\to} H^k(F)$$

for all $k \geq 0$, where R is the Poincaré residue. Since $\mathbb{C}^{n+1}\backslash F$ is a smooth affine variety, we can use the Grothendieck theorem (1.21) and write

$$H^*(\mathbb{C}^{n+1}\backslash F) = H^*(C_f^*),$$

where $C_f^* = H^0(\mathbb{C}^{n+1}\backslash F, \Omega^*_{\mathbb{C}^{n+1}\backslash F})$ is the algebraic de Rham complex of the complement $\mathbb{C}^{n+1}\backslash F$. The map $\alpha: (\Omega^*, D_f) \to C_f^*$, given by

(2.12)
$$\alpha(\omega) = \omega - (df \wedge \Delta(\omega))/N(f-1),$$

is a morphism of differential graded algebras.

Moreover, for $[\omega] \in H^{k+1}(\mathbb{C}^{n+1}\backslash F)$ it follows that

$$R\alpha(\omega) = -N^{-1}\delta(\omega).$$

Hence α induces an isomorphism of *algebras*

$$\alpha^*: H^*(\Omega^*, D_f) \to H^*(\mathbb{C}^{n+1}\backslash F).$$

This should be compared to the main result in [OS], where all isomorphisms depend on some choices.

(2.13) **Example.** Consider a weighted homogeneous polynomial $f \in S_N$ such that $0 \in \mathbb{C}^{n+1}$ is an *isolated* singularity for f. Then the sequence $\partial f/\partial x_0, \dots, \partial f/\partial x_n$ is a *regular sequence* in the polynomial ring S, see, for instance, [D4], p. 111. It follows that the only nontrivial cohomology group in the corresponding Koszul complex K_f^* is the top-dimensional one, namely, $H^{n+1}(K_f^*)$, see, for instance, [Ms], p. 135, where the result is stated in terms of homology groups. (Sometimes this result is called the *de Rham lemma*, see, for instance, [Dg1], (4.1), or [Gl], (1.7).) For any weighted homogeneous polynomial $f \in S$ we consider the graded Milnor algebra

(2.14)
$$M(f) = S\bigg/\left(\frac{\partial f}{\partial x_0}, \dots, \frac{\partial f}{\partial x_n}\right).$$

Let
$$w = w_0 + \cdots + w_n = \deg(dx_0 \wedge \cdots \wedge dx_n)$$

be the sum of all the weights. Then we have the following obvious identification of homogeneous components

(2.15)
$$H^{n+1}(K_f^*)_s = M(f)_{s-w},$$
$$x^a \, dx_0 \wedge \cdots \wedge dx_n \leftrightarrow x^a,$$

where $x^a = x_0^{a_0} \cdots x_n^{a_n}$.

It is well known that $M(f)_s = 0$ for $s > (n+1)N - 2w$, see, for instance, [D4], p. 113. Hence the nonzero terms in the E_1-term of the spectral sequence $'E_r(f)_0$ can be pictured as follows:

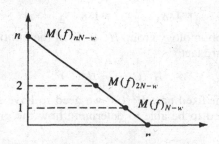

It follows that the spectral sequence $'E_r(f)_0$ degenerates at the E_1-term, i.e.,

$$'E_1(f)_0 = 'E_2(f)_0 = \cdots = 'E_\infty(f)_0.$$

For a monomial $x^a \in M(f)_{sN-w}$, the corresponding cohomology class in $H^n(U)$ is given by the following differential form:

$$(2.16) \qquad \omega(x^a) = \frac{x^a \Delta(dx_0 \wedge \cdots \wedge dx_n)}{f^s}.$$

Using Steenbrink's results in [S2] it follows that the inclusion in (1.31) is, in this case, an equality, i.e., we have

$$(2.17) \qquad P^s H^n(U) = F^s H^n(U).$$

In explicit terms, it follows that the differential forms $\omega(x^a)$, for the monomials x^a providing a basis for $M(f)_{tN-w}$ for $t = 1, \ldots, n - s + 1$, give a basis of the subspace $F^s H^n(U)$. We now consider two even more concrete examples.

(2.18) **Examples.** (i) Let $n = 2m - 1$ be an odd number and let $f = x_0^2 + \cdots + x_{2m-1}^2$ be a nondegenerate quadratic form. Then dim $H^n(U) = 1$ and a generator is given by

$$\omega(1) = \frac{\Delta(dx_0 \wedge \cdots \wedge dx_{2m-1})}{f^m}.$$

By the above remark, it follows that the only nontrivial mixed Hodge number of the cohomology group $H^n(U)$ is, in this case,

$$h^{m,m} H^{2m-1}(U) = 1.$$

(ii) Let $n = 2$ and consider the cubic polynomial $f = x_0^3 + x_1^3 + x_2^3$. Then dim $H^2(U) = 2$ and a basis is given by

$$\omega(1) = \frac{\Delta(dx_0 \wedge dx_1 \wedge dx_2)}{f}, \qquad \omega(x_0 x_1 x_2) = \frac{x_0 x_1 x_2 \omega(1)}{f}.$$

Moreover, $\omega(1)$ spans the subspace $F^2 H^2(U)$. The only nontrivial mixed Hodge numbers are $h^{2,1} = h^{1,2} = 1$. As an easy application of these explicit computations consider the following situation.

Let G be the group $\{\lambda \in \mathbb{C}^*; \lambda^3 = 1\}$ and consider the action of G on the complement U of the above cubic curve defined by

$$\lambda \cdot (x_0 : x_1 : x_2) = (x_0 : x_1 : \lambda x_2).$$

To compute the cohomology group $H^2(U/G)$ of the quotient variety U/G we may use the standard fact

$$H^2(U/G) = H^2(U)^G.$$

But to compute the fixed part $H^2(U)^G$ we need to have an explicit basis for $H^2(U)$, and we need to be able to determine how the group G acts on this basis. It is easy to check that

$$\lambda^*(\omega(1)) = \lambda\omega(1), \qquad \lambda^*(\omega(x_0 x_1 x_2)) = \lambda^2\omega(x_0 x_1 x_2).$$

It follows that

(2.19) $$H^2(U/G) = H^2(U)^G = 0.$$

(2.20) **Exercise.** Show that the eigenspace $H^n(F)_a$ of the cohomology of the Milnor fiber of an *isolated* weighted homogeneous singularity can be identified with the direct sum

$$\bigoplus_{s \geq 1} M(f)_{sN-w+a}.$$

Going beyond the realm of isolated singularities, let $m = \dim f^{-1}(0)_{\text{sing}}$ be the dimension of the singular locus of the polynomial f. Then the size (more precisely, the width) of the spectral sequence $'E_r(f)$ is given by the following Greuel–Saito generalized version of the de Rham lemma, see, for instance, [GI], (1.7).

(2.21) **Proposition.** *If* $m = \dim f^{-1}(0)_{\text{sing}}$, *then*

$$H^s(K_f^*) = 0 \qquad for \quad s \leq n - m.$$

(2.22) **Corollary.** *If* $m = \dim f^{-1}(0)_{\text{sing}}$, *then* $\tilde{H}^s(F) = 0$ *for all* $s < n - m$.

Note that this result is a weaker version of the result (3.22) due to Kato–Matsumoto.

(2.23) **Remark.** As soon as $\dim f^{-1}(0)_{\text{sing}} > 0$, the first term of the spectral sequence $'E_r(f)$ is *infinite*, namely, there are infinitely many pairs (s, t) such that

$$'E_1^{s,t}(f) \neq 0.$$

This remark is related to the fact that, in this case,

$$\dim M(f) = \infty.$$

Hence it makes sense to ask if the spectral sequence $'E_r(f)$ degenerates after a finite number of steps, i.e., if there exists an integer m such that

$$'E_m(f) = 'E_{m+1}(f) = \cdots = 'E_\infty(f).$$

This question is answered affirmatively in our paper [D9]. It is however an *open* problem to find good estimates for the integer m in terms of the geometry of the hypersurface $V: f = 0$.

We end this section by explaining a *reduction to the projective setting*, i.e., a way to compute the dimensions of the eigenspaces $H^*(F)_a$ for $a \neq 0$ in terms of some eigenspaces $H^*(\tilde{U})_a$ associated with the complement $\tilde{U} = \tilde{\mathbb{P}} \setminus \tilde{V}$, where

$\tilde{V}$ is the hypersurface given by the equation

(2.24) $$\tilde{f}(x, t) = f(x) + t^N = 0$$

in the weighted projective space $\tilde{\mathbb{P}} = \mathbb{P}(w_0, \ldots, w_n, 1)$.

Let $\tilde{F}$ be the Milnor fiber associated to this new polynomial and let F_N be the Milnor fiber of the singularity t^N. Using the Thom–Sebastiani formula (3.3.21) it follows that

(2.25) $$H^k(\tilde{U}) = H^k(\tilde{F})_0 = \bigoplus_{a \neq 0} \tilde{H}^{k-1}(F)_a \otimes \mathbb{C}\langle \delta_{-a} \rangle,$$

where δ_{-a} is the eigenvector of the monodromy operator h_N^* acting on $\tilde{H}^0(F_N)$ corresponding to the eigenvalue $\exp(-2\pi i a N^{-1})$. Consider the cyclic group of order N

$$G = \{\lambda \in \mathbb{C}^*; \lambda^N = 1\}$$

and the following action of G on the complement $\tilde{U}$

(2.26) $$\lambda \cdot (x_0 : \cdots : x_n : t) = (x_0 : \cdots : x_n : \lambda^{-1} t).$$

Since multiplication by λ^{-1} is related to the *inverse* of the monodromy operator h_N^*, it follows from (2.25) that we have the following identification of eigenspaces:

(2.27) $$\tilde{H}^{k-1}(F)_a \simeq H^k(\tilde{U})_a$$

for all $k \geq 1, a \neq 0$.

As shown in a special case in Example (2.18(ii)) above, the computation of the eigenspaces $H^k(\tilde{U})_a$ is very easy once we have an explicit basis for the cohomology groups $H^k(\tilde{U})$ in terms of algebraic differential forms. That is why, for the most part in the following sections, we shall be concerned with the computations of the cohomology $H^*(U)$ of the hypersurface complement $U = \mathbb{P} \setminus V$.

§3. Singularities with a One-Dimensional Critical Locus

The topology associated with a weighted homogeneous polynomial f with $\dim f^{-1}(0)_{\text{sing}} = 1$ is related to some classical results by Zariski [Z1], as well as to some recent investigations in Singularity Theory, see, for instance, [Si2], [Si3], [Si4], [My], [Sr]. In this section we construct a setting for efficiently doing computations related to this class of singularities using algebraic differential forms. We follow the presentation given in [D6] rather closely with some improvements.

Since the critical locus $f^{-1}(0)_{\text{sing}}$ is one dimensional, we can write

(3.1) $$f^{-1}(0)_{\text{sing}} = \{0\} \cup \left(\bigcup_{i=1}^{p} \mathbb{C}^* a_i \right)$$

for some points $a_i \in \mathbb{C}^{n+1}$, one in each irreducible component of this critical locus. Let G_i be the *isotropy group* of the point a_i with respect to the $\mathbb{C}^*$-action on $\mathbb{C}^{n+1}$ associated to the weights **w**. This group G_i is precisely the multiplicative group of the d_i-roots of unity, where

(3.2) $$d_i = \text{g.c.d.}\{w_j; a_i^j \neq 0\},$$

where $a_i = (a_i^0, \ldots, a_i^n)$. We assume that the weights **w** satisfy the condition

(3.3) $$\text{g.c.d.}(w_0, \ldots, w_n) = 1,$$

and hence a generic point in $\mathbb{C}^{n+1}$ has a trivial isotropy group.

Let H_i be a small G_i-invariant transversal to the orbit $\mathbb{C}^* \cdot a_i$ at the point a_i. The isolated hypersurface singularity

(3.4) $$(Y_i, a_i) = (H_i \cap f^{-1}(0), a_i)$$

is called the *transversal singularity* of f along the branch $\overline{\mathbb{C}^* \cdot a_i}$ of the critical locus. Note that this singularity (Y_i, a_i) is G_i-invariant.

(3.5) **Example.** Consider the polynomial

$$f = x_0^{265} + x_0 x_1^{11} + x_0 x_2^8 + x_2 x_3^4$$

of degree $N = 265$ with respect to the weights $\mathbf{w} = (1, 24, 33, 58)$.

This homogeneity type $(\mathbf{w}; N)$ is "famous," since the associated Poincaré series is a polynomial in spite of the fact that there is no isolated singularity with this homogeneity type, see [AGV1], p. 201. The critical locus is the irreducible curve

$$f^{-1}(0)_{\text{sing}}: x_0 = x_3 = x_1^{11} + x_2^8 = 0.$$

Hence we can take $a_1 = (0, -1, 1, 0)$. The corresponding isotropy group G_1 is cyclic of order 3, since

$$d_1 = \text{g.c.d.}(24, 33) = 3.$$

We can take as a G_1-invariant transversal the hyperplane

$$H_1: x_2 = 1.$$

To move the point a_1 to the origin, we use the following coordinates on H_1:

$$x = x_0, \qquad y = x_1 + 1, \qquad z = x_3.$$

In these new coordinates, the transversal singularity (Y_1, a_1) has the equation

$$g = x^{265} + x(y-1)^{11} + x + z^4 = 0.$$

Using the weights

$$\text{wt}(x) = \text{wt}(y) = \tfrac{1}{2}, \qquad \text{wt}(z) = \tfrac{1}{4},$$

and basic facts about semiweighted homogeneous singularities, see, for instance, [D4], p. 117, we infer that the singularity (Y_1, a_1) has type A_3. Note also that the isotropy group G_1 acts on the transversal H_1 via the formula

$$\lambda \cdot (x, y, z) = (\lambda x, y, \lambda z).$$

As a result, we have

$$g(\lambda \cdot (x, y, z)) = \lambda g(x, y, z),$$

i.e., the equation g is *not* G_1-invariant, only its zero set $g^{-1}(0) = Y_1$ is so.

The weighted projective hypersurface V defined by a polynomial f in $\mathbb{P} = \mathbb{P}(\mathbf{w})$ is a V-manifold (i.e., has only quotient singularities, see Appendix B) at all points, except the points a_i, $i = 1, \ldots, p$. Here a_i denotes both a point in $\mathbb{C}^{n+1} \backslash \{0\}$ and its class in $\mathbb{P} = \mathbb{C}^{n+1} \backslash \{0\}/\mathbb{C}^*$. At such a point a_i the hypersurface V has a *hyperquotient singularity* Y_i/G_i in the sense of M. Reid [Rd].

We are mainly interested in finding a way to compute the Betti numbers

$$b_j(V) = \dim H^j(V)$$

of this hypersurface. The reader can extend without any difficulty the result (5.2.11) to the weighted homogeneous case (see also Appendix B). As a consequence, only two Betti numbers, namely, $b_{n-1}(V)$ and $b_n(V)$, are to be computed. Note also that the Euler characteristic $\chi(V)$ can be computed by a formula (which is in general a conjecture, but which holds in most interesting cases) involving the weights $\mathbf{w}$, the degree N, and some local invariants of the G_i-singularities (Y_i, a_i), see [D5], Prop. (3.19), and the related facts in [D8]. It follows that it is enough to compute the Betti number $b_n(V)$.

On the other hand, it is known since the striking example of Zariski, involving sextic curves in $\mathbb{P}^2$ with six cusps as singularities situated (or not) on a conic [Z1], that this Betti number $b_n(V)$ depends not only on the data listed above for $\chi(V)$ but also on the *position* of the singularities of V in $\mathbb{P}$.

Let Σ be the set $\{a_1, \ldots, a_p\}$ in $\mathbb{P}$ and set $V^* = V \backslash \Sigma$, $\mathbb{P}^* = \mathbb{P} \backslash \Sigma$. In spite of the fact that $\mathbb{P}^*$ and V^* are only V-manifolds, we have the following result similar to (2.2.14).

(3.6) Lemma. *There is a natural exact "Gysin" sequence*

$$\to H^k(\mathbb{P}^*) \xrightarrow{i^*} H^k(\mathbb{P}^* \backslash V^*) \xrightarrow{R} H^{k-1}(V^*) \xrightarrow{\delta} H^{k+1}(\mathbb{P}^*) \to.$$

Proof. Such a "Gysin" sequence would be an obvious consequence of the existence of a "Thom" isomorphism

$$(3.7) \qquad \theta^*: H^{k+1}(\mathbb{P}^*, \mathbb{P}^* \backslash V^*) \to H^{k-1}(V^*)$$

(recall the proof of (2.2.13), (2.2.14)). Assume for the moment that we are in the usual projective case, i.e., $\mathbb{P} = \mathbb{P}^n$. Then we can *construct* the homology Thom isomorphism

$$\theta: H_{k-1}(V \backslash \Sigma) \to H_{k+1}(\mathbb{P}^n \backslash \Sigma, \mathbb{P}^n \backslash V) \ .$$

from the following commutative diagram:

$$
\begin{array}{ccc}
H_{k-1}(V\backslash\Sigma) & \xrightarrow{\;\;\theta\;\;} & H_{k+1}(\mathbb{P}^n\backslash\Sigma, \mathbb{P}^n\backslash V) \\
& & \\
& H^{2n-k-1}(V, \Sigma) & \\
\end{array}
$$

(3.8) $\cap 0$ $\cap 0$

see [Do2], pp. 314 and 321.

In the general case, $\mathbb{P}$ is a $\mathbb{C}$-homology manifold and (V, Σ) is a relative $\mathbb{C}$-cohomology manifold. As a result, the ascending arrows in (3.8) are still isomorphisms (i.e., the corresponding Alexander and Lefschetz duality theorems still hold with $\mathbb{C}$-coefficients) and we can *define* the morphism θ using this diagram (3.8). It follows that the "Thom" morphism (3.7) (simply defined as the dual of θ) is itself an isomorphism. It is not difficult to see that the exact sequence (3.6) is in fact an exact sequence of MHS, and the corresponding "residue" morphism

$$R = \theta^* \circ \partial$$

is a MHS morphism of type $(-1, -1)$.   □

(3.9) **Exercise.** Show that:

(i) The inclusion $\mathbb{P}^* \to \mathbb{P}$ induces isomorphisms

$$H^k(\mathbb{P}) \cong H^k(\mathbb{P}^*) \qquad \text{for} \quad 0 \leq k \leq 2n - 2.$$

(ii) The inclusion $\mathbb{P}\backslash V \to \mathbb{P}^*$ induces trivial morphisms

$$H^k(\mathbb{P}^*) \to H^k(\mathbb{P}\backslash V) \qquad \text{for} \quad 0 \leq k \leq 2n - 2.$$

(iii) The inclusion $V^* \to \mathbb{P}^*$ induces a monomorphism

$$H^2(\mathbb{P}^*) \to H^2(V^*) \qquad \text{for} \quad n \geq 3.$$

(iv) The composition

$$H^{2k}(\mathbb{P}^*) \to H^{2k}(V^*) \xrightarrow{(\theta^*)^{-1}} H^{2k+2}(\mathbb{P}^*, \mathbb{P}\backslash V) \to H^{2k+2}(\mathbb{P}^*)$$

is nontrivial for $0 \leq k < n - 1$, where the first and last morphisms are induced by the corresponding inclusions.

Hint (for (iv) in the homogeneous case). This composition is just the cup product with $N\alpha$ where α is the "canonical" generator of $H^2(\mathbb{P}^*)$. To see this, use the formula in [Do2], p. 321, (with $(\theta^*)^{-1} = e$) and note that the element in $H^2(V^*)$ corresponding to $N\alpha$ is just the first Chern class (or Euler class) of the normal bundle of V^* in $\mathbb{P}^*$. Compare with (5.1.10).

Consider now the diagram of inclusions:

$$
\begin{array}{ccc}
V^* & \xrightarrow{\;\;i\;\;} & V \\
& & \\
& \mathbb{P} & \\
\end{array}
$$

We can define, in analogy to (5.2.18), the *primitive* cohomology groups of V and V^* by the formulas

(3.10)
$$H_0^k(V) = \text{coker}(H^k(\mathbb{P}) \to H^k(V)),$$
$$H_0^k(V^*) = \text{coker}(H^k(\mathbb{P}) \to H^k(V^*)),$$

(see also (B25)). The inclusion i induces a homomorphism

$$H_0^k(V) \xrightarrow{i_0^*} H_0^k(V^*).$$

Moreover, using (3.9) (ii) and (iv), it follows that the composition

$$\text{im } R \hookrightarrow H^{k-1}(V^*) \to H_0^{k-1}(V^*)$$

is an isomorphism for $0 < k < 2n - 2$. In other words, we have proved the following.

(3.11) Corollary.

(i) *The "residue" morphism R induces a $(-1, -1)$-type isomorphism*

$$\bar{R}: H^k(U) \to H_0^{k-1}(V^*)$$

of MHS for $0 < k < 2n - 2$.

(ii) *When $\dim V = 1$, there is a MHS exact sequence*

$$0 \to H^2(U) \xrightarrow{R} H^1(V^*) \xrightarrow{\delta} H^3(\mathbb{P}^*) \to 0.$$

(iii) *When $\dim V = 2$, there is an isomorphism $\delta: H^3(V^*) \to H^5(\mathbb{P}^*)$.*

Consider now, as in Steenbrink [S3], the following long exact sequence of MHS:

(3.12)
$$\to H_\Sigma^k(V) \to H^k(V) \xrightarrow{i^*} H^k(V^*) \xrightarrow{\delta} H_\Sigma^{k+1}(V) \to .$$

For k in the middle dimensions, this yields the following long exact sequence

(3.12')
$$\to H_\Sigma^k(V) \to H_0^k(V) \xrightarrow{i_0^*} H_0^k(V^*) \xrightarrow{\delta} H_\Sigma^{k+1}(V) \to .$$

There are three distinct cases to discuss.

Case 1. $\dim V > 2$.

Using (3.11(i)) and the fact that $\mathbb{P}\backslash V$ is an affine variety, we deduce that the right-hand end of the sequence (3.12') looks like

(3.12'')
$$H^n(U) \xrightarrow{\theta} H_\Sigma^n(V) \to H_0^n(V) \to 0,$$

where $\theta = \delta \circ \bar{R}$ is a MHS morphism of type $(-1, -1)$.

This apparently complicated morphism θ can be described quite explicitly as follows. Let B_i be a small G_i-invariant open ball in the transversal H_i, centered at the point a_i. The quotient $\bar{B}_i = B_i/G_i$ can be regarded as a contractible neighborhood of the point a_i in $\mathbb{P}$. By excision, it follows that

$$H_\Sigma^n(V) = H^n(V, V\backslash\Sigma) = \bigoplus H^n(\bar{B}_i \cap V, \bar{B}_i \cap V\backslash\{a_i\})$$

Since $\bar{B}_i \cap V$ is contractible, we get

$$H^n(\bar{B}_i \cap V, \bar{B}_i \cap V\{a_i\}) \overset{\delta}{\underset{\sim}{\leftarrow}} H^{n-1}(\bar{B}_i \cap V\setminus\{a_i\}).$$

Consider now the Gysin sequence associated to the smooth hypersurface $Y_i \setminus \{a_i\}$ in the punctured ball $B_i \setminus \{a_i\} = B_i^*$. The residue morphism

$$H^n(B_i \setminus Y_i) \overset{R_i}{\underset{\sim}{\rightarrow}} H^{n-1}(B_i^* \cap Y_i)$$

is an isomorphism, compatible with the G_i-actions which exist on both sides. It follows that

$$H^{n-1}(\bar{B}_i \cap V\setminus\{a_i\}) = H^{n-1}(B_i^* \cap Y_i)^{G_i} \overset{R_i}{\underset{\sim}{\leftarrow}} H^n(B_i \setminus Y_i)^{G_i}.$$

Putting all these isomorphisms together we get from (3.12″) the exact sequence

$$(3.13) \qquad H^n(U) \overset{\rho}{\rightarrow} \bigoplus H^n(B_i \setminus Y_i)^{G_i} \overset{\delta}{\rightarrow} H_0^n(V) \rightarrow 0,$$

where the morphism ρ is induced by the restriction of differential forms on U to the open sets $\bar{B}_i \setminus V$ for $i = 1, \ldots, p$. It is important to note that δ is a MHS morphism of type $(-1, -1)$.

Case 2. $\dim V = 1$.

In this case, we have to use (2.2.20) and (3.11(ii)) to get the following huge commutative diagram:

$$(3.14) \qquad
\begin{array}{ccccccc}
& 0 & & 0 & & & \\
& \uparrow & & \uparrow & & & \\
0 \rightarrow & H^3(\mathbb{P}^*) & \overset{\rho''}{\rightarrow} & \bigoplus H^3(\bar{B}_i^*) & & \rightarrow H^4(\mathbb{P}) \rightarrow 0 \\
& \uparrow & & \uparrow & & & \\
& H^1(V^*) & \overset{\rho'}{\rightarrow} & \bigoplus H^1(V \cap \bar{B}_i^*) & \rightarrow H^2(V) \rightarrow 0 \\
& \uparrow & & \uparrow & & & \\
& H^2(U) & \overset{\rho}{\rightarrow} & \bigoplus H^2(B_i \setminus Y_i)^{G_i} & & & \\
& \uparrow & & \uparrow & & & \\
& 0 & & 0 & & &
\end{array}$$

Here the vertical morphisms come from Gysin sequences, while the rows come from the cohomology sequences of various pairs. It follows that

$$\dim \operatorname{coker} \rho' = \dim \operatorname{coker} \rho + \dim \operatorname{coker} \rho'',$$

in other words,

$$\dim \operatorname{coker} \rho = \dim H_0^2(V).$$

A little more care shows that in fact we have an isomorphism of MHS

$$\operatorname{coker} \rho \cong H_0^2(V)(-1),$$

i.e., the exact sequence (3.13) is also valid in this second case.

Case 3. dim $V = 2$.

This case can be treated similarly and the exact sequence (3.13) is seen to hold again.

Note that since MHS morphisms are *strictly* compatible with the Hodge filtrations (see Appendix C), we can derive a whole bunch of exact sequences from (3.13), namely,

$$(3.15) \qquad F^s H^n(U) \xrightarrow{\rho} \bigoplus F^s H^n(B_i \backslash Y_i)^{G_i} \xrightarrow{\delta} F^{s-1} H_0^n(V) \to 0.$$

Using Theorem (1.31) it follows that any element in $F^s H^n(U)$ can be represented by a form

$$\omega(h) = \frac{h\Omega}{f^{n-s+1}},$$

where $\Omega = \Delta(dx_0 \wedge \cdots \wedge dx_n)$ and $h \in S_{(n-s+1)N-w}$. Suppose now that for all the transversal singularities (Y_i, a_i) we have the following equality:

$$F^s H^n(B_i \backslash Y_i) = P^s H^n(B_i \backslash Y_i)$$

(this is true, for instance, when all these singularities (Y_i, a_i) are weighted homogeneous, see [S2] and (3.17) below). Then it follows that the image of the map

$$F^s H^n(U) \xrightarrow{\rho} \bigoplus F^s H^n(B_i \backslash Y_i)^{G_i}$$

is the same as the image of the map e given by the composition

$$(3.16) \qquad S_{(n-s+1)N-\omega} \to P^s H^n(U) \to \bigoplus F^{s-1} H^n(B_i \backslash Y_i)^{G_i},$$

$$e(h) = \rho(\omega(h)).$$

We turn now to the problem of computing the local cohomology groups $H^n(B_i \backslash Y_i)^{G_i}$ in terms of differential forms. We discuss first this problem without a group action. Let B be a small open ball centered at the origin of $\mathbb{C}^n$, Y a hypersurface in B with defining equation $g = 0$. In this situation, we have a local de Rham complex A_∞^* consisting of germs at the origin of meromorphic differential forms with polar singularities along Y, recall (1.35), (1.36). There is a polar filtration P on this complex and we can consider the corresponding spectral squence $(E_r(g), d_r)$, similar to the global case treated in Section 2 of this chapter. When $(Y, 0)$ is an *isolated* hypersurface singularity, the first terms $E_1(g), E_2(g), \ldots$ of this spectral sequence can be quite explicitly computed, see for details [D5] and [D8].

This local spectral sequence $(E_r(g), d_r)$ degenerates at the second term (i.e., $E_2(g) = E_\infty(g)$) if and only if the isolated singularity $(Y, 0)$ is weighted homogeneous. In such a case an explicit basis for the group $H^n(B \backslash Y)$ can be described as follows. Assume that g is weighted homogeneous of type $(v_1, \ldots, v_n; M)$ and let $M(g)$ be the corresponding graded Milnor algebra. Let $\{y^\alpha, \alpha \in A_k\}$ be a monomial basis for the homogeneous component $M(g)_{kM-v}$ where $v = v_1 + \cdots + v_n$. Then the differential forms

$$(3.17) \qquad \omega(y^\alpha) = y^\alpha g^{-k} \omega_n \qquad \text{with} \quad \omega_n = dy_1 \wedge \cdots \wedge dy_n$$

for $\alpha \in A_k$, $k = 1, \ldots, n - s + 1$, give a basis for the subspace $F^s H^n(B \backslash Y)$, see [S2]. Note also that by [S2], the group $H^n(B \backslash Y)$ has in this case a pure Hodge structure of weight $n + 2$. Indeed, the formula for μ_0, in terms of the mixed Hodge numbers for F given in (C26), is related to the fact that the cohomology group $H^{n-1}(Y \backslash \{0\})$ has in this case a pure Hodge structure of weight n. See also (C28). As the residue isomorphism

$$H^n(B \backslash Y) \to H^{n-1}(Y \backslash \{0\})$$

is a $(-1, -1)$-type morphism, it follows that the cohomology group $H^n(B \backslash Y)$ has a pure Hodge structure of weight $n + 2$ when $(Y, 0)$ is a weighted homogeneous singularity.

(3.18) **Examples.** (i) *A_1-singularities with even embedding dimension $(n = 2m)$.*
A local equation in this case is given by

$$g(y) = y_1^2 + y_2^2 + \cdots + y_{2m}^2,$$

and has the obvious weights $v_1 = \cdots = v_{2m} = 1$ and degree $M = 2$. A generator for the cohomology group $H^{2m}(B \backslash Y)$ is given by the differential form

$$\omega(1) = g^{-m} \omega_{2m}.$$

The corresponding cohomology class has type $(m + 1, m + 1)$. It is important to note that the class of a differential form

$$\gamma = h g^{-m} \omega_{2m} \quad \text{with} \quad h \in \mathcal{O}_{C^n, 0}$$

in the cohomology group $H^{2m}(B \backslash Y)$ is given by

$$[\gamma] = h(0) [\omega(1)],$$

i.e., it is given by the evaluation of the function germ h at the origin of C^n.

(ii) *Simple-elliptic surface singularities.*
Let $n = 3$ and $Y: g = 0$ be one of the simply-elliptic singularities $\tilde{E}_6$, $\tilde{E}_7$, $\tilde{E}_8$ as in (2.4.9).
In all of these three cases, the hessian of g

(3.19) $$\text{hess}(g) = \det\left(\frac{\partial^2 g}{\partial y_i \, \partial y_j}\right)$$

has degree $3M - 2v = M$. It follows that $H^3(B \backslash Y)$ is two dimensional with a basis given by

$$\omega(1) = g^{-1} \omega_3 \quad \text{and} \quad \omega(\text{hess}(g)) = \text{hess}(g) g^{-2} \omega_3.$$

The element $[\omega(1)]$ has type $(3, 2)$, while the element $[\omega(\text{hess}(g))]$ has type $(2, 3)$. Again it is easy to see that the class of a differential form

$$\gamma = h g^{-1} \omega_3 \quad \text{with} \quad h \in \mathcal{O}_{C^3, 0}$$

in the cohomology group $H^3(B \backslash Y)$ is given by

$$[\gamma] = h(0) [\omega(1)].$$

When we have a G-singularity $(Y, 0)$, then the cohomology group

$$H^n(B \backslash Y)^G,$$

i.e., the G-invariant subgroup in $H^n(B \backslash Y)$, can easily be computed when we have an explicit basis for this group $H^n(B \backslash Y)$ on which the G-action is clearly seen.

(3.20) Example. Let $(Y, 0)$ be the A_1-singularity from (3.18(i)). Consider the group $G = \{\pm 1\}$ and the G-action on $\mathbb{C}^{2m}$ given by

$$-1 \cdot (y_1, \dots, y_{2m}) = (-y_1, y_2, \dots, y_{2m}).$$

Then it is clear that the induced action on the cohomology group $H^{2m}(B \backslash Y)$ satisfies

$$-1 \cdot \omega(1) = -\omega(1)$$

and hence

$$H^{2m}(B \backslash Y)^G = 0.$$

If we go back to Example (3.5), we see that it may happen that the coordinates in which the G-action on $\mathbb{C}^n$ is linear are not necessarily coordinates in which the singularity g is weighted homogeneous (here we assume that the given singularity $Y: g = 0$ is known to be weighted homogeneous in some coordinate system.

There are two ways to remedy this unpleasant situation. The first one is to try to find a better coordinate system, i.e., a system in which the G-action is linear *and* the equation g is weighted homogeneous. The existence of such a coordinate system follows from a general result due to G. Müller [Mü].

(3.21) Example. In (3.5) we can make the coordinate change

$$\bar{x} = x, \qquad \bar{y} = x^{264} + (y - 1)^{11} + 1, \qquad \bar{z} = z.$$

In this new coordinate system $(\bar{x}, \bar{y}, \bar{z})$ the equation for Y_1 becomes

$$g_1 = \bar{x}\bar{y} + \bar{z}^4 = 0$$

and the G_1-action remains linear

$$\lambda(\bar{x}, \bar{y}, \bar{z}) = (\lambda \bar{x}, \bar{y}, \lambda \bar{z}) \qquad \text{for all} \quad \lambda \in G_1.$$

The second method is useful when in the given coordinate system the G-action is linear and the equation g is semiweighted homogeneous, exactly as in (3.5).

More precisely, let us assume that $g = g_0 + g'$ where g_0 is a weighted homogeneous IHS of type $(v_1, \dots, v_n; M)$ and all the monomials in the germ g' have degrees $> M$ with respect to the weights v. Consider the family of singularities

(3.22) $$g_t(y) = t^{-M} g(t \cdot y) = g_0(y) + g'_t(y)$$

for $t \in [0, 1]$. This family is μ-constant, see, for instance, [D4], p. 116.

Let $K_t = \partial \bar{B} \cap Y_t$ be the link associated to the singularity Y_t: $g_t = 0$. Then, using the conic structure of singularities (1.5.1), we have

$$H^n(B \backslash Y_t) = H^n(S_\varepsilon^{2n-1} \backslash K_t),$$

where $S_\varepsilon^{2n-1} = \partial \bar{B}$.

By the cohomology Wang sequence (3.1.18′) associated to the Milnor fibration of the singularity $(Y_{t,0})$ we have

$$H^{n-1}(F_t) \xrightarrow{T_t - I} H^{n-1}(F_t) \to H^n(S_\varepsilon^{2n-1} \backslash K_t) \to 0.$$

Since in a μ-constant family the monodromy operators T_t form a constant family of linear maps, it follows that we get a trivial vector bundle

$$\pi: V \to [0, 1], \qquad \pi^{-1}(t) = H^n(B \backslash Y_t).$$

Consider now the differential forms

(3.23) $$\omega_t(y^\alpha) = y^\alpha g_t^{-k} \omega_n$$

for all $k = 1, \ldots, n$, where the monomials y^α form a basis for $M(g_0)_{kM-v}$ as in (3.17). These forms can be regarded as continuous sections of the bundle π. Since for $t = 0$ they form a basis in the fiber $\pi^{-1}(0) = (B \backslash Y_0)$, it follows that they also form a basis in $\pi^{-1}(t)$ for t small enough. But the singularity Y_t is equivalent to $Y_1 = Y$ by a homothety as in formula (3.22). It follows that the differential forms $\omega_t(y^\alpha)$ form a basis in $H^n(B \backslash Y_t)$ for any $t \in [0, 1]$.

(3.23′) **Exercise.** Consider the following argument. Let Y be the hypersurface in $B \times [0, 1]$ corresponding to the union of the hypersurfaces Y_t above. Consider the inclusions

$$B \backslash Y_0 \xrightarrow{i_0} B \times [0, 1] \backslash Y \xleftarrow{i_1} B \backslash Y_1.$$

Both inclusions are homotopy equivalences and hence

$$H^n(B \backslash Y_0) \xleftarrow[\sim]{i_0^*} H^n(B \times [0, 1] \backslash Y) \xrightarrow[\sim]{i_1^*} H^n(B \backslash Y_1).$$

Consider the cohomology classes

$$[\omega_t(y^\alpha)]$$

in the middle cohomology group. Since $i_0^*[\omega_t(y^\alpha)]$ form a basis in $H^n(B \backslash Y_0)$, it follows that $i_1^*[\omega_t(y^d)]$ also form a basis in $H^n(B \backslash Y_1)$. What is wrong with this argument?

From the exact sequence (3.13) we see that not all the singularities $a_i \in \Sigma$ play a role in the computation of $H_0^n(V)$. Namely, it is natural to define the subset $\Sigma' \subset \Sigma$ by the following property

$$a_i \in \Sigma' \iff H^n(B_i \backslash Y_i)^{G_i} \neq 0.$$

We call the points in Σ' the *essential singularities* of the weighted hypersurface V. When $\Sigma' = \varnothing$, then $H_0^n(V) = 0$ and we are done.

(3.24) **Example.** Consider the surface $V : f = 0$ in $\mathbb{P}(1, 24, 33, 58)$ where f is the polynomial from (3.5). It follows that $\Sigma' = \varnothing$ and hence $H^3(V) = 0$. Using the computation of the Euler characteristic $\chi(V)$ given in [D5], we get

$$b_2(V) = \chi(V) - 2 = 252.$$

We end this section with some remarks on how to compute the Alexander polynomial Δ_V of a hypersurface V in $\mathbb{P}^n$ having only isolated singularities via the exact sequence (3.13). If $F : f - 1 = 0$ is the associated Milnor fiber, then $\Delta_V = \Delta_V^1$ is just the characteristic polynomial of the monodromy operator

$$h^* : H^{n-1}(F) \to H^{n-1}(F),$$

see (4.1.14).

Since $h^N = 1$, it follows that all the roots of the Alexander polynomial Δ_V are N-roots of unity. The multiplicity of 1 as a root in Δ_V is clearly

$$(3.25) \qquad m(1) = \dim H^{n-1}(F)^{h^*} = \dim H^{n-1}(U) = \dim H_0^n(V).$$

It is clear that this multiplicity $m(1)$ can be calculated from the exact sequence (3.13). For a root $\lambda = \exp(2\pi i a N^{-1})$ different from 1 we have, in the notation of (2.27),

$$(3.26) \qquad m(\lambda) = \dim H^{n-1}(F)_a = \dim H^n(\tilde{U})_a = \dim H_0^{n+1}(\tilde{V})_a.$$

The multiplicative group $G = \{\lambda; \lambda^N = 1\}$ acts on the exact sequence (3.13) corresponding to the hypersurface $\tilde{V}$. Passing to eigenspaces, we get the following exact sequence:

$$(3.27) \qquad H^{n+1}(\tilde{U})_a \overset{\rho}{\to} \bigoplus H^{n+1}(\tilde{B}_i \setminus \tilde{Y}_i)_a \to H_0^{n+1}(\tilde{V})_a \to 0$$

where the notation is obvious. Several examples of concrete computations using this exact sequence (or a slightly modified version of it) are given in the next section.

To limit the possible roots of the Alexander polynomial Δ_V, we can proceed as follows. For each transversal singularity (Y_i, a_i) we can define a *local Alexander polynomial* Δ_{Y_i} as the characteristic polynomial of the corresponding local monodromy operator

$$\Delta_{Y_i}(t) = \det(tI - h_{Y_i}^*).$$

We can be more careful and define a *reduced* local Alexander polynomial $\tilde{\Delta}_{Y_i}$ by the formula

$$(3.28) \qquad \tilde{\Delta}_{Y_i}(t) = \prod (t - \lambda)^{a(\lambda)} \qquad \text{with} \quad a(\lambda) = \dim \ker(\lambda I - h_{Y_i}^*).$$

It is easy to see that the polynomial $\tilde{\Delta}_{Y_i}$ divides Δ_{Y_i} and that we have equality $\tilde{\Delta}_{Y_i} = \Delta_{Y_i}$ if and only if the monodromy operator $h_{Y_i}^*$ is diagonalizable. From the exact sequence (3.27) we then get the following result (see Thom–Sebastiani (3.3.21) and compare with [Li4]).

(3.29) **Corollary.** *The Alexander polynomial* Δ_V *divides the product* $\tilde{\Delta}_{Y_1} \dots \tilde{\Delta}_{Y_p}$ *of all the reduced local Alexander polynomials associated to the singularities of the hypersurface V.*

§4. Alexander Polynomials and Defects of Linear Systems

In this section we consider some more explicit cases of the setting discussed in the previous section, namely, that V is a hypersurface in $\mathbb{P}^n$ having only isolated singularities. Our approach to compute the associated Alexander polynomial Δ_V using algebraic differential forms should be compared with the other possible approach based on constructing an explicit resolution of singularities $\hat{V} \to V$, see, for instance, [Es], [Cl], [Sch], [Wn], [LV]. Such a resolution $\hat{V}$ is of course difficult to construct as soon as dim $V \geq 3$ and the singularities of V are worse than nodes.

Sometimes we can determine the Betti number $b_n(V)$ directly from the exact sequence (3.13) with no additional computations, just taking into account the MHS properties of this sequence.

(4.1) **Example.** Assume that V is a surface having only singularities of type $T_{p,q,r}$. Then it follows from (C29), and our discussion following (3.17), that the cohomology groups $H^3(B_i \setminus Y_i)$ have weights equal to 6 in this case. Since the cohomology group $H_0^3(V)$ has pure weight equal to 3 according to [S3], it follows that the $(-1, -1)$-type morphism δ in (3.13) is trivial. Hence

$$H_0^3(V) = 0,$$

in spite of the fact that all the $T_{p,q,r}$ surface singularities are essential.

Now we pass to the discussion of the really interesting examples, where the Betti number $b_n(V)$ and the Alexander polynomial Δ_V may depend on the *position* of the singularities of the hypersurface V. In the statements below we use the following notation. Let S_k denote the homogeneous component of degree k in the polynomial ring

$$S = \mathbb{C}[x_0, \dots, x_n].$$

When no other weights are mentioned, we use the usual weights $w_0 = \dots = w_n = 1$. Let Σ be a finite set of points in $\mathbb{P}^n$ and consider the linear system

(4.2) $S_k(\Sigma) = \{h \in S_k; \, h|\Sigma = 0\}.$

The difference

(4.3) $\text{def } S_k(\Sigma) = \#\Sigma - \text{codim } S_k(\Sigma)$

is called the *defect* of the linear system $S_k(\Sigma)$. Here $\#\Sigma$ denotes the number of points in Σ and the codimension is taken with respect to the vector space S_k. The number def $S_k(\Sigma)$ measures the *degree of independence* of the points in Σ with respect to the polynomials in S_k.

(4.4) **Exercise.** (i) Consider the evaluation map

$$\theta: S_k \to \mathbb{C}^p, \qquad h \mapsto (h(a_1), \ldots, h(a_p)),$$

where $\Sigma = \{a_1, \ldots, a_p\}$. Show that

$$\mathrm{def}\, S_k(\Sigma) = \dim(\mathrm{coker}\,\theta).$$

(ii) Take $n = 2$ and discuss the possible values of $\mathrm{def}\, S_1(\Sigma)$ in terms of $\#\Sigma = 1, 2, 3$.

(iii) Take $n = 2$ and discuss the possible values of $\mathrm{def}\, S_2(\Sigma)$ in terms of $\#\Sigma = 1, 2, \ldots, 6$.

Our first result deals with *nodal hypersurfaces* (i.e., hypersurfaces having only A_1-singularities) and is related to results in [O3], [Cl], [Sch], [Wn].

(4.5) **Theorem.** *Let V be a nodal hypersurface in $\mathbb{P}^n$ of degree N. Then:*

(i) $\Delta_V = 1$ *if nN is odd;*
(ii) $\Delta_V(\lambda) = [\lambda + (-1)^{n+1}]^{\mathrm{def}\,S_k(\Sigma)}$ *if nN is even, where Σ is the singular locus of V and $k = nN/2 - n - 1$.*

Proof. In case (i) both n and N are odd integers. As a result, the set of essential singularities $\Sigma'(V)$ for the hypersurface V is empty and hence $H_0^n(V) = 0$ by (3.13). If we look at the associated hypersurface

$$\tilde{V}: f(x) + t^N = 0 \quad \text{in } \mathbb{P}^{n+1}$$

we see again that $\Sigma'(\tilde{V}) = \varnothing$ and hence $H_0^{n+1}(\tilde{V}) = 0$. Using (3.25) and (3.26) we get our claim (i).

The proof for case (ii) has to be subdivided into two subcases.

Subcase 1 (*n odd, N even*)

In this case $\Sigma'(V) = \varnothing$ and $H_0^n(V) = 0$ as above. Identifying a point $a \in \mathbb{P}^n$ with the point $(a : 0) \in \mathbb{P}^{n+1}$, we see that

$$\Sigma'(\tilde{V}) = \Sigma.$$

The local Alexander polynomials for the singularities in $\Sigma'(V)$ are all equal to

$$\Delta_Y(\lambda) = \lambda + 1.$$

Hence by (3.29) the only possible root of the Alexander polynomial Δ_V is -1. The multiplicity $m(-1)$ of this root can be computed as explained in (B27). Namely, let

$$V^2: f(x) + t^2 = 0$$

be the *double covering* of $\mathbb{P}^n$ ramified along the hypersurface V. Note that V^2 can be regarded as a hypersurface in the weighted projective space

$$\mathbb{P} = \mathbb{P}(1, 1, \ldots, 1, N/2).$$

As in (B27) we have

$$m(-1) = \dim H_0^{n+1}(V^2).$$

To compute this number we consider the exact sequence (3.15) associated with the hypersurface V^2. In this case all the isotropy groups G_i are trivial. Next, each of the local cohomology groups $H^{n+1}(\tilde{B}_i \setminus \tilde{Y}_i)$ are one-dimensional of Hodge type $(m+1, m+1)$ where $2m = n+1$ as in (3.18(i)). It follows that the cohomology group $H_0^{2m}(V^2)$ has Hodge type (m, m) if nonzero. Hence $H_0^{2m}(V^2)$ is the cokernel of the "evaluation" map e in (3.16) with $s = m+1$ and $w = n + 1 + N/2$. This map e is indeed given by evaluation by (3.18(i)) and hence we are done by (4.4(i)).

Subcase 2 (n even, N arbitrary)
In this situation $\Sigma'(V) = \Sigma$ and $\Sigma'(\tilde{V}) = \varnothing$. Then the only possible root of the Alexander polynomial is 1 and its multiplicity $m(1)$ is given by (3.25).
The proof ends by using (3.16) as above. ☐

(4.6) **Example.** The maximal number of nodes a cubic hypersurface in $\mathbb{P}^4$ can have is ten and such a hypersurface is projectively equivalent to

$$V: x_1^3 + x_2^3 + x_0 x_1 x_2 + x_3^3 + x_4^3 + x_0 x_3 x_4 = 0,$$

see for details [Wn]. It is easy to see (by explicit computation) that the ten nodes of V are not situated in any hyperplane of $\mathbb{P}^4$. In other words,

$$\text{def } S_1(\Sigma) = 10 - 5 = 5.$$

Hence using (4.5) we get the following formula for the corresponding Alexander polynomial

$$\Delta_V(\lambda) = (\lambda - 1)^5.$$

Now let V be a nodal hypersurface of odd dimension, i.e., we are in Subcase 2 above. Then the corresponding defect

$$\delta = \text{def } S_k(\Sigma)$$

is just the primitive Betti number $b_n^0(V)$.

The following result gives a *lower bound* for the defect δ in terms of $p = \#\Sigma$ and the Betti number $b_{n-1}(V(N))$ of a *smooth* hypersurface of degree N in $\mathbb{P}^n$.

(4.7) **Exercise.**

$$\delta \geq p - b_{n-1}(V(N))/2.$$

Hint. Use just the exact sequence (3.13).

(4.8) **Examples.** (i) For the cubic with ten nodes from (4.6) we have

$$p = b_3(V(3)) = 10.$$

The lower bound given by (4.7) is strict, since $\delta = 5$.

(ii) Hirzebruch [Hz3] has constructed a quintic three-fold in $\mathbb{P}^4$ with $p = 126$ nodes and $\delta = 25$. In this case $b_3(V(5)) = 204$, hence the lower bound given by (4.7) is 24.

(iii) According to van Geemen–Werner [GW] there exist quintics in $\mathbb{P}^4$ with $p = 100$ nodes and any defect δ between 4 and 16.

We consider now plane sextic curves with six cusps A_2 as singularities. Zariski has already shown in [Z1] that these sextics fall into two distinct families. We say that a sextic curve V with six cusps is of *type* (i) when its cusps are situated on a conic in $\mathbb{P}^2$. When the cusps are not situated on a conic we say that V is a sextic of *type* (ii).

(4.9) **Theorem.** *The Alexander polynomial of a sextic curve V in $\mathbb{P}^2$ with six cusps is given by the following formula:*

$$\Delta_V(\lambda) = \begin{cases} \lambda^2 - \lambda + 1 & \text{if } V \text{ is of type (i),} \\ 1 & \text{if } V \text{ is of type (ii).} \end{cases}$$

Proof. We clearly have $\Sigma'(V) = \varnothing$ and hence $H_0^2(V) = 0$ (this is obvious anyway since V is irreducible). All the singularities in $\Sigma'(\tilde{V})$ are simple-elliptic of type $\tilde{E}_8$, namely,

$$\tilde{Y}_i: x^2 + y^3 + t^6 = 0.$$

The corresponding local cohomology groups $H^3(\tilde{B}_i \setminus \tilde{Y}_i)$ can be described as in (3.18(ii)). Hence the exact sequence (3.16) applied to $\tilde{V}$ with $s = 3$ looks like

$$e: S_2 \to \bigoplus F^3 H^3(\tilde{B}_i \setminus \tilde{Y}_i) \cong \mathbb{C}^6.$$

Moreover, the morphism e is, by (3.18(ii)), just the evaluation of a polynomial h in $\mathbb{C}[x, y, z, t]_2$ at the set Σ in $\mathbb{P}^3$. If we write

$$h = h_2 + h_1 t + h_0 t^2 \quad \text{with} \quad h_j \in \mathbb{C}[x, y, z]_j$$

it follows that the condition $h(a_i : 0) = 0$ is equivalent to the condition $h_2(a_i) = 0$. When the sextic curve V is of type (i), the defect of the linear system $S_2(\Sigma)$ is 1.

Since $H_0^3(\tilde{V})$ has a pure Hodge structure of weight 3 by [S3], it follows that the only nonzero mixed Hodge numbers for $H_0^3(\tilde{V})$ are $h^{2,1} = h^{1,2} = 1$. Using the explicit description of the generator for $F^3 H^3(\tilde{B}_i \setminus \tilde{Y}_i)$ coming from (3.18(ii)), it follows that

$$F^3 H^3(\tilde{B}_i \setminus \tilde{Y}_i) = F^3 H^3(\tilde{B}_i \setminus Y_i)_1,$$

where the notation is as in (3.27). Moreover, since the group action on $H_0^3(V)$ comes from a real action on $H_0^3(V, \mathbb{R})$, it follows that the only eigenvalues that occur are

$$\lambda_1 = \exp(2\pi i/6), \qquad \lambda_5 = \exp(10\pi i/6).$$

In other words,

$$\Delta_V(\lambda) = (\lambda - \lambda_1)(\lambda - \lambda_5) = \lambda^2 - \lambda + 1.$$

When the sextic curve V is of type (ii) it follows by definition that the defect of the linear system $S_2(\Sigma)$ is 0, and hence

$$\Delta_V = 1. \qquad\qquad \square$$

(4.10) **Corollary.** *Two sextic curves with six cusps and having distinct types cannot be equisingularly deformed one into the other.*

Note that an equation for a sextic curve of type (i) is given by

(4.11) $$V: (x^2 + y^2)^3 + (y^3 + z^3)^2 = 0.$$

This equation has been considered in (4.4.16) where we have computed the corresponding fundamental group $\pi_1(\mathbb{P}^2 \backslash V)$.

(4.12) **Remark.** An equation for a sextic curve of type (ii) can be found in [O6]. Among other facts, Oka proves there that the group $\pi_1(\mathbb{P}^2 \backslash V)$ is abelian in this case.

A generalization of equation (4.11) to higher dimensions is given by the following type of equations, considered first by Libgober [Li4] (in a more restrictive setting than ours). Let $f_i \in \mathbb{C}[x_0, \ldots, x_n]$ be a homogeneous polynomial of degree $d_i \geq 1$ for $i = 1, \ldots, n$ and let N be a common multiple of the integers d_i. Assume that these data satisfy the following conditions:

(i) $\Sigma_0 = \{x \in \mathbb{P}^n; f_1(x) = \cdots = f_n(x) = 0\}$ consists of exactly $d_1 \cdots d_n$ points;
(ii) $e_i = Nd_i^{-1} > 1$ for all $i = 1, \ldots, n$;
(iii) the hypersurface in $\mathbb{P}^n$ defined by

(4.13) $$V = V(f_1, \ldots, f_n; N): f = f_1^{e_1} + \cdots + f_n^{e_n} = 0$$

has only isolated singularities.

Note that $\Sigma = V_{\text{sing}} \supset \Sigma_0$.

(4.14) **Theorem.** *With the above assumptions, the hypersurface $V = V(f_1, \ldots, f_n; N)$ has a nontrivial Alexander polynomial.*

Proof. There is a unique pair of nonnegative integers (k, t) such that $k = tN - d_1 - \cdots - d_n$ and $k < N$. There are two cases to consider.

Case 1 ($k = 0$)
At any point $a \in \Sigma_0$, the singularity (V, a) is of Brieskorn–Pham type, and is given in local coordinates by the equation

$$(V, a): g = u_1^{e_1} + \cdots + u_n^{e_n} = 0.$$

In each cohomology group $H^n(B_a \backslash V)$ for $a \in \Sigma_0$ and B_a, a small open ball in $\mathbb{P}^n$ centered at a, we have a nonzero element given by the meromorphic form

$$\omega_a = \frac{du_1 \wedge \cdots \wedge du_n}{g^t}$$

(use the weights $\mathrm{wt}(u_i) = d_i$ and note that $\deg(g) = N$ with respect to these weights).

It is enough to show that the map ρ in (3.13) is not surjective. Indeed, then 1 is a root of the polynomial Δ_V and hence $\Delta_V \neq 1$ as claimed. But since (3.13) is a MHS exact sequence, it follows from (C22) that it is enough to show that the "evaluation" map e in (3.16) is not surjective for $s = n - t + 1$. Hence we have to deal with homogeneous polynomials h of degree

$$sN - w = d_1 + \cdots + d_n - n - 1.$$

It follows from the generalized version of the Cayley–Bacharach theorem, see [GH], p. 671, that the composition

$$S_{sN-w} \to \bigoplus_{a \in \Sigma} F^{n-t+1} H^n(B_a \backslash V) \xrightarrow{\mathrm{pr}} \bigoplus_{a \in \Sigma_0} F^{n-t+1} H^n(B_a \backslash V)$$

is not surjective, where pr is just the projection corresponding to the inclusion $\Sigma_0 \subset \Sigma$. This clearly ends the proof in this case.

Case 2 ($k \neq 0$)

Consider now the hypersurface $\tilde{V}$ in $\mathbb{P}^{n+1}$. For any point $a \in \Sigma_0$, the germ $(\tilde{V}, a)$ is given by an equation

$$(\tilde{V}, a): \tilde{g}(u, v) = u_1^{e_1} + \cdots + u_n^{e_n} + v^N = 0.$$

Consider the element in $H^{n+1}(\tilde{B}_a \backslash \tilde{V})$ given by

$$\tilde{\omega}_a = \frac{v^{k-1} du_1 \wedge \cdots \wedge du_n \wedge dv}{\tilde{g}^t}.$$

Using a similar argument, we can apply the generalized Cayley–Bacharach theorem and deduce that the Alexander polynomial Δ_V is divisible by the product

$$\left(\lambda - \exp\left(\frac{2\pi i k}{N}\right)\right)\left(\lambda - \exp\left(-\frac{2\pi i k}{N}\right)\right).$$

Thus the second case is also proved. $\square$

The example which we discuss next, obtained recently by Artal–Bartolo [AB], is perhaps the simplest one showing the dependence of the topology (reflected in the Alexander polynomial) on the position of the singularities.

(4.15) **Theorem.** *Let $V \subset \mathbb{P}^2$ be a curve consisting of nine lines, such that at any intersection point the number of lines meeting there is at most five. Let Σ' be the set of those intersection points where exactly three lines come together. Then the Alexander polynomial of the curve V is given by the formula*

$$\Delta_V(\lambda) = (\lambda - 1)^8 (\lambda^2 + \lambda + 1)^{\mathrm{def} \, S_3(\Sigma')}.$$

Proof. The multiplicity of 1 as a root in Δ_V is given by

$$m(1) = \dim H_0^2(V) = 8$$

since V has nine irreducible components. The essential singularities of the associated surface $\tilde{V}$ correspond exactly to the triple points in Σ'. They are given in suitable local coordinates by the equation

$$\tilde{Y}_i: \tilde{g} = x^3 + y^3 + t^9 = 0.$$

The local cohomology group $H^3(\tilde{B}_i \setminus \tilde{Y}_i)$ is generated by the forms

$$\omega(t^2) = t^2 g^{-1}\, dx \wedge dy \wedge dt, \qquad \omega(xyt^5) = xyt^5 g^{-2}\, dx \wedge dy \wedge dt.$$

The form $\omega(t^2)$ generates the subspace $F^3 H^3(\tilde{B}_i \setminus \tilde{Y}_i)$.

The corresponding "evaluation" map from (3.16) is, in this case,

$$e: \mathbb{C}[x, y, t]_5 \to \bigoplus F^3 H^3(\tilde{B}_i \setminus \tilde{Y}_i).$$

Writing

$$h = \Sigma h_j(x, y) t^{5-j}, \qquad h_j \in \mathbb{C}[x, y]_j,$$

as in the proof of (4.9), it follows that only the coefficient $h_3(x, y)$ of t^2 plays a role. In other words,

$$h^{2,1} H_0^3(\tilde{V}) = \operatorname{def} S_3(\Sigma').$$

Note that the action of the generator $\lambda_1 = \exp(2\pi i/9)$ of the corresponding group G on the vector space $\bigoplus F^3 H^3(\tilde{B}_i \setminus \tilde{Y}_i)$ is just multiplication by λ_1^3. Since the minimal polynomial of λ_1^3 over $\mathbb{Q}$ is exactly $\lambda^2 + \lambda + 1$, we are done.   □

(4.16) Example. The following two examples of line arrangements as in (4.15) are given in [AB] in order to exhibit a change in the first Betti number

$$b_1(F) = \deg \Delta_V$$

of the corresponding Milnor fiber. The first arrangement is given by

$$V_1: xyz(x - y)(y + bz)(x - y - z)(ax + y + z)(ax + y + bz)$$

$$\times (abx + (a - ab + 1)y + bz) = 0,$$

where $ab(a + 1)(b - 1)(a + b)(a - ab + 1)(ab + b - 1) \neq 0$. In this case, there are nine triple intersection points in Σ' and there is a pencil of cubics passing through all these nine points. In other words, in this case,

$$\operatorname{def} S_3(\Sigma') = 1.$$

The second arrangement is given by

$$V_2: xyz(x + y)(x + az)(y + bz)(cx + (c + 1)y + bz)$$

$$\times (c(a + b)x + a(c + 1)y + ab(c + 1)z)(cx + (c + 1)y + acz) = 0,$$

where $abc(c + 1)(ac - b)(a + b) \neq 0$ and $c(a - bc) - b(c + 1) = 0$. In this second arrangement there are again nine triple points in Σ', but there is only one cubic curve passing through all these nine points. In other words, in this case,

$$\operatorname{def} S_3(\Sigma') = 0.$$

Before proving the last result in this section, some comments are in order. The lower bound for the defect δ given in (4.7) may be interpreted as saying that when a hypersurface V (of odd dimension) has lots of singularities (in our case nodes), then the Betti number $b_n^0(V)$ is nonzero. On the other hand, the hypersurface W_{n-1}^d from (5.4.21(i)) has just one "bad" singularity and still it has a nonzero Betti number $b_n^0(W_{n-1}^d)$. The result (4.3.11) tells us that an irreducible curve C having just a "good" singularity has an abelian fundamental group $\pi_1(\mathbb{P}^2 \backslash C)$ for its complement.

Since the Alexander polynomial Δ_C should be regarded as a weaker analog of the fundamental group $\pi_1(\mathbb{P}^2 \backslash C)$ as explained in (4.1.20), the following result is a natural extension of (4.3.11) to higher dimensions.

(4.17) **Theorem.** *Let V be a hypersurface of degree N in $\mathbb{P}^n$ having just one singular point p. Assume that*

$$\mu - \det(V, p) < N.$$

Then

$$\Delta_V = 1.$$

Proof. The claim in (4.17) is equivalent to

$$H_0^n(V) = H_0^{n+1}(\tilde{V}) = 0.$$

Choose a linear system of coordinates on $\mathbb{P}^n$ such that $p = (1 : 0 : \cdots : 0)$ and the hyperplane $H: x_0 = 0$ has a transversal intersection with the hypersurface V. In the affine chart $x_0 = 1$ the equation for V is given by

(4.18) $$g = g_m + \cdots + g_{N-1} + g_N = 0,$$

where $g_k \in \mathbb{C}[x_1, \ldots, x_n]_k$ is the homogeneous component of degree k in g. Consider now the family of hypersurfaces

$$g_h = g_m + \cdots + g_{N-2} + (g_{N-1} + h) + g_N = 0$$

parametrized by $h \in \mathbb{C}[x_1, \ldots, x_n]_{N-1}$. The assumption on the singularity (V, p) implies that for polynomials h as above, with small enough coefficients, we have

$$\mu(g_h, 0) = \mu(g, 0) = \mu(V, p).$$

On the other hand, for h generic, the projective hypersurface corresponding to the affine equation g_h will have just one singularity, namely, the point p. In this way, we can find a family of hypersurfaces V_t for $t \in [0, \varepsilon]$ such that:

(i) $V_0 = V$;

(ii) all the hypersurfaces V_t have just one singular point, namely p, and

$$\mu(V_t, p) = \text{const.};$$

(iii) the affine equation (4.18) corresponding to the hypersurface V_ε has a non-degenerate term g_{N-1}^ε (i.e., $g_{N-1}^\varepsilon = 0$ defines an IHS at the origin of $\mathbb{C}^n$)

and the hypersurface

$$W: x_0 g_{N-1}^\varepsilon + g_N = 0$$

in $\mathbb{P}^n$ is smooth outside p.

The corresponding family of hypersurfaces $\tilde{V}_t$ in $\mathbb{P}^{n+1}$ is again a μ-constant family. Using (3.1.8) and (5.4.3) we get

$$H_0^n(V) = H_0^n(V_t), \qquad H_0^{n+1}(\tilde{V}) = H_0^{n+1}(\tilde{V}_t),$$

for any $t \in [0, \varepsilon]$. In other words, we may assume from the beginning that the term g_{N-1} satisfies the conditions in (iii) above. Consider now a new family of hypersurfaces V_t in $\mathbb{P}^n$, namely, the family associated with the following affine equation:

$$g_t = t(g_m + \cdots + g_{N-2}) + g_{N-1} + g_N.$$

As in the proof of (4.3.11) we get a degeneration of hypersurfaces, i.e., a family of projective hypersurfaces V_s of $s \in [0, 1]$ with:

(α) $V_0 = W: x_0 g_{N-1} + g_N = 0$;
(β) the family (V_s, p) is μ-constant for $s \in (0, 1]$ and $V_1 = V$.

Using (5.4.3) we get an epimorphism

$$H_0^n(V_0) \to H_0^n(V_s) \qquad \text{for} \quad s > 0.$$

Indeed, with the notation used in the proof of (5.4.3), the lattice homomorphism $\varphi_{V_s}: L_1^s \to \bar{L}$ for $s > 0$ small enough factorizes as

$$L_1^s \hookrightarrow L_1^0 \xrightarrow{\varphi_{V_0}} \bar{L},$$

where L_1^s is the Milnor lattice corresponding to the singularity (V_s, p). Hence

$$H_n^0(V) = H_n^0(V_s) = \ker \varphi_{V_s} \subset \ker \varphi_{V_0} = H_n^0(V_0).$$

It remains to apply the duality between homology and cohomology with $\mathbb{C}$-coefficients. A similar argument can be applied to the corresponding family $\tilde{V}_s$ of hypersurfaces in $\mathbb{P}^{n+1}$. Hence we get again an epimorphism

$$H_0^{n+1}(\tilde{V}_0) \to H_0^{n+1}(\tilde{V}_s).$$

Thus it remains to prove that

$$H_0^n(W) = H_0^{n+1}(\tilde{W}) = 0.$$

The result is clear for $\tilde{W}$, since the only singular point of this hypersurface is not essential (use $(N - 1, N) = 1$). We show now that $H_0^n(W) = 0$ using the corresponding exact sequence (3.13).   □

The only singularity (W, p) is given by the affine equation

$$(W, p): g = g_{N-1}(x_1, \ldots, x_n) + g_N(x_1, \ldots, x_n) = 0,$$

i.e., it is a semiweighted homogeneous singularity by our assumption. Using (3.23), it follows that a basis for the cohomology space $H^n(B \backslash W)$, with B a small open ball centered at p, is given by the differential forms

$$\omega_1(x^\alpha) = x^\alpha g^{-k} \, dx_1 \wedge \cdots \wedge dx_n$$

for all $k = 1, \ldots, n$ and $x^\alpha = x_1^{\alpha_1} \cdots x_n^{\alpha_n}$, a basis for $M(g_{N-1})_{k(N-1)-n}$. Consider the form

(4.19)
$$\overline{\omega}(x^\alpha) = \frac{x_0^{k-1} x^\alpha \Omega}{(x_0 g_{N-1} + g_N)^k},$$

where $\Omega = \Delta(dx_0 \wedge \cdots \wedge dx_n)$. It is clear that

$$\rho(\overline{\omega}(x^\alpha)) = \omega_1(x^\alpha).$$

Hence the morphism ρ in (3.13) is surjective and $H_0^n(W) = 0$.

(4.20) **Remark.** Consider a curve

$$W: z g_{N-1}(x, y) + g_N(x, y) = 0$$

having $p = (0:0:1)$ as its unique singular point and assume that (W, p) is an ordinary $(N-1)$-multiple point (i.e., g_{N-1} has no repeated factors).

In the exact sequence (3.13) for such a curve W we have

$$\dim H^2(B \backslash W) = N - 2 = \dim H^2(U)$$

(use the formula (5.4.4(ii))). Using the proof above it follows that the morphism ρ is in this special case an isomorphism. In other words, the differential forms in (4.19) give a basis for the cohomology space $H^2(U)$. As a concrete example, the differential form

$$\frac{\Delta(dx \wedge dy \wedge dz)}{xyz + x^3 + y^3}$$

generates the space $H^2(\mathbb{P}^2 \backslash C)$ where C is the nodal cubic $xyz + x^3 + y^3 = 0$. It seems a difficult problem to find such explicit basis for $H^n(\mathbb{P}^n \backslash V)$ in the case of a general hypersurface V. Compare with (2.18).

(4.21) **Remark.** Let $(Y, 0)$ be an IHS at the origin of $\mathbb{C}^n$. Since any such singularity is $\mathscr{K}$-finitely determined, see, for instance, [D4], p. 81, it follows that the number $\mu\text{-det}(Y, 0)$ is always finite and well defined. Now choose an integer $N > \mu\text{-det}(Y, 0)$ such that singularity $(Y, 0)$ has a *globalization* to a hypersurface V in $\mathbb{P}^n$ of degree N. This means that V is smooth outside a point $p \in V$ and that the singularity (V, p) is analytically isomorphic to $(Y, 0)$. The existence of such globalizations is standard, see, for instance, Brieskorn [B3]. Identifying (V, p) with $(Y, 0)$, the proof of (4.17) gives us an epimorphism

$$H^n(\mathbb{P}^n \backslash V) \xrightarrow{\rho} H^n(B \backslash Y) \to 0,$$

where B is a small open ball in $\mathbb{C}^n$ centered at the origin. Using Theorem (1.31) it follows that the Hodge filtration F and the polar filtration P on this local cohomology group satisfy

$$P^s H^n(B \backslash Y) \supset F^s H^n(B \backslash Y)$$

for all s. Compare with (1.39).

(4.22) **Remark.** Let V be a hypersurface in $\mathbb{P}^n$ having only isolated singularities at the points $a_1, \ldots, a_p$. Set $V^* = V \backslash V_{\text{sing}}$. An interesting topological invariant of V is the (middle perversity) *intersection cohomology* with $\mathbb{C}$-coefficients. We have the following description of these intersection cohomology groups in terms of usual cohomology groups:

$$IH^k(V) = \begin{cases} H^k(V) & \text{for } k > n - 1, \\ H^k(V^*) & \text{for } k < n - 1, \\ \text{im}(H^{n-1}(V) \to H^{n-1}(V^*)) & \text{for } k = n - 1, \end{cases}$$

see [S3] and [Kw], p. 48.

We can compute the intersection cohomology Betti number $\dim IH^{n-1}(V)$ from the exact sequences (3.12), (3.13). Namely, we get

$$\dim IH^{n-1}(V) = b_{n-1}(\mathbb{P}^n) + \dim \ker \rho,$$

where ρ is the morphism from (3.13). Let $\mu_0(V, a_i)$ be the $(n-1)$th Betti number of the link of the singularity (V, a_i).
   Note that

$$\mu_0(V, a_i) = \dim H^n(B_i \backslash V).$$

Using this relation and the exact sequence (3.13) again, we get the following:

(4.23) **Corollary.**

(i) *The middle intersection cohomology Betti number of the hypersurface V is given by the formula*

$$\dim IH^{n-1}(V) = b_{n-1}(V) + b_n^0(V) - \sum \mu_0(V, a_i).$$

In particular, this topological invariant of the hypersurface V depends on the position of the singularities in the ambient projective space $\mathbb{P}^n$ in general.

(ii) *The intersection cohomology Euler characteristic*

$$\chi(IH^*(V)) = \sum (-1)^k \dim IH^k(V)$$

satisfies the formula

$$\chi(IH^*(V)) = \chi(V) + (-1)^n \sum \mu_0(V, a_i).$$

In particular, this topological invariant $\chi(IH^*(V))$ does not depend on the position of the singularities of the hypersurface V.

(4.24) **Remark.** Let V be a hypersurface as above with singular points $a_1, \ldots,$ a_p. Let $\pi: \hat{V} \to V$ be a resolution of singularities for V. It is natural to ask what information we can derive on the topological invariants (e.g., Betti numbers) of the smooth variety $\hat{V}$ using the results in this book.

When dim $V \geq 3$, the resolution $\hat{V}$ is in general quite difficult to handle (i.e., we have no uniqueness or minimality results as in the surface case). See, however, [Rd]. For any resolution $\hat{V}$ we have

$$h^{p,q}(H^n(\hat{V})) \geq h^{p,q}(H^n(V)).$$

Indeed, $H^n(V)$ has a pure Hodge structure of weight n by [S3] and hence $H^n(\pi)$ is *injective*. *Hint.* In general ker $H^n(\pi) = W_{n-1} H^n(V)$ as in [Df5].

In the case of surfaces we can do much better. Then $H^3(\pi)$ is known to be an *isomorphism*, see [BaK], p. 122. Hence $b_3(\hat{V}) = b_1(\hat{V}) = b_3(V)$. The remaining Betti number $b_2(\hat{V})$ can be computed from the Euler characteristic $\chi(\hat{V})$. This invariant in turn can be computed from the Euler characteristic $\chi(V)$ and the data associated with the exceptional divisors $E_i = \pi^{-1}(a_i)$. Assume that E_i has r_i irreducible components E_{ij} and set

$$g_i = \sum_j g(E_{ij}),$$

$b_i =$ the numbers of loops in the dual graph associated to E_i.

Using (2.3.1) and assuming that π is a very good resolution for each of the singularities (V, a_i) we get

$$\chi(\hat{V}) = \chi(V) - \sum_{i=1, p} (2g_i + b_i - r_i).$$

Similar formulas can be obtained in the case of nodal threefolds when $\pi: \hat{V} \to V$ is a so-called "small" resolution. For details we refer to [Wn].

APPENDIX A

Integral Bilinear Forms and Dynkin Diagrams

(A1) **Definition.** A *lattice* $(M, (,))$ is a pair consisting of a finitely generated free abelian group M together with a bilinear form

$$(,): M \times M \to \mathbb{Z}$$

which is either:

(i) symmetric, i.e., $(x, y) = (y, x)$ for all $x, y \in M$;
(ii) skew-symmetric, i.e. $(x, y) = -(y, x)$ for all $x, y \in M$.

For simplicity, we refer to M as a lattice when the bilinear form $(,)$ is clear. The main example which we have in mind is the Milnor lattice $L_X = (\tilde{H}_n(F), \langle , \rangle)$ of an n-dimensional IHS $X: f = 0$ as defined in (3.3.6).

(A2) **Definition.** A symmetric lattice M is *even* if $(x, x) \equiv 0 \pmod 2$ for any $x \in M$.

A symmetric lattic M is *odd* if it is not even.

Note that the Milnor lattice L_X of an even dimensional IHS X is an even lattice by (3.3.7).

(A3) **Definition.** The lattice M is *nondegenerate* if it satisfies the following two equivalent conditions:

(i) $$\mathrm{Rad}(M) := \{x \in M; (x, y) = 0 \text{ for all } y \in M\} = 0;$$

(ii) the natural group homomorphism

$$i_M: M \to M' := \mathrm{Hom}(M, \mathbb{Z}), x \mapsto (x, \cdot)$$

is injective.

The lattice M is *unimodular* if i_M is an isomorphism.

When the lattice M is nondegenerate, we call the finite group

$$D(M) = \mathrm{coker}(i_M)$$

the *discriminant group* of M and denote its order $|D(M)|$ by $\det(M)$. The quotient lattice $\overline{M} = M/\operatorname{Rad} M$ is called the *reduced* lattice associated to M.

(A4) **Exercise.** (i) Show that

$$D(M) = D(-M), D(M_1 \oplus M_2) = D(M_1) \oplus D(M_2).$$

Here $-M$ is the lattice $(M, -(\ ,\ ))$ obtained by changing to signs of all the products (x, y) and $M_1 \oplus M_2$ is the direct sum lattice, i.e.,

$$(x_1 + x_2, y_1 + y_2) = (x_1, y_1) + (x_2, y_2)$$

for all $x_1, y_1 \in M_1$ and $x_2, y_2 \in M_2$.

(ii) Show that $\det(L_X) = |\Delta(1)|$, for a nondegenerate Milnor lattice L_X, where Δ denotes the characteristic polynomial of the monodromy operator of X. *Hint.* Recall the proof of (3.4.7).

(A5) **Lemma.** *If $N \subset M$ is a sublattice in the nondegenerate lattice M such that* $\operatorname{rk} M = \operatorname{rk} M$, *then the quotient group M/N is finite and its order satisfies the relation*

$$|M/N|^2 \det M = \det N.$$

Proof. Use the structure of the subgroup N as described in [La], p. 393. $\qquad\square$

(A6) **Definition.** Let M and N be two lattices. A group homomorphism $\varphi: M \to N$ is called a *lattice morphism* if $(x, y) = (\varphi(x), \varphi(y))$ for all $x, y \in M$.

A lattice morphism φ is called an *embedding* (resp. an *isomorphism*) if φ is a group monomorphism (resp. isomorphism).

(A7) **Theorem** (Structure of Skew-Symmetric Lattices). *Any skew-symmetric lattice M is isomorphic to a direct sum of "elementary" skew-symmetric lattices*

$$(\mathbb{Z}^2, (\ )_{d_1}) \oplus \cdots \oplus (\mathbb{Z}^2, (\ )_{d_k}) \oplus (\mathbb{Z}^m, (\ )_0)$$

where $\mathbb{Z}^2 = \mathbb{Z}e_1 + \mathbb{Z}e_2$ and $(e_1, e_2)_{d_i} = d_i$ for some integers $d_i > 0$ and $(x, y)_0 = 0$ for all $x, y \in \mathbb{Z}^m$. Moreover the positive integers d_i are uniquely determined if we ask in addition that $d_1 | d_2 \cdots | d_k$.

For a proof, see [La], p. 380. In particular, M is nondegenerate if and only if $m = 0$ and then $\det M = d_1^2 \cdots d_k^2$. And M is unimodular if and only if $m = 0$ and $d_1 = \cdots = d_k = 1$.

We use the *notation* I_{2k} for this unimodular skew-symmetric lattice of rank $2k$.

(A8) **Exercise.** Let $X: f = 0$ be a reduced plane curve singularity with Milnor number μ and number of irreducible branches r. Then the Milnor lattice L_X is isomorphic to the direct sum

$$I_{\mu-r+1} \oplus (\mathbb{Z}^{r-1}, (\ )_0).$$

Hint. Use the Wang exact sequence (3.1.18) associated to the Milnor fibration for X.

When M is a symmetric lattice we can obtain a real bilinear form tensoring by $\mathbb{R}$. All the usual terminology for the latter applies to M and hence we can speak about the *signature* sign $M = (m_-, m_0, m_+)$. For instance, M is *negative definite* when $m_0 = m_+ = 0$, and M is *indefinite* when $m_- > 0$ and $m_+ > 0$. The difference $m_+ - m_-$ is called the *index* of the lattice.

The classification of the indefinite unimodular lattices is due to Milnor [M2] and we do not recall it here since it is more complicated than (A7). Moreover, the Milnor lattices L_X for the most familiar singularities X are neither indefinite nor unimodular, so the best way to introduce them to the reader is just by listing them. As remarked in (3.3.23), to each IHS X there are two naturally associated Milnor lattices, a symmetric one L_X^s and a skew-symmetric one L_X^{ss}. In what follows, we list the Dynkin diagrams for several important classes of singularities, namely for all simple and unimodular singularities in Arnold's lists [AGV1]. These Dynkin diagrams determine at once the symmetric lattices L_X^s and they determine also the skew-symmetric lattices L_X^{ss} via Gabrielov's result (3.3.22'). Recall that each vertex in a Dynkin diagram D corresponds to a vanishing cycle Δ_i with $\langle \Delta_i, \Delta_i \rangle = -2$. Two distinct vertices, corresponding to vanishing cycles Δ_i and Δ_j, respectively, are joined by k edges (resp. k dotted edges) if their intersection number (Δ_i, Δ_j) is k (resp. $-k$). Hence

$$\overset{i}{\bullet}\!\!-\!\!-\!\!-\!\!-\!\!\overset{j}{\bullet} \quad \text{means } (\Delta_i, \Delta_j) = 1 \text{ and}$$

$$\overset{i}{\bullet}\!\!-\!\!-\!\!-\!\!-\!\!\overset{j}{\bullet} \quad \text{means } (\Delta_i, \Delta_j) = -2.$$

Note also that the vertices in a Dynkin diagram are numbered (corresponding to the order of the vanishing cycles $\Delta_1 \ldots, \Delta_\mu$ in the associated distinguished basis Δ), *unless* any order is good (i.e., any order of Δ_i's corresponds to a distinguished basis).

All the singularities discussed in what follows are stably equivalent to surface singularities in $\mathbb{C}^3$. The reader has already encountered the notations for (and the equations of) these singularities in Chapter 2, §4, so here we describe only their Dynkin diagrams and some of their properties.

(A9) Dynkin Diagrams for the Simple Singularities.

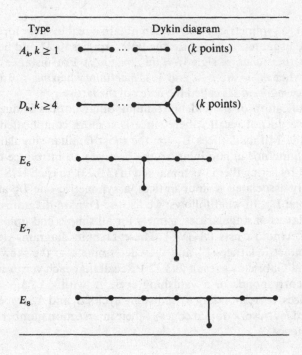

Type	Dykin diagram
$A_k, k \geq 1$	(k points)
$D_k, k \geq 4$	(k points)
E_6	
E_7	
E_8	

We let A_k denote the (symmetric) Milnor lattice of the singularity A_k and use the same convention for all the other singularities discussed.

A first striking fact about the Dynkin diagrams of the simple singularities is that they coincide with their resolution graphs, see (2.4.3). This remark gives us the first part of the following result (for a complete proof we refer to [Df4]).

(A10) Proposition.

(i) *The Milnor lattices A_k, D_k, E_6, E_7 and E_8 are negative definite.*

(ii) *Any IHS X whose symmetric Milnor lattice is negative definite is a simple singularity.*

(iii) *$\det A_k = k + 1, \det D_k = 4, \det E_l = 9 - l$ for $l = 6, 7, 8$.*

(A11) Dynkin Diagrams for the Simple-Elliptic Singularities $\tilde{E}_6$, $\tilde{E}_7$, and $\tilde{E}_8$.
First we associate to each of these simple-elliptic singularities $\tilde{E}$ a triple of positive integers (p, q, r) as follows

$$\tilde{E}_6 \mapsto (3, 3, 3), \tilde{E}_7 \mapsto (2, 4, 4), \tilde{E}_8 \mapsto (2, 3, 6).$$

A look at the equations for these singularities given in (2.4.9) will explain to the reader where these triples come from! See also (A13). Next to each triple

(p, q, r) we associate the following diagram:

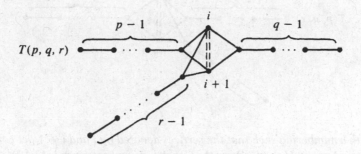

with a numbering such that the vertices denoted by i and $i + 1$ get consecutive indices.

(A12) Proposition.

(i) *The correspondence $\tilde{E} \mapsto (p, q, r) \mapsto \tilde{T}(p, q, r)$ associates to each simple-elliptic singularity $\tilde{E}_k (k = 6, 7, 8)$ a Dynkin diagram.*

(ii) *The lattice $\tilde{E}_k$ is negative semidefinite, its radical has rank 2 and the corresponding reduced Milnor lattice $\tilde{\tilde{E}}_k$ is exactly the Milnor lattice E_k.*

(iii) *Any IHS X whose symmetric Milnor lattice is negative semidefinite is a simple-elliptic singularity.*

For a proof, see, for instance, [Df4]. For a proof of the following more complicated results we refer to Ebeling [E1] and [E2].

(A13) Proposition.

(i) *A Dynkin diagram for the cusp singularity $T_{p,q,r}$ is given by the diagram $\tilde{T}(p, q, r)$ described above.*

(ii) $\mathrm{sign}(T_{p,q,r}) = (p + q + r - 3, 1, 1)$.

(iii) $\det(\bar{T}_{p,q,r}) = pqr (1 - 1/p - 1/q - 1/r)$.

This is not a surprise, since we can regard the simple-elliptic singularities $\tilde{E}_k$ as special cases of the $T_{p,q,r}$-singularities for

$$\frac{1}{p} + \frac{1}{q} + \frac{1}{r} = 1.$$

(A14) Proposition.

(i) *A Dynkin diagram for the triangle singularity $D_{p,q,r}$ is given by the following diagram:*

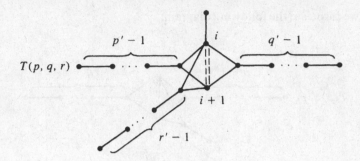

with a numbering such that the vertices denoted by i and $i + 1$ get consecutive indices and (p', q', r') are the Gabrielov numbers given in (2.4.7).

(ii) $\text{sign}(D_{p,q,r}) = (p' + q' + r' - 2, 0, 2)$.

(iii) $\det(D_{p,q,r}) = p'q'r'(1 - 1/p' - 1/q' - 1/r')$.

Recall that $D_{p,q,r}$ denotes in fact two singularities: one weighted homogeneous and the other one semiweighted homogeneous. By our discussion in (3.1.19) it follows that the two singularities have the same topological invariants, in particular, the same Dynkin diagrams.

(A15) **Exercise.** Check that the Dolgacev numbers (p, q, r) and the Gabrielov numbers (p', q', r') associated to a given triangle singularity $D_{p,q,r}$ in (2.4.7) satisfy the following relation

$$pqr\left(1 - \frac{1}{p} - \frac{1}{q} - \frac{1}{r}\right) = p'q'r'\left(1 - \frac{1}{p'} - \frac{1}{q'} - \frac{1}{r'}\right).$$

This is part of a "strange duality" which is explained in [EW].

(A16) **Exercise.** Show that the skew-symmetric Milnor lattice L_X^{ss} associated to the triangle singularity

$$S_{11}: x^4 + y^2z + xz^2 = 0$$

is isomorphic to the lattic $I_{10} \oplus (\mathbb{Z}, (\ )_0)$. *Hint.* Use (A14) and (3.3.22').

Recall now the notion of an embedding of lattices $\varphi: M \to N$ from (A6). The lattice N is then called a *supralattice* of the lattice M. The embedding φ is called *primitive* if coker φ has no torsion. Assume from now on that M is a nondegenerate lattice. We would like to have a control over the set of all the possible torsion groups arising from various embeddings of a given lattice M. In other words, we have to consider the set

(A17) $T(M) = \{\text{Tors}(N/M); N \text{ a supralattice of } M\}$

where $\text{Tors}(G)$ denotes the torsion part of a finitely generated abelian group G.

(A18) Exercise. Show that

$$T(M) = \{\text{Tors}(N/M); N \text{ is a supralattice of } M \text{ and rk } N = \text{rk } M\}.$$

A first restriction on the finite groups F in $T(M)$ is given by the next

(A19) Lemma.

$$F \in T(M) \Rightarrow |F|^2 \text{ divides } \det M.$$

In particular, $T(M) = \{0\}$ when M is a unimodular lattice.

Proof. Use (A5).   □

To get a finer description of the groups in $T(M)$ we proceed as follows. The bilinear form (,) on M has a natural extension to a bilinear form on $M' = \text{Hom}(M, \mathbb{Z})$ with values in $\mathbb{Q}$ which can be defined in one of the following two equivalent ways. Identify M with the image of the monomorphism $i_M: M \to M', x \mapsto (x, \cdot)$ and let $u, v \in M'$.

(i) Take an integer k such that $ku = x$ is in M and define

$$(u, v) = \frac{1}{k} v(x).$$

(ii) Take two integers k, l such that $ku = x$ and $lv = y$ are in M and define

$$(u, v) = \frac{1}{kl}(x, y).$$

This new bilinear form induces by passing to the quotient a bilinear form

(A20) $$b_M: D(M) \times D(M) \to \mathbb{Q}/\mathbb{Z}$$

called the bilinear *discriminant form*.

(A21) Lemma. *The bilinear discriminant form b_M is nondegneerate in the sense that*

$$b_M(u, v) = 0 \quad \text{for all} \quad u \in D(M) \Rightarrow v = 0.$$

Proof. Let k be the smallest positive integer such that ku belongs to M. There is an element $v \in M'$ such that $v(ku) = 1$ and hence $(u, v) = 1/k$. Since $b_M(u, v) = 0$, it follows that $k = 1$, i.e., $u \in M$ and hence $u = 0$ in $D(M)$.   □

(A22) Definition. A subgroup $H \subset D(M)$ is called *isotropic* if $b_M | H \times H = 0$.

(A23) Proposition. *There is a bijection between the set $T(M)$ and the set of isotropic subgroups in $D(M)$.*

Proof. Let $F = N/M$ be a quotient in $T(M)$. Then we have the obvious inclusions

$$M \subset N \subset M'$$

since any element $n \in N$ induces an element in M' by $m \mapsto (n, m) \in \mathbb{Z}$. The subgroup

$$F = N/M \subset M'/M = D(M)$$

is clearly isotropic. This argument can be reversed and hence gives rise to a one-to-one correspondence between the set $T(M)$ and the set of isotropic subgroups in $D(M)$. $\square$

(A24) **Exercise.** (i) Show that $b_{M_1 \oplus M_2} = b_{M_1} \oplus b_{M_2}$.

(ii) Let $M_d = (\mathbb{Z}^2, (\ )_d)$ be the "elementary" skew-symmetric lattice considered in (A7). Then show that

$$D(M_d) = (\mathbb{Z}/d\mathbb{Z})^2$$

and

$$b_{M_d}((\hat{a}_1, \hat{a}_2), (\hat{b}_1, \hat{b}_2)) = \frac{a_1 b_2 - a_2 b_1}{d} \in \mathbb{Q}/\mathbb{Z}.$$

Show that $T(M_d) = \{0\}$ if and only if the lattice M_d is unimodular.

In the symmetric case the results are much more interesting. Assume from now on that M is an even nondegenerate symmetric lattice. Then its discriminant bilinear form b_M is determined by its *quadratic form* q_M, defined as follows:

(A25) $q_M : D(M) \to \mathbb{Q}/2\mathbb{Z}, \qquad q_M(x + M) = (x, x) + 2\mathbb{Z}.$

Moreover, a subgroup $F \subset D(M)$ is isotropic if and only if $q_M|F = 0$.

Using the description of the Milnor lattices A_k, D_k, E_k given in (A9), we can prove the following.

(A26) (i) $D(A_k) = \mathbb{Z}/(k + 1)\mathbb{Z}, q(\hat{1}) = -k(k + 1)^{-1}$.
       (ii) $T(A_k) = \{\mathbb{Z}/e\mathbb{Z}; e^2 | k + 1 \text{ and } k(k + 1)e^{-2} \in 2\mathbb{Z}\}$.

In particular, $T(A_k) = \{0\}$ for $k = 1, \dots, 6$.

(A27) (i)   For k even, $D(D_k) = (\mathbb{Z}/2\mathbb{Z})^2$ and the generators u_1, u_2 of $D(D_k)$ can be chosen such that $q(u_1) = 1$ and $q(u_2) = -k/4$.
       (ii) For k odd, $D(D_k) = \mathbb{Z}/4\mathbb{Z}$ and $q(\hat{1}) = -k/4$.

In particular, $T(D_k) = \{0, \mathbb{Z}/2\mathbb{Z}\}$ for $k \equiv 0 \pmod 8$ and $T(D_k) = \{0\}$ otherwise.

(A28) (i) $D(E_6) = \mathbb{Z}/3\mathbb{Z}$ and $q(\hat{1}) = 2/3$.
       (ii) $D(E_7) = \mathbb{Z}/2\mathbb{Z}$ and $q(\hat{1}) = 1/2$.
       (iii) $D(E_8) = 0$.

In particular, $T(E_k) = \{0\}$ for $k = 6, 7, 8$. Other explicit computations of quadratic forms for Milnor lattices of singularities can be found in [EW].

Two important results on lattices depending on the notions introduced above are the following ones, due to Nikulin [Nk], see also [Dg2].

(A29) Theorem (Uniqueness of a Primitive Embedding). *Let $i: M \to N$ be a primitive embedding of an even nondegenerate lattice M of signature (m_-, m_+) into an even nondegenerate lattice N of signature (n_-, n_+). Then this embedding is unique up to an automorphism of N provided the following conditions are satisfied:*

(i) $n_- \geq m_-, n_+ \geq m_+$;
(ii) $\operatorname{rk}(N) - \operatorname{rk}(M) \geq l(D(M)) + 2$, *where $l(F)$ denotes the minimal number of generators of a finite abelian group F.*

(A30) Theorem (Uniqueness of a Milnor Lattice). *The symmetric Milnor lattice $L = L^s(X)$ of an IHS X is determined uniquely by its signature (l_-, l_0, l_+) and the discriminant form $q: D(\bar{L}) \to \mathbb{Q}/2\mathbb{Z}$ associated to its reduced lattice $\bar{L}$.*

This result (A30) can be used to determine the Milnor lattices for large classes of singularities (but not their Dynkin diagrams since it does not provide us with distinguished bases in L). Consider the class of weighted homogeneous surface singularities in $\mathbb{C}^3$. Then the signature of the Milnor lattice $L = L^s(X)$ of such a singularity $(X, 0)$ can be determined using results of Steenbrink, [S2], see Appendix C. To determine the remaining ingredient, namely, the discriminant form one can proceed as follows [Dg2], [LW]. We use integer coefficients for (co)homology and the subscript t stands for the "torsion part" of a finitely generated abelian group. If $\alpha: G \to H$ is a group homomorphism between two such groups, then the restricted homomorphism $\alpha_t: G_t \to H_t$ is well defined.

Let M be a compact closed oriented $(2n - 1)$-manifold and recall the definition of the *linking form* of M

$$\operatorname{lk}: H_{n-1}(M)_t \times H_{n-1}(M)_t \to \mathbb{Q}/\mathbb{Z}.$$

Given two classes $v_1, v_2 \in H_{n-1}(M)_t$, we can represent them by two disjoint cycles V_1 and V_2 (use transversality and the fact that $\dim V_1 + \dim V_2 < \dim M$). Some integral multiple kV_1 of V_1 bounds an n-chain C_1 in M and we set

$$\operatorname{lk}(v_1, v_2) = \frac{1}{k} C_1 \cdot V_2 (\operatorname{mod} \mathbb{Z})$$

(compare to Definition (3.3.12)). Suppose now that M bounds an oriented compact $2n$-manifold N such that

$$H_{n-1}(N)_t = 0.$$

The composite

$$H_n(N) \to H_n(N, \partial N) \xrightarrow{D} H^n(N) \xrightarrow{\sim} H_n(N)'$$

is the adjoint of the intersection product $(\ ,\ )$ on $H_n(N)$, recall (2.3.6). It is obvious that $H_n(N)_t \subset \mathrm{Rad}(\ ,\ )$ and hence

$$\overline{H}_n(N) := H_n(N)/\mathrm{Rad}(\ ,\ )$$

is a nondegenerate lattice. Let (D_N, b_N) be the corresponding discriminant bilinear form.

(A31) **Proposition.** *There is a natural isomorphism*

$$(D_N, b_N) \simeq (H_{n-1}(M)_t, -\mathrm{lk}).$$

Proof. Let w_1, w_2 be two cycles in $H_n(N, M)$ and choose representatives W_1, W_2 in (N, M) for them which are transversal and $\partial W_1 \cap \partial W_2 = \varnothing$. Under these conditions, the intersection number $W_1 \cdot W_2$ is a well-defined integer, but not an invariant of the classes w_1, w_2 as remarked in [AGV2], p. 12. Let $\overline{w}_1, \overline{w}_2$ be the elements in D_N corresponding to the cycles w_1 and w_2 respectively. Let $v_i = \partial w_i$ be the associated cycles in $H_{n-1}(M)$. We consider only classes w_1, w_2 such that the classes v_1, v_2 are torsion elements. Then there exists an integer k such that $kv_1 = 0$, i.e., $kv_1 = \partial C_1$ for some n-cycle C_1 in M. But then $kW_1 - C_1$ is an absolute n-cycle in N and hence we have by definition

$$b_N(\overline{w}_1, \overset{\ast}{w}_2) = \frac{1}{k}(kW_1 - C_1, W_2).$$

Indeed, note that $kW_1 - C_1 \sim kW_1$ in $H_n(N, \partial N)$ and we use here the intersection number of an absolute cycle with a relative cycle, exactly like in [AGV2], p. 11. Again by definition we have

$$\mathrm{lk}(v_1, v_2) = \frac{1}{k}(C_1, \partial W_2) = \frac{1}{k}(C_1, W_2).$$

Hence

$$b_N(\overline{w}_1, \overline{w}_2) + \mathrm{lk}(v_1, v_2) = (W_1, W_2) \in \mathbb{Z}$$

which ends the proof. $\qquad\square$

(A32) **Example.** Let $M = K$ be the link of an n-dimensional IHS X and let $N = F$ be the corresponding Milnor fiber. The the condition $H_{n-1}(F)_t = 0$ follows from (3.2.1). The discriminant bilinear form (D_N, b_N) is in this case exactly the discriminant bilinear form $(D(\overline{L}), b_{\overline{L}})$ associated with the reduced Milnor lattice $\overline{L}$ of the singularity X. Proposition (A31) says that the linking form of the link K is essentially the same as the discriminant bilinear form $b_{\overline{L}}$.

(A33) **Example.** Let $M = L(X, 0)$ be the link of a normal surface singularity $(X, 0)$ and N the total space of a resolution of the singularity $(X, 0)$. Then $H_1(M)_t = 0$ by (2.3.1). Proposition (A31) says in this case that the linking form of the link M can be computed from the intersection matrix of the resolution (i.e., from the corresponding dual resolution graph).

Combining (A32) and (A33) we deduce that the discriminant bilinear form (and hence also the quadratic form) of the reduced Milnor lattice $\bar{L}$ of an isolated surface singularity in $\mathbb{C}^3$ can be computed from its dual resolution graph. In particular, using the result by Orlik–Wagreich (2.4.21), the formula for the signature in Steenbrink [S2], and Nikulin's result (A30), we can determine the Milnor lattices of all the weighted homogeneous surface singularities in $\mathbb{C}^3$.

A last lattice-theoretic fact that we need in our book is the following.

(A34) **Lemma.** *Let N be a unimodular lattice and $M \subset N$ be a nondegenrate sublattice. Consider the orthogonal complement of M in N*

$$M^\perp = \{x \in N; (x, y) = 0 \text{ for all } y \in M\}.$$

Then $M^\perp$ is a nondegenerate lattice such that

$$\det M^\perp = \det M \cdot |\mathrm{Tors}(N/M)|^{-2}.$$

Proof. Let $\bar{M} = \{x \in N; kx \in M \text{ for some integer } k\}$. Then the embedding $\bar{M} \subset N$ is primitive and $\mathrm{Tors}(N/M) = \bar{M}/M$. Moreover, we have by (A5)

$$\det(\bar{M}) = \det(M) \cdot |\bar{M}/M|^{-2}.$$

Using this, we can clearly assume that the embedding $M \subset N$ is primitive.

Let $i_M: M \to M'$ and $i_N: N \xrightarrow{\sim} N'$ be the usual embeddings. Then one has

$$\det M \cdot \det M^\perp = \det(M + M^\perp) = |N/(M + M^\perp)|^2$$

and

$$\frac{N}{M + M^\perp} \cong \frac{N'}{i_N(M + M^\perp)} \cong \frac{M'}{i_M(M)} \cong D(M).$$

Hence $\det M^\perp = \det(M)$ as claimed in (A34). $\qquad \square$

APPENDIX B

Weighted Projective Varieties

Weighted projective varieties are increasingly important in all branches of algebraic geometry and even in mathematical physics, see, for instance, [Dg1], [S2], and [Ro]. In this appendix we survey some of their basic properties, with special attention to their *topological properties*. The proofs for the statements below which are not included here can be found in [Dg1], if the reader is not explicitly referred to a different source.

Let w_0, w_1, ..., w_n be a set of strictly positive integers and consider the associated $\mathbb{C}^*$ action on the affine space $\mathbb{C}^{n+1}$, namely,

(B1) $$t \cdot (x_0, \ldots, x_n) = (t^{w_0} x_0, \ldots, t^{w_n} x_n)$$

as in (3.1.10). The integers w_i are called *weights* and we denote by $\mathbf{w} = (w_0, \ldots, w_n)$ the set of all these weights. The *weighted projective space* of type $\mathbf{w}$, denoted by $\mathbb{P}(\mathbf{w})$ can be defined either geometrically, as the quotient

(B2) $$\mathbb{P}(\mathbf{w}) = \mathbb{C}^{n+1} \setminus \{0\} / \mathbb{C}^*$$

or algebraically, as

(B3) $$\mathbb{P}(\mathbf{w}) = \text{Proj } \mathbb{C}[x_0, \ldots, x_n]$$

where the grading for the polynomial ring comes from the relations $\deg(x_i) = \text{wt}(x_i) = w_i$. For the general definition of the scheme Proj S, for any graded ring S, we refer to [Hn], p. 76.

We can regard the weighted projective space $\mathbb{P}(\mathbf{w})$ as a quotient space in two other useful ways. Let S denote the unit sphere in $\mathbb{C}^{n+1}$ and note that the subgroup $S^1 \subset \mathbb{C}^*$ has a naturally induced action on the sphere S. It is also clear that

(B4) $$\mathbb{P}(\mathbf{w}) = S/S^1.$$

Let $G(m)$ be the multiplicative group of all the mth roots of unity. Consider the product group

(B5) $$G(\mathbf{w}) = G(w_0) \times \cdots \times G(w_n).$$

If we let $G(\mathbf{w})$ act on the projective space $\mathbb{P}^n$ coordinate wise, then it is clear that

(B6) $$\mathbb{P}(\mathbf{w}) = \mathbb{P}^n/G(\mathbf{w}).$$

Moreover, the quotient map $\pi\colon \mathbb{P}^n \to \mathbb{P}(\mathbf{w})$ is given explicitly by the following formula

(B7) $$\pi(x_0 : \cdots : x_n) = (x_0^{w_0} : \cdots : x_n^{w_n}).$$

Being a quotient variety of $\mathbb{P}^n$ under a finite group action, it follows that $\mathbb{P}(\mathbf{w})$ is a *complete normal* variety, having only quotient singularities, see (2.3.14).

We can obviously assume that the weights $\mathbf{w}$ satisfy the following condition

(B8) $$\text{g.c.d.}(w_0, \ldots, w_n) = 1$$

which means that the action (B1) is effective. It is often possible to assume a stronger condition on the weights $\mathbf{w}$, namely

(B9) $$\text{g.c.d.}(w_0, \ldots, \hat{w}_i, \ldots, w_n) = 1 \qquad \text{for all} \quad i = 0, \ldots, n.$$

This stronger restriction is useful in many cases. One example of such a situation is the following description of the singular locus $\mathbb{P}(\mathbf{w})_{\text{sing}}$ of a weighted projective space, see [DDv].

(B10) **Proposition.** *Assume that the weights* $\mathbf{w}$ *satisfy the condition* (B9). *Then*

$$x \in \mathbb{P}(\mathbf{w})_{\text{sing}} \Leftrightarrow \text{g.c.d.} \{w_j; x_j \neq 0\} > 1.$$

(B11) **Exercise.** (i) Identify the type of the Hirzebruch–Jung singularities in $\mathbb{P}(2, 3, 5)_{\text{sing}}$.

(ii) Why is the statement (B10) is obviously *false* when the weights $\mathbf{w}$ do not satisfy (B9)?

A first basic topological property of the weighted projective spaces is the following.

(B12) **Proposition.**
$$\pi_1(\mathbb{P}(\mathbf{w})) = 0.$$

Proof. Use the definition (B4) for the weighted projective space $\mathbb{P}(\mathbf{w})$, the fact that $\pi_1(S) = 0$ and [Ar]. $\square$

The projection map

$$S \to S/S^1 = \mathbb{P}(\mathbf{w})$$

is the analog of the Hopf map, but it is no longer an S^1-bundle map. Hence there is no associated Gysin sequence with $\mathbb{Z}$-coefficients. However, since all the isotropy groups of this S^1-action on S are finite, there is a *Smith–Gysin sequence* with $\mathbb{Q}$- or $\mathbb{C}$-coefficients, see [Bd], p. 162. It follows that Proposition (5.1.6) has the following analog in the weighted case:

(B13) Proposition. *The rational cohomology algebra* $H^*(\mathbb{P}(\mathbf{w}); \mathbb{Q})$ *for* $n =$ $\dim \mathbb{P}(\mathbf{w}) \geq 1$ *is a truncated polynomial algebra*

$$\mathbb{Q}[\alpha]/(\alpha^{n+1})$$

generated by an element α *of degree 2.*

(B14) Remark. See [Kws] for a method to compute the *integral* cohomology algebra $H^*(\mathbb{P}(\mathbf{w}); \mathbb{Z})$ in terms of the weights $\mathbf{w}$.

Now let $f_i \in \mathbb{C}[x_0, \ldots, x_n]$ be weighted homogeneous polynomials of degree d_i with respect to the weights $\mathbf{w}$, for $i = 1, \ldots, c$. Consider the affine variety in $\mathbb{C}^{n+1}$ (a weighted cone)

(B15) $$CV: f_1 = \cdots = f_c = 0$$

and the subvariety V in $\mathbb{P}(\mathbf{w})$ defined by these equations, i.e.,

(B16) $$V = CV \setminus \{0\}/\mathbb{C}^* = \operatorname{Proj} \mathbb{C}[x_0, \ldots, x_n]/(f_1, \ldots, f_c).$$

When $\dim V = n - c$, we refer to V as being a *weighted complete intersection*. When $c = 1$, we call V a *weighted hypersurface*.

(B17) Definition. A weighted complete intersection V is called *quasismooth* if the associated weighted affine cone CV has an isolated singularity at the origin.

We would like to emphasize that a quasismooth weighted complete intersection V is *not* smooth as an algebraic variety. A discussion of the singular locus V_{sing} of such a variety V, in particular sufficient conditions for the relation

$$V_{\text{sing}} = V \cap \mathbb{P}(\mathbf{w})_{\text{sing}}$$

to hold, can be found in [D3]. What *is* true, is that all the singularities of a quasismooth variety are quotient singularities. Such varieties are sometimes called *V*-manifolds, see for instance, [S2], [S6].

One of the main topological properties of the *V*-manifolds is the following.

(B18) Proposition. *Any V-manifold M is a* $\mathbb{Q}$*-homology manifold of dimension* $2 \dim M$.

Proof. If $m = \dim M$, we have to show that

$$H_x^*(M; \mathbb{Q}) = H_0^*(D^{2m}; \mathbb{Q}),$$

where $x \in M$ is an arbitrary point and D^{2m} is the closed $2m$-dimensional disc. The problem being a local one on M, we can assume that the germ(M, x) is a quotient singularity $(\mathbb{C}^m/G, 0)$ as in (2.3.14). Moreover we can assume that $G \subset U(m)$, i.e., the unit sphere S^{2m-1} in $\mathbb{C}^m$ is G-invariant. Then the link K of the singularity $(\mathbb{C}^m/G, 0)$ can be identified with the quotient S^{2m-1}/G, a generalized lens space. $\square$

At the $\mathbb{Q}$-cohomology group level, we have the equalities

$$H_x^s(M; \mathbb{Q}) = H_0^s(\mathbb{C}^m/G; \mathbb{Q}) = H^{s-1}(S^{2m-1}/G; \mathbb{Q}) \overset{\alpha}{=} H^{s-1}(S^{2m-1}; \mathbb{Q})^G$$

$$= H^{s-1}(S^{2m-1}; \mathbb{Q}) = H_0(D^{2m}; \mathbb{Q}).$$

The equality α comes from a general result in transformation groups, see [Bd], p. 120. Moreover, the fact that the induced action of G on the cohomology $H^{\cdot}(S^{2m-1}; \mathbb{Q})$ is trivial follows from the fact that the group $U(m)$ is connected (any element $h \in U(m)$ can be continuously deformed into the identity).

(B19) **Corollary.** *The rational cohomology of a weighted projective space $\mathbb{P}(\mathbf{w})$ or of a quasismooth weighted complete intersection $V \subset \mathbb{P}(\mathbf{w})$ satisfies Poincaré duality.*

Let $K_V = S \cap CV$ be the link associated to a weighted complete intersection. In analogy to (B4) we have

(B20) $$V = K_V/S^1.$$

(B21) **Corollary.** *A weighted complete intersection V is connected for dim $V \geq$ 1 and simply-connected for dim $V \geq 2$.*

Proof. Exactly as in (B12), using the relation $\pi_1(K_V) = 0$ from (3.2.12). $\qquad \square$

(B22) **Lefschetz Theorem over $\mathbb{Q}$.** *If V is a weighted complete intersection in $\mathbb{P}(\mathbf{w})$, then the morphism*

$$j^k: H^k(\mathbb{P}(\mathbf{w}); \mathbb{Q}) \to H^k(V; \mathbb{Q})$$

induced by the inclusion is an isomorphism for $k < $ dim V and a monomorphism for $k = $ dim V.

Proof. The proof given for (5.2.6) can be applied to this situation, since:

(i) As shown in (B18), the weighted projective space $\mathbb{P}(\mathbf{w})$ is a $\mathbb{Q}$-homology manifold, hence all the duality theorems (Poincaré, Alexander, Lefschetz) can be used with $\mathbb{Q}$-coefficients.

(ii) The complement in $\mathbb{P}(\mathbf{w})$ of a weighted hypersurface is still an affine variety. This follows from the description (B6) and the fact that the quotient of an affine variety under a finite group is affine, see for instance [KK], p. 314, where the corresponding result for Stein spaces is proved.

Using the ramified covering $\pi: \mathbb{P}^n \to \mathbb{P}(\mathbf{w})$ from (B7) we can associate to each weighted variety $V \subset \mathbb{P}(\mathbf{w})$ the subvariety $\tilde{V} = \pi^{-1}(V)$ in $\mathbb{P}^n$. Since $V = \tilde{V}/G(\mathbf{w})$, it follows from [Bd], p. 120, that we have

(B23) $$H^*(V; \mathbb{Q}) = H^*(\tilde{V}; \mathbb{Q})^{G(\mathbf{w})}.$$

However this formula is not very useful in practice, since the dimension of the singular locus of $\tilde{V}$ can be high, even if we start with a quasismooth complete intersection V. □

(B24) Exercise. Consider the weights $\mathbf{w} = (3, 2, 3, 2)$ and the weighted hypersurface

$$V: x^3 + xy^3 + z^3 + zt^3 = 0$$

in $\mathbb{P}(\mathbf{w})$. Show that V is quasismooth, but dim $\tilde{V}_{\text{sing}} = 1$, i.e., the surface $\tilde{V}$ is not even normal.

We can use (B23) (the version with $\mathbb{C}$-coefficients) to prove the following.

(B25) Lemma/Definition ($A = \mathbb{R}, \mathbb{C}$).

(i) *The morphism* $j^k: H^k(\mathbb{P}(\mathbf{w}); A) \to H^k(V; A)$ *induced by the inclusion* $j: V \to \mathbb{P}(\mathbf{w})$ *is a monomorphism for all* $k \leq 2 \dim V$.
(ii) $H_0^k(V; A) \doteq \operatorname{coker} j^k$ *is called the primitive cohomology of the variety* V.
(iii) $H_0^k(V; A) = H_{2n-k-1}(U; A)$, *where* $U = \mathbb{P}(\mathbf{w}) \backslash V$.

When V is a weighted hypersurface in $\mathbb{P}(\mathbf{w})$ associated to a weighted homogeneous polynomial f of degree N with respect to the weights $\mathbf{w}$, we have the following additional facts. Let $F: f - 1 = 0$ be the corresponding affine Milnor fiber and let $h: F \to F$ be the monodromy homeomorphism as in (3.1.19). Let $H^*(F)_0$ be the fixed part cohomology in $H^*(F; \mathbb{C})$ under the monodromy operator. Since $U = F/\langle h \rangle$, where $\langle h \rangle$ denotes the finite group generated by the transformation h, it follows as above that

$$\text{(B26)} \qquad\qquad H^*(U) = H^*(F)_0.$$

(B27) Exercise. (i) Use the previous relation to obtain an analog of Theorem (5.2.11) in the weighted hypersurface case.

(ii) Let d be a divisor of $N = \deg f$, $d > 1$. Consider the weighted hypersurface $V^d: f^d(x_0, \ldots, x_n, t) = f(x_0, \ldots, x_n) + t^d = 0$ in the weighted projective space $\mathbb{P}(w_0, \ldots, w_n, Nd^{-1})$. This hypersurface is the d-fold covering of $\mathbb{P}(\mathbf{w})$ ramifed along the hypersurface V, see [DDv].

Let U^d be the complement of the weighted hypersurface V^d. Show that dim $H^s(U^d)$ is equal to the number of eigenvalues (counted with multiplicities) of the monodromy operator h^* on $H^{s-1}(F; \mathbb{C})$ which are dth roots of unity and different from 1. *Hint.* Use the Thom–Sebastiani formula.

Let us consider now in more detail quasismooth weighed complete intersections. Their topology is completely determined by their weights $\mathbf{w} = (w_0, \ldots, w_n)$ and multidegree $\mathbf{d} = (d_1, \ldots, d_c)$.

(B28) Proposition. *Let* V, V' *be two quasismooth weighted complete intersection in* $\mathbb{P}(\mathbf{w})$ *having the same multidegree* $\mathbf{d}$. *Then the varieties* V *and* V' *are homeomorphic.*

For a proof of this result, based on an S^1-equivariant Ehresmann Fibration theorem we refer to [D1]. Let V be a quasismooth complete intersection in $\mathbb{P}(\mathbf{w})$ and let K_V be the associated link. Then K_V is a smooth manifold with an induced S^1-action, recall (B20). For a point $x \in K_V$, the isotropy group S_x^1 is the multiplicative group of $w(x)$-roots of unity, where

(B29) $$w(x) = \text{g.c.d.}\{w_i; x_i \neq 0\}$$

(compare with (B10)).

(B30) **Definition.** The weighted complete intersection V is called *strongly smooth* if it is quasismooth and if all the isotropy groups S_x^1 for $x \in K_V$ are the same.

In such a case the projection $K_V \to V$ is an $S^1 = S^1/S_x^1$-bundle map. In particular V gets in this way the structure of a smooth manifold.

(B31) **Example.** Consider a set of weights $\mathbf{w} = (w_0, \ldots, w_n)$ such that g.c.d.$(w_i, w_j) = 1$ for all $i \neq j$. Let N be a common multiple of these weights w_i and consider the weighted hypersurface

$$V: f(x) = x_0^{N/w_0} + \cdots + x_n^{N/w_n} = 0.$$

Then the hypersurface V is strongly smooth.

The main properties of this important class of weighted complete intersections are contained in the following result, see [D1].

(B32) **Proposition.** *Let V be a weighted complete intersection of weights $\mathbf{w}$ and multidegree $\mathbf{d}$ which is strongly smooth.*

(i) *Any other quasismooth weighted complete intersection with the same weights $\mathbf{w}$ and multidegree $\mathbf{d}$ is also strongly smooth and diffeomorphic to V.*
(ii) *The integral cohomology algebra $H^*(V)$ is torsion free.*

Now let V be a quasismooth weighted complete intersection in $\mathbb{P}(\mathbf{w})$ of multidegree $\mathbf{d}$. Using (B19), (B22), and (B13), it follows that

(B33) $$H^i(V, \mathbb{Q}) = H^i(\mathbb{P}^m; \mathbb{Q}) \qquad \text{for} \quad i \neq m,$$

where $m = \dim V$. In view of (B28) it is natural to try to compute in terms of $\mathbf{w}$ and $\mathbf{d}$ the following topological invariants for V:

(i) the middle Betti number $b_m(V)$ or, alternatively, the Euler characteristic $\chi(V)$ of V;
(ii) (when $m = \dim V$ is even) the index $\tau(V)$ of the cup product

$$H^m(V; \mathbb{Q}) \times H^m(V; \mathbb{Q}) \to H^{2m}(V; \mathbb{Q}) = \mathbb{Q}.$$

It turns out that the simplest way to compute these *topological invariants* is by computing some *analytical invariants*, namely the mixed Hodge numbers $h^{p,q}(V)$, see Appendix C.

The general case is treated in [H2] and [S6]. Here we present the result only for the hypersurface case, see [S2], [Dg1]. Assume that the hypersurface V is defined by a weighted homogeneous polynomial f of type $(w_0, \ldots, w_n; N)$. Let $w = w_0 + \cdots + w_n$ be the sum of all these weights. Note that the Milnor algebra

$$M(f) = \frac{\mathbb{C}[x_0, \ldots, x_n]}{(\partial f/\partial x_0, \ldots, \partial f/\partial x_n)}$$

(see [D4], p. 111) is a graded algebra, using $\deg x_i = \operatorname{wt} x_i = w_i$ for $i = 0, \ldots, n$.

Let $M(f)_s$ denote the homogeneous component of degree s in $M(f)$ with respect to this grading.

(B34) **Theorem** (Steenbrink). *The mixed Hodge numbers of the primitive cohomology group $H_0^{n-1}(V)$ of a quasismooth weighted hypersurface V in $\mathbb{P}(\mathbf{w})$ are given by*

$$h_0^{i, n-i-1}(V) = \dim_{\mathbb{C}} M(f)_{(i+1)N-w}.$$

(B35) **Examples.** (i) $\dim V = 1$.

Then V is a smooth connected curve (there are no quotient singularities in dimension 1) of genus

$$g(V) = h_0^{0, 1}(V) = \dim M(f)_{N-w}.$$

To have a numerical example, consider the curve $V: x_0^2 + x_1^3 + x_2^5 = 0$. Here $\mathbf{w} = (15, 10, 6)$, $N = 30$. Since $N - w = -1 < 0$, it follows that $g(V) = 0$, i.e., $V \simeq \mathbb{P}^1$, the projective line.

(ii) $\dim V = 2$.

Then V is a normal surface and its index $\tau(V)$ is given by the formula

$$\tau(V) = 1 + 2 \dim M(f)_{N-w} - M(f)_{2N-w}.$$

Here we have used the general formula for $\tau(V)$ in terms of mixed Hodge numbers $h^{p,q}(V)$, see Appendix C as well as the equality

$$\dim M(f)_{(i+1)N-w} = \dim M(f)_{(n-i)N-w}$$

holding for any i, which is a well-known property of this Milnor algebra, see [D4], p. 113. To have again a numerical example, let $C: g(x, y, z) = 0$ be a smooth sextic curve in $\mathbb{P}^2$. Let V be the double covering of $\mathbb{P}^2$ ramified along the curve C. Then an equation for V is

$$V: f(x, y, z, t) = g(x, y, z) + t^2 = 0$$

in the weighted projective space $\mathbb{P}(1, 1, 1, 3)$, i.e., $w_0 = w_1 = w_2 = 1$, $w_3 = 3$, $N = 6$. A simple computation (if you need take $g(x, y, z) = x^6 + y^6 + z^6$!) shows that

$$\tau(V) = -16.$$

This should be compared to (5.3.33) since the double covering V is also a classical model for K3 surfaces.

(iii) *Closures of Affine Milnor Fibers.*
Let

$$F: f(x_1, \ldots, x_n) - 1 = 0$$

be the affine Milnor fiber of the weighted homogeneous polynomial f of type $(w_1, \ldots, w_n; N)$, having an isolated singularity at the origin. Assume that n is odd and let (μ_-, μ_0, μ_+) be the signature of the Milnor lattice $H_{n-1}(F)$.

Consider the compactification $\mathbb{C}^n \subset \mathbb{P}(\mathbf{w})$, where $\mathbf{w} = (1, w_1, \ldots, w_n)$. The closure $\bar{F}$ corresponding to this compactification is the weighted hypersurface given by

$$\bar{f}(x_0, \ldots, x_n) = x_0^N - f(x_1, \ldots, x_n) = 0.$$

The Milnor algebras $M(f)$ and $M(\bar{f})$ are related by the following obvious relation

$$M(\bar{f}) = M(f) \otimes \mathbb{C}[x_0]/(x_0^{N-1}).$$

Using this it follows that

$$h_0^{i,n-i-1}(\bar{F}) = \dim M(\bar{f})_{(i+1)N-w-1} = \sum_{s=1, N-1} \dim M(f)_{iN-w+s},$$

where $w = w_1 + \cdots + w_n$. Using [S2] (see also (C26)) it turns out that

$$\tau(\bar{F}) = 1 + \mu_+ - \mu_-.$$

This formula can be regarded as an analog of (5.3.24).

Beyond the case of quasismooth complete intersections, one knows very little about the topology of weighted varieties.

Assume, for instance, that we are again in the hypersurface case and

$$\dim CV_{\text{sing}} = 1$$

(this is the analog of a hypersurfaces with isolated singularities in the usual projective space $\mathbb{P}^n$, the case considered in Chapter 5, §4). In the weighted case, even the formulas for the Euler characteristics $\chi(V)$ and $\chi(F)$ are not completely clear. However, there are conjectural formulas which are known to hold in many special cases, see [D5].

APPENDIX C

Mixed Hodge Structures

In this appendix we survey some of the basic definitions and properties of mixed Hodge structures. A similar introduction to mixed Hodge structures can be found in Durfee [Df5], while for details and complete proofs we refer to the original papers by Deligne [De1] and to the forthcoming book by Steenbrink [S6].

For a smooth manifold X the famous *de Rham theorem* says that we have an isomorphism

$$(C1) \qquad H^*(X; \mathbb{R}) = H_{DR}^*(X; \mathbb{R}),$$

i.e., any real singular cohomology class $[c]$ for X can be represented by a closed differential form ω on X, see, for instance, [GH], p. 44. However, the differential form ω is by no means uniquely determined by the class $[c]$. Assume from now on that X is an oriented compact smooth manifold which has a Riemannian metric g. Then we can define the classical *-operator*

$$(C2) \qquad *: E^s(X) \to E^{n-s}(X),$$

where $n = \dim X$, and $E^s(X)$ denotes the vector space of all differential s-forms on X, see [We], p. 158. Using this operator, we define

$$\delta: E^s(X) \to E^{s-1}(X) \qquad \delta = (-1)^s (*)^{-1} \circ d \circ *,$$

where $d: E^s(X) \to E^{s+1}(X)$ is the exterior differentiation of forms. Let

$$(C3) \qquad \Delta = d\delta + \delta d$$

be the corresponding *Laplace operator*.

A differential form $\omega \in E^s(X)$ is called *harmonic* if it satisfies one of the following equivalent conditions:

(i) $\Delta\omega = 0$;
(ii) $d\omega = \delta\omega = 0$.

Let $H^s(X)$ be the vector space of all harmonic forms in $E^s(X)$.

(C4) **Theorem** (Hodge).
$$H^*(X; \mathbb{R}) = H^*(X),$$

i.e., each cohomology class $[c] \in H^s(X; \mathbb{R})$ has a unique harmonic representative $\omega \in H^s(X)$.

For a proof of this famous result we refer to [We].

Assume from now on that X is a compact complex manifold (which also implies a natural orientation on X). Let $E^s(X; \mathbb{C}) = E^s(X) \otimes \mathbb{C}$ be the vector space of all the differentiable s-forms on X with $\mathbb{C}$-values. Recall that there is a natural decomposition

(C5)
$$E^m(X; \mathbb{C}) = \bigoplus_{p+q=m} E^{p,q}(X, \mathbb{C}),$$

the so-called decomposition into (p, q)-types.

Note that the Laplace operator Δ extends to this new space $E^m(X; \mathbb{C})$ and so does the definition of harmonic forms. Moreover, we clearly have

$$H^m(X; \mathbb{C}) = H^m(X) \otimes \mathbb{C},$$

where the left-hand side denotes the space of all harmonic complex m-forms on X. Let h be a Hermitian metric on X. Then the real part $g = \text{Re}(h)$ of this metric is a Riemannian metric on X, while its imaginary part $\Omega = \text{Im}(h)$ is a 2-form of type $(1, 1)$, i.e., $\Omega \in E^{1,1}(X; \mathbb{C})$.

(C6) **Definition.** The Hermitian metric h is called *Kähler* if $d\Omega = 0$.

A complex manifold X is called *Kähler* if X has a Hermitian metric h which is Kähler.

(C7) **Example.** (i) The Fubini–Study metric on $\mathbb{P}^n$ is Kähler, see for details [We], p. 190.

(ii) If X is a Kähler manifold and $Y \subset X$ is a closed complex submanifold, then Y is again Kähler. In particular, all the smooth projective varieties are Kähler.

Assume from now on that X is a compact Kähler manifold.

(C8) **Proposition.** *Let $\omega = \sum_{p+q=m} \omega_{p,q}$ be the decomposition of an m-form on X according to (p, q)-types. Then the form ω is harmonic if and only if all the components $\omega_{p,q}$ are harmonic.*

In other words, we have a decomposition

(C9)
$$H^m(X; \mathbb{C}) = \bigoplus_{p+q=m} H^{p,q}(X),$$

where $H^{p,q}(X)$ denotes the $\mathbb{C}$-vector space of all harmonic forms of type (p, q) on X. Note that the vector space $E^m(X, \mathbb{C})$ has a natural conjugation coming from the complex conjugation on $\mathbb{C}$ and denoted by $\omega \mapsto \bar{\omega}$. Moreover, we obviously have

(C10) $$\overline{H^{p,q}(X)} = H^{q,p}(X).$$

Let us define the *Hodge numbers* of the manifold X by the formula

(C11) $$h^{p,q}(X) = \dim H^{p,q}(X).$$

Note that (C10) implies $h^{pq}(X) = h^{qp}(X)$. The decomposition (C9) for the cohomology of X has some direct *topological consequences*, see, for instance, [We], pp. 198–208, or [GH], pp. 117–126.

(C12) **Corollary.** *Let X be a compact Kähler manifold. Then:*

(i)   *the odd Betti numbers $b_{2k+1}(X)$ are even;*
(ii) *the even Betti numbers $b_{2k}(X)$ for $0 \le k \le \dim X$ are nonzero;*
(iii) *assume that $n = \dim X$ is even; then the index $\tau(X)$ of the manifold X can be computed in terms of the Hodge numbers of X via the following formula:*

$$\tau(X) = \sum_{\substack{1 \le p,q \le n \\ p \equiv q \,(\mathrm{mod}\,2)}} (-1)^p h^{p,q}(X).$$

To go further, it is convenient to define abstractly the structure we have obtained in (C9), (C10).

(C13) **Definition.** A (pure) *Hodge structure of weight m* is a pair (H, F), where H is a finite dimensional $\mathbb{R}$-vector space and F is a decreasing filtration on $H_{\mathbb{C}} = H \otimes \mathbb{C}$ (called the *Hodge filtration*) such that:

(i) F is a finite filtration, i.e., there exist $s, t \in \mathbb{Z}$ with $F^s H_{\mathbb{C}} = H_{\mathbb{C}}$ and $F^t H_{\mathbb{C}} = 0$;
(ii) $H_{\mathbb{C}} = F^p H_{\mathbb{C}} \oplus \overline{(F^{m-p+1} H_{\mathbb{C}})}$ for all $p \in \mathbb{Z}$, where the conjugation on $H_{\mathbb{C}}$ is induced from the complex conjugation on $\mathbb{C}$.

If we define $H^{p,q} = F^p H_{\mathbb{C}} \cap \overline{(F^q H_{\mathbb{C}})}$ for any pair (p, q) with $p + q = m$, then we have the following relations:

(α) $H_{\mathbb{C}} = \bigoplus_{p+q=m} H^{p,q}$;
(β) $\overline{H^{p,q}} = H^{q,p}$;

i.e., the abstract analog of (C9) and (C10) holds true. Conversely, starting with a finite direct sum decomposition of $H_{\mathbb{C}}$ satisfying the properties (α) and (β), we can define the Hodge filtration by the formula

(C14) $$F^p H_{\mathbb{C}} = \bigoplus_{s \ge p} H^{s, m-s}.$$

It is easy to check that the properties (i) and (ii) in Definition (C13) are then fulfilled. When there is no danger of confusion, we denote the Hodge structure (H, F) simply by H.

(C15) **Definition.** Let H and H' be pure Hodge structures of the same weight m. An $\mathbb{R}$-linear map $\varphi: H \to H'$ is called a *morphism of Hodge structures* if $\varphi_{\mathbb{C}} = \varphi \otimes 1_{\mathbb{C}}: H_{\mathbb{C}} \to H'_{\mathbb{C}}$ satisfies $\varphi_{\mathbb{C}}(F^p H_{\mathbb{C}}) \subset F^p H'_{\mathbb{C}}$ for all p.

(C16) **Remarks.** (i) When $\varphi: H \to H'$ is a morphism of Hodge structures, then the vector spaces $\ker \varphi$, $\operatorname{im} \varphi$, and $\operatorname{coker} \varphi$ have canonically induced Hodge structures (of the same weight).

(ii) If φ is a morphism of Hodge structures, then $\varphi_{\mathbb{C}}$ is *strictly compatible* with the Hodge filtrations, i.e.,

$$\varphi_{\mathbb{C}}(F^p H_{\mathbb{C}}) = F^p H'_{\mathbb{C}} \cap \operatorname{im} \varphi_{\mathbb{C}} \qquad \text{for all } p.$$

(C17) **Example.** (i) If X is a compact Kähler manifold it follows from (C9), (C10), and our discussion in Definition (C13), that the space $H^m(X; \mathbb{R})$ has a pure Hodge structure of weight m.

(ii) Let $f: X \to X'$ be a complex analytic map between the compact Kähler manifolds X and X'. Then the induced morphism

$$f^*: H^m(X'; \mathbb{R}) \to H^m(X; \mathbb{R})$$

is a morphism of Hodge structures. This property comes from the obvious fact that $f^*: E^m(X'; \mathbb{C}) \to E^m(X; \mathbb{C})$ preserves the decompositions into (p, q)-types.

We can now define the central concept of this appendix.

(C18) **Definition** (Deligne [De1]). A *mixed Hodge structure* (MHS) is a triple (H, W, F) where:

 (i) H is a finite dimensional $\mathbb{R}$-vector space;
 (ii) W is a finite increasing filtration on H called the *weight filtration*;
(iii) F is a finite decreasing filtration on $H_{\mathbb{C}}$ called the *Hodge filtration*, such that $(\operatorname{Gr}_k^W H, F)$ is a Hodge structure of weight k for all k.

More explicitly, we look at the graded pieces

$$\operatorname{Gr}_k^W H = W_k H / W_{k-1} H$$

with respect to the weight filtration and ask that the filtration induced by F on $(\operatorname{Gr}_k^W H)_{\mathbb{C}}$ (denoted again by F) is a Hodge filtration as in (C13). This induced Hodge filtration is explicitly given by

$$F^p(\operatorname{Gr}_k^W H)_{\mathbb{C}} = F^p H_{\mathbb{C}} \cap W_k H_{\mathbb{C}} + W_{k-1} H_{\mathbb{C}} / W_{k-1} H_{\mathbb{C}}.$$

When (H, W, F) is a MHS (usually denoted simply by H) we can define the associated *mixed Hodge numbers* $h^{p,q}(H)$ by the formula

(C19) $$h^{p,q}(H) = \dim \operatorname{Gr}_F^p \operatorname{Gr}_{p+q}^W H_{\mathbb{C}}.$$

(C20) **Exercise.** Let (H, W, F) be a MHS and let $k \in \mathbb{Z}$ be any integer. Show that the following data $(H(k), W, F)$ where $H(k) = H$, $W_m H(k) = W_{m+2k} H$, $F^p H(k)_{\mathbb{C}} = F^{p+k} H_{\mathbb{C}}$ for all $m, p \in \mathbb{Z}$ again define a MHS. Show that

$$h^{p,q}(H(k)) = h^{p+k, q+k}(H).$$

(C21) **Definition.** (i) Let H and H' be two MHS and let $\varphi: H \to H'$ be an $\mathbb{R}$-linear map. Then φ is called a MHS *morphism* if φ is compatible with both filtrations, i.e.,

$$\varphi(W_m H) \subset W_m H' \qquad \text{for all } m,$$

$$\varphi(F^p H_{\mathbb{C}}) \subset F^p H'_{\mathbb{C}} \qquad \text{for all } p.$$

(ii) An $\mathbb{R}$-linear map $\varphi: H \to H'$ is called a MHS *morphism of type* (k, k) for some $k \in \mathbb{Z}$ if the induced map $\varphi: H \to H'(k)$ is a MHS morphism Equivalently, we can ask that the map $\varphi: H(-k) \to H'$ is a MHS morphism.

(C22) **Proposition** (Deligne [De1]). *If φ is a MHS morphism, then φ is strictly compatible with both filtrations W and F.*

The following simple example shows that MHS do arise quite naturally on the cohomology of algebraic varieties.

(C23) **Example.** Let X_1 and X_2 be two smooth projective varieties such that $X_1 \cap X_2$ is again smooth and let $X = X_1 \cup X_2$ be their union (e.g., take X_1 and X_2 to be smooth curves in $\mathbb{P}^2$). The Mayer–Vietoris sequence in cohomology with $\mathbb{R}$-coefficients (which are not written explicitly in order to simplify the notation) associated with the closed covering $X = X_1 \cup X_2$ looks like

$$\to H^{k-1}(X_1) \oplus H^{k-1}(X_2) \xrightarrow{\alpha} H^{k-1}(X_1 \cap X_2) \xrightarrow{\delta} H^k(X) \xrightarrow{\beta} H^k(X_1) \oplus H^k(X_2) \to.$$

The morphisms α and β here are essentially induced by inclusions and as a result they are morphisms of Hodge structures. Using (C16(i)) it follows that:

(i) coker α is a Hodge structure of weight $k - 1$;
(ii) im β is a Hodge structure of weight k.

The exact sequence

$$0 \to \text{coker } \alpha \to H^k(X) \to \text{im } \beta \to 0$$

shows that the quotients $\mathrm{Gr}_s^W H^k(X)$ have a natural Hodge structure of weight s if we set

$$0 = W_{k-2} H^k(X) \subset W_{k-1} H^k(X) = \text{coker } \alpha \subset W_k H^k(X) = H^k(X).$$

The fundamental fact about MHS is that the cohomology of any algebraic variety has in a canonical way such a structure. More precisely, we have the following result due to Deligne [De1].

(C24) Theorem. *There is a functorial* MHS *on $H^*(X; \mathbb{R})$ for any algebraic variety X such that the following properties hold for all $m \geq 0$:*

(i) *the weight filtration W on $H^m(X; \mathbb{R})$ satisfies*

$$0 = W_{-1} \subset W_0 \subset \cdots \subset W_{2m} = H^m(X; \mathbb{R});$$

for $m \geq n = \dim X$, we also have $W_{2n} = \cdots = W_{2m}$;

(ii) *the Hodge filtration F on $H^m(X; \mathbb{C})$ satisfies $H^m(X; \mathbb{C}) = F^0 \supset \cdots \supset F^{m+1} = 0$. For $m \geq n = \dim X$, we also have $F^{n+1} = 0$;*

(iii) *if X is a smooth variety, then $W_{m-1} = 0$ (i.e., all weights on $H^m(X; \mathbb{R})$ are $\geq m$) and $W_m = j^* H^m(\bar{X})$ for any compactifications*

$$j: X \hookrightarrow \bar{X};$$

(iv) *If X is a projective variety, then $W_m = H^m(X)$ (i.e., all weights on $H^m(X; \mathbb{R})$ are $\leq m$) and $W_{m-1} = \ker p^*$ for any proper map*

$$p: \tilde{X} \to X \quad \text{with } \tilde{X} \text{ smooth.}$$

(C25) Remarks. (i) The word "functorial" above means that any morphism $f: X \to X'$ of algebraic varieties induces a MHS morphism

$$f^*: H^m(X'; \mathbb{R}) \to H^m(X; \mathbb{R}) \qquad \text{for any } m.$$

(ii) Using (iii) and (iv) in (C24), it follows that $H^m(X; \mathbb{R})$ has a pure Hodge structure when X is smooth *and* projective. This structure coincides precisely with the Hodge structure coming from (C9).

(iii) The statement (C24(iii)) holds for the larger class of varieties X which are $\mathbb{Q}$-homology manifolds. Using (B18) this covers the class of V-manifolds, in particular, the weighted projective spaces $\mathbb{P}(\mathbf{w})$ and the quasismooth weighted complete intersections. Since these latter varieties are also projective, it follow as in (ii) that their cohomology even has a pure Hodge structure. Using (C16(i)), it follows that the primitive cohomology $H_0^m(X; \mathbb{R})$ has a natural MHS too. The corresponding mixed Hodge numbers $h^{p,q}(H_0^m(X; \mathbb{R}))$ are simply denoted by $h_0^{p,q}(X)$. This notation has already been used in (B24). Note also that for a projective $\mathbb{Q}$-homology manifold X all the results in (C12) hold true.

(C26) Example (Steenbrink [S2]). Let f be a weighted homogeneous polynomial of type $(w_0, \ldots, w_n; N)$ having an isolated singularity at the origin of $\mathbb{C}^{n+1}$. Let $F: f - 1 = 0$ be the corresponding affine Milnor fiber. From (C24) it follows that the middle cohomology group $H^n(F; \mathbb{R})$ has a MHS, such that all the weights are $\geq n$. More precisely, we have

$$0 = W_{n-1} \subset W_n \subset W_{n+1} = H^n(F; \mathbb{R}),$$

and the corresponding mixed Hodge numbers are given by the following formulas:

$$h^{p,n-p}(F) = \sum_{i=1,N-1} \dim M(f)_{pN-w+i},$$

$$h^{p,n+1-p}(F) = \dim M(f)_{pN-w},$$

where $M(f)$ is the graded Milnor algebra

$$\frac{\mathbb{C}[x_0,\ldots,x_n]}{(\partial f/\partial x_0,\ldots,\partial f/\partial x_n)}$$

as in (B34) and $w = w_0 + \cdots + w_n$, see [S2]. Assume now that n is even and let (μ_-,μ_0,μ_+) be the signature of the Milnor lattice. Then we have

$$\mu_0 = \sum_p h^{p,n+1-p}(F),$$

$$\mu_- = \sum_{p\,\text{even}} h^{p,n-p}(F),$$

$$\mu_+ = \sum_{p\,\text{odd}} h^{p,n-p}(F).$$

Note that the monodromy homeomorphism $h: F \to F$ is in fact an algebraic morphism by (3.1.19) and hence $h^*: H^n(F; \mathbb{R}) \to H^n(F; \mathbb{R})$ is a MHS morphism.

(C27) **Remark.** There are some other classes of cohomology groups of objects familiar in Algebraic Geometry (or even in Analytic Geometry) which carry natural MHS, see, for instance, [AGV], [De1], [Df7], [Na], [S1], [S3].

We mention here explicitly the following situations:

(i) The *cohomology of the Milnor fiber* of a hypersurface singularity $(X, 0)$. It is important to point out that the monodromy operator h^* is no longer a MHS morphism in general as it is in the special case (C26). However, the semisimple part h_a^* of the monodromy operator h^* is a MHS morphism and using it we can define the *spectrum* of the singularity $(X, 0)$. This spectrum has very interesting semicontinuity properties, see [S4] and [AGV2], the latter for many interesting applications also.

(ii) *Relative cohomology* groups $H^*(X, Y; \mathbb{R})$ where (X, Y) is a pair of algebraic varieties. These MHS are functorial with respect to algebraic maps of pairs. A special case of this situation is the *cohomology with supports*

$$H_Y^*(X; \mathbb{R}) = H^*(X, X\setminus Y; \mathbb{R}),$$

where Y is a closed subvariety in X.

(iii) *Local cohomology* of analytic spaces $H_x^*(X; \mathbb{R})$ where X is a complex analytic space and $x \in X$ is a point. It is clear that these cohomology groups depend only on the singularity (X, x). The corresponding MHS are functorial with respect to analytic map germs $f: (X, x) \to (X', x')$.

Assume now that (X, x) is an isolated singularity and let $K = L(X, x)$ be the corresponding link. The isomorphisms

$$H_x^m(X) = H^m(X, X\setminus\{x\}) = H^{m-1}(X\setminus\{x\}) = H^{m-1}(K)$$

show that it is natural to consider also MHS on the cohomology of *links* of isolated singularities. These MHS have some useful properties described in the following result.

(C28) **Proposition.** *Let K be the link of an n-dimensional isolated singularity (X, x). Then the natural MHS on $H^m(K)$ has all the weights $>m$ (resp. $\leq m$) for $m \geq n$ (resp. for $m < n$).*

(C29) **Example** (Links of Normal Surface Singularities). Let (X, x) be a normal surface singularity and let $(\tilde{X}, D) \to (X, x)$ be a very good resolution, as in Chapter 2, §3. Let K be the link corresponding to (X, x). Then the MHS on $H^2(K; \mathbb{R})$ has all the weights >2 according to (C28). It can be shown, as in [Df7], that the weight filtration here looks like

$$0 = W_2 \subset W_3 \subset W_4 = H^2(K; \mathbb{R}).$$

Moreover, we have

$$\dim W_3 = 2h^{1,2} = 2\sum g(D_i),$$

where D_i are the irreducible components of the exceptional divisor D, and $\dim(W_4/W_3) = h^{2,2}$ is equal to the number of cycles in the dual graph of the resolution. Hence these two numbers are independent of the resolution $(\tilde{X}, D)$ chosen for the singularity (X, x), see also (2.3.2).

To have a concrete example, consider a $T_{p,q,r}$-singularity

$$X: xyz + x^p + y^q + z^r = 0 \quad \text{with} \quad \frac{1}{p} + \frac{1}{q} + \frac{1}{r} < 1.$$

Using the description of the very good resolution given in (2.4.5), it follows that the MHS on $H^2(K; \mathbb{R})$ has the following mixed Hodge numbers:

$$h^{1,2} = h^{2,1} = 0, \qquad h^{2,2} = 1.$$

(C30) **Remark.** Most of the morphisms occurring in the usual exact sequences are MHS morphisms in an obvious way. Here are two such examples:

 (i) In the exact sequence

$$\xrightarrow{\delta} H_Y^m(X) \to H^m(X) \to H^m(X\setminus Y) \xrightarrow{\delta} H_Y^{m+1}(X) \to$$

all the morphisms (including δ) are MHS morphisms.

(ii) Consider the Gysin sequence from (2.2.14)

$$\to H^m(M) \xrightarrow{j^*} H^m(M\setminus D) \xrightarrow{R} H^{m-1}(D) \xrightarrow{\delta} H^{m+1}(M) \to,$$

where M is a smooth algebraic variety and D is a smooth hypersurface. Then the residue morphism R is a MHS morphism of type $(-1, -1)$, while the connecting homomorphism δ (which is, in fact, Poincaré dual to the morphism $H^*(M) \to H^*(D)$ induced by the inclusion $D \subset M$) is a MHS morphism of type $(1, 1)$.

In case we prefer to work only with MHS morphisms, we can write the above exact sequence in the following form (recall (C20)):

$$H^m(M) \xrightarrow{j^*} H^m(M \backslash D) \xrightarrow{R} H^{m-1}(D)(-1) \xrightarrow{\delta} H^{m+1}(M) \to .$$

(C31) **Remark.** The MHS on the cohomology of algebraic varieties behave in a nice way with respect to Poincaré, Alexander, or Lefschetz duality. For this we refer to Fujiki [Fj] and mention here only the following result.

Let V be a weighted hypersurface in a weighted projective space $\mathbb{P}(\mathbf{w})$, with $\dim \mathbb{P}(\mathbf{w}) = n$. Then we have the following relation between the mixed Hodge numbers for $U = \mathbb{P}(\mathbf{w}) \backslash V$ and the mixed Hodge numbers for the primitive cohomology of V (recall also (B25(iii))

$$h^{p,q}(H^m(U)) = h^{n-p,n-q}(H_0^{2n-m-1}(V)).$$

In the final part of this appendix we discuss some nice topological applications of the existence of MHS on the cohomology of algebraic varieties and links. The first result is so crystal clear that we just state it and refer for a proof to Durfee [Df8].

(C32) **Proposition.** *Let X and Y be smooth projective varieties and suppose that*

$$X = X_1 \cup \cdots \cup X_n, \qquad Y = Y_1 \cup \cdots \cup Y_n,$$

are disjoint decompositions of X and Y into quasiprojective varieties X_i, Y_j. Assume, moreover, that the variety X_i is algebraically isomorphic to the variety Y_i for all $i = 1, \ldots, n$. Then

$$h^{p,q}(X) = h^{p,q}(Y) \qquad \text{for all pairs } (p, q).$$

In particular, the Betti numbers $b_k(X)$ and $b_k(Y)$ coincide for all k.

(C33) **Remark.** It might be interesting to compare (C32) with our example (5.4.29).

Now let $(X, 0)$ be an ICIS at the origin of $\mathbb{C}^N$ with $\dim(X, 0) = n + 1$. Let $f: (X, 0) \to (\mathbb{C}, 0)$ be an analytic function germ such that $(X_0, 0) = (f^{-1}(0), 0)$ is again an ICIS. Let $K = K(X, 0)$ and $K_0 = K(X_0, 0)$ be the corresponding links. The morphism

$$j^m: H^m(K; \mathbb{R}) \to H^m(K_0; \mathbb{R})$$

induced by the inclusion $j: K_0 \to K$ is obviously trivial for $m \neq n$. Indeed, in this range, at least one of the groups $H^m(K; \mathbb{R})$ or $H^m(K_0; \mathbb{R})$ is trivial by

(3.2.12). On the other hand, the following statement is not at all obvious from a purely topological point of view.

(C34) **Proposition.** *The morphism j^n is trivial.*

Proof. Using (C28) it follows that the MHS on $H^n(K_0)$ has weights $\leq n$, while the MHS on $H^n(K)$ has weights $>n$. Since j^n is a MHS morphism it should preserve the weight filtrations. But this is possible only when $j^n = 0$.   □

(C35) **Remark.** R. Randell has introduced in [Ra] the concept of a d-negligible plane curve singularity as follows. Let $C: g = 0$ be an isolated plane curve singularity and let $d \geq 2$ be an integer. Consider the surface singularity (the so-called "d-suspension of $(C, 0)$") defined by

$$S: g(x, y) + t^d = 0 \quad \text{in } \mathbb{C}^3.$$

Let K_C and K_S be the corresponding links and let $j: K_C \to K_S$ be the inclusion $(x, y) \mapsto (x, y, 0)$. The singularity $(C, 0)$ is called d-negligible if the morphism

$$j_*: H_1(K_C; \mathbb{Z}) \to H_1(K_S; \mathbb{Z})$$

is surjective.

(C36) **Corollary.** *If the singularity $(C, 0)$ is d-negligible, then $H^1(K_S; \mathbb{R}) = 0$.*

Proof. If the above morphism j_* is surjective, it follows by duality that the morphism

$$j^*: H^1(K_S; \mathbb{R}) \to H^1(K_C; \mathbb{R})$$

is injective. According to (C34) this can happen only if $H^1(K_S; \mathbb{R}) = 0$.   □

References

[AC1]   A'Campo, N.: Sur la monodromie des singularités isolées d'hypersurfaces complexes, *Invent. Math.* **20** (1973), 147–169.

[AC2]   A'Campo, N.: Le nombre de Lefschetz d'une monodromie, *Indag. Math.* **35** (1973), 113–118.

[AC3]   A'Campo, N.: La fonction zeta d'une monodromie, *Comm. Math. Helv.* **50** (1975), 233–248.

[AM]   Akbulut, S., McCarthy, J.D.: *Casson's Invariant for Oriented Homology 3-Spheres, an Exposition*, Mathematical Notes 36, Princeton University Press, Princeton, 1990.

[Ar]   Armstrong, M.A.: Calculating the fundamental group of an orbit space, *Proc. Amer. Math. Soc.* **84** (1982), 267–271.

[AGV1]   Arnold, V.I., Gusein-Zade, S.M., Varchenko, A.N.: *Singularities of Differ-*
and   *entiable Maps*, Vols. 1 and 2, Monographs in Mathematics 82 and 83,
[AGV2]   Birkhäuser, Boston, 1985 and 1988.

[AB]   Artal-Bartolo, E.: Sur le premier nombre de Betti de la fibre de Milnor du cone sur une courbe projective plane et son rapport avec la position des points singuliers, preprint 1990, University of Wisconsin, Madison.

[ABG]   Atiyah, M.F., Bott, R., Garding, L.: Lacunas for hyperbolic differential operators with constant coefficients II, *Acta Math.* **131** (1973), 145–206.

[BS]   Bănică, C., Stănăsilă, O.: *Méthodes Algébriques dans la Théorie Globale des Espaces Complexes*, Gauthier-Villars, Paris, 1977.

[BPV]   Barth, W., Peters, C., Van de Ven, A.: *Compact Complex Surfaces*, Erg. der Math., 3. Folge, Band 4, Springer-Verlag, Berlin, 1984.

[BD]   Barthel, G., Dimca, A.: On complex projective hypersurfaces which are homology $\mathbb{P}^n$'s (preprint 1989).

[BaK]   Barthel, G., Kaup, L.: *Topologie des Surfaces Complexes Compactes Singulières*, Sém. Math. Sup. 80, Les Presses de l'Université de Montreal, 1982.

[BO]   Bauer, S., Okonek, C.: The algebraic geometry of representation spaces associated to Seifert fibered homology 3-spheres, *Math. Ann.* **286** (1990), 45–76.

[Bi]   Birman, J.S.: *Braids, links, and Mapping Class Groups*, Annals of Mathematical Studies 82, Princeton, 1974.

[BT]   Bott, R., Tu, L.W.: *Differential Forms in Algebraic Topology*, Graduate Texts in Mathematics 82, Springer-Verlag, New York, 1982.

[Bd]   Bredon, G.E.: *Introduction to Compact Transformation Groups*, Academic Press, New York, 1972.

[BS1]   Briançon, J., Speder, J.P.: La trivialité topologique n'implique pas les condition de Whitney, *C. R. Acad. Sci. Paris, Sér. A* **280** (1975), 365–367.

[BS2]   Briançon, J., Speder, J.P.: Les conditions de Whitney impliquent μ^* constant, *Ann. Inst. Fourier (Grenoble)*, **26** (1976), 153–163.

[B1]   Brieskorn, E.: Beispiele zur Differentialtopologie von Singularitäten, *Invent. Math.* **2** (1966), 1–14.

[B2]   Brieskorn, E.: *Singular Elements of Semi-Simple Algebraic Groups*, Actes Congrés Int. Math. Nice 1970, Vol. 2, 279–284.

[B3]   Brieskorn, E.: Die Monodromie der isolierten Singularitäten von Hyperflächen, *Manuscripta Math.* **2** (1970), 103–161.

[B4]   Brieskorn, E.: Sur les groupes de tresses (d'apres V.I. Arnold), *Sém. Bourbaki 1971/72, Exp. 401*, Lecture Notes in Mathematics 317, Springer-Verlag, Berlin, 1973.

[B5]   Brieskorn, E.: The unfolding of exceptional singularities, *Nova Acta Leopoldina (NF 52)*, **240** (1981), 65–93.

[BK]   Brieskorn, E., Knörrer, H.: *Plane Algebraic Curves*, Birkhäuser, Boston, 1986.

[Bt1]   Broughton, S.A.: On the topology of polynomial hypersurfaces, *Proc. Symp. Pure Math. 40, Part I (Arcata Singularities Conference)*, American Mathematical Society, 1983, pp. 167–178.

[Bt2]   Broughton, S.A.: Milnor numbers and the topology of polynomial hypersurfaces, *Invent. Math.* **92** (1988), 217–241.

[Bc]   Bruce, J.W.: Stratifying jet spaces—some special cases, *Math. Proc. Cambridge Philos. Soc.* **85** (1979) 439–444.

[BG]   Bruce, J.W., Giblin, P.J.: A stratification of the space of plane quartic curves, *Proc. London Math. Soc.* **42** (1981), 270–298.

[BW]   Bruce, J.W., Wall, C.T.C.: On the classification of cubic surfaces, *J. London Math. Soc.* **19** (1979), 245–256.

[BV]   Burghelea, D., Verona, A.: Local homological properties of analytic sets, *Manuscripta Math.* **7** (1972), 55–66.

[Ch1]   Cheniot, D.: Le théorème de Van Kampen sur le groupe fondamental du complémentaire d'une courbe algébrique plane, *Sém. F. Norguet 1971/72*, Lecture Notes in Mathematics 409, Springer-Verlag, Berlin, 1974.

[Ch2]   Cheniot, D.: Une démonstration du théorème de Zariski sur les sections hyperplanes d'une hypersurface projective et du théorème du Van Kampen sur le group fondamental du complémentaire d'une courbe projective plane, *Compositio Math.* **27** (1973), 141–158.

[Cd]   Choudary, A.D.R.: On the resultant hypersurface, *Pacific J. Math.* **142** (1990), 259–263.

[Cl]   Clemens, C.H.: Double solids, *Adv. in Math.* **47** (1983), 107–230.

[CM]   Coxeter, H.S.M., Moser, W.O.J.: *Generators and Relations for Discrete Groups*, 2nd edn., Springer-Verlag, Berlin, 1965.

[CF]   Crowell, R., Fox, R.: *Knot Theory*, Ginn, Boston, 1963.

[De1]   Deligne, P.: Theorie de Hodge, II and III, *Publ. Math. IHES* **40** (1971), 5–58 and **44** (1974), 5–77.

[De2]   Deligne, P.: Le groupe fondamental du complement d'une courbe plane n'ayant que des points double ordinaires est abelian, *Sém. Bourbaki, 1979/*

80, Lecture Notes in Mathematics 842, Springer-Verlag, Berlin 1981, pp. 1–10.

[DD]   Deligne, P., Dimca, A.: Filtrations de Hodge et par l'ordre du pôle pour les hypersurfaces singulières, *Ann. Sci. École Norm. Sup.* **23** (1990), 645–656.

[tDP]   tom Dieck, T., Petrie, T.: Contractible affine surfaces of Kodaira dimension one, *Japan J. Math.* **16** (1990), 147–169.

[D1]   Dimca, A.: Monodromy and Betti numbers of weighted complete intersections, *Topology* **24** (1985), 369–374.

[D2]   Dimca, A.: On the homology and cohomology of complete intersections with isolated singularities, *Compositio Math.* **58** (1986), 321–339.

[D3]   Dimca, A.: Singularities and coverings of weighted complete intersections, *J. Reine Angew. Math.* **366** (1986), 184–193.

[D4]   Dimca, A.: *Topics on Real and Complex Singularities*, Vieweg Advanced Lectures in Mathematics, Vieweg, Braunschweig/Wiesbaden 1987.

[D5]   Dimca, A.: On the Milnor fibrations of weighted homogeneous polynomials, *Compositio Math.* **76** (1990), 19–47.

[D6]   Dimca, A.: Betti numbers of hypersurfaces and defects of linear systems, *Duke Math. J.* **60** (1990), 285–298.

[D7]   Dimca, A.: On the connectivity of affine hypersurfaces, *Topology* **29** (1990), 511–514.

[D8]   Dimca, A.: Differential forms and hypersurface singularities, in: *Singularity Theory and its Applications, Warwick 1989, Proceedings*, Lecture Notes in Mathematics 1462, Springer-Verlag, Berlin, 1991, pp. 122–153.

[D9]   Dimca, A.: On the de Rham cohomology of a hypersurface complement, *Amer. J. Math.* **113** (1991), 763–771.

[DDv]   Dimca, A., Dimiev, S.: On analytic coverings of weighted projective spaces, *Bull. London Math. Soc.* **17** (1985), 234–238.

[Do1]   Dold, A.: Partitions of unity in the theory of fibrations, *Ann. of Math.* **78** (1963), 223–255.

[Do2]   Dold, A.: *Lectures on Algebraic Topology*, 2nd ed., Springer-Verlag, New York, 1980.

[Dg1]   Dolgachev, I.: Weighted projective varieties, in: *Group Actions and Vector Fields, Proceedings 1981*, Lecture Notes in Mathematics 956, Springer-Verlag, Berlin, 1982, pp. 34–71.

[Dg2]   Dolgachev, I.: Integral quadratic forms: applications to algebraic geometry (after V. Nikulin). *Sem. Bourbaki 1982/83, Exp. 611, Astérisque 105–106*, Soc. Math. de France, 1983.

[DgLi]   Dolgachev, I., Libgober, A.: On the fundamental group of the complement to a discriminant variety, in: *Algebraic Geometry, Chicago 1980*, Lecture Notes in Mathematics 862, Springer-Verlag, New York, 1981, pp. 1–25.

[DS]   van Doorn, M.G.M., Steenbrink, J.H.M.: A supplement to the monodromy theorem, *Abh. Math. Sem. Univ. Hamburg* **59** (1989), 225–233.

[Df1]   Durfee, A.H.: Foliations of odd-dimensional spheres, *Ann. of Math.* **96** (1972), 407–411.

[Df2]   Durfee, A.H.: The characteristic polynomial of the monodromy, *Pacific J. Math.* **59** (1975), 21–26.

[Df3]   Durfee, A.H.: The signature of smoothings of complex surface singularities, *Math. Ann.* **232** (1978), 85–98.

[Df4]   Durfee, A.H.: Fifteen characterizations of rational double points and simple critical points, *Enseign. Math.* **25** (1979), 131–163.

[Df5]   Durfee, A.H.: A naive guide to mixed Hodge theory, *Proc. Symp. Pure Math. 40, Part I (Arcata Singularities Conferences)*, American Mathematical Society, 1983, pp. 313–320.

[Df6]   Durfee, A.H.: Neighborhoods of algebraic sets. *Trans. Amer. Math. Soc.* **276** (1983), 517–530.

[Df7]   Durfee, A.H.: Mixed Hodge structures on punctured neighborhoods, *Duke Math. J.* **50** (1983), 1017–1040.

[Df8]   Durfee, A.H.: Algebraic varieties which are disjoint union of subvarieties, in: *Geometry and Topology: Manifolds, Varieties, and Knots*, Marcel Dekker, New York, 1987, pp. 99–102.

[DfK]   Durfee, A.H., Kauffman, L.: Periodicity of branched cyclic covers, *Math. Ann.* **218** (1975), 157–174.

[E1]   Ebeling, W.: Quadratische Formen und Monodromiegruppen von Singularitäten, *Math. Ann.* **255** (1981), 463–498.

[E2]   Ebeling, W.: Milnor lattices and geometric bases of some special singularities, in: *Noeuds, Tresses et Singularités*, Monographie No. 31 de l'Enseignement Math., Université de Genève, 1983, pp. 129–146; *Enseign. Math.* **29** (1983), 263–280.

[E3]   Ebeling, W.: *The Monodromy Groups of Isolated Singularities of Complete Intersections*, Lecture Notes in Mathematics 1293, Springer-Verlag, Berlin, 1987.

[E4]   Ebeling, W.: An example of two homeomorphic, nondiffeomorphic complete intersection surfaces, *Invent. Math* **99** (1990), 651–654.

[EO]   Ebeling, W., Okonek, C.: Donaldson invariants, monodromy groups and singularities, *Internat. J. Math.* **1** (1990), 233–250.

[EW]   Ebeling, W., Wall, C.T.C.: Kodaira singularities and an extension of Arnold's strange duality, *Compositio Math.* **56** (1985) 3–77.

[Eh]   Ehresmann, C.: Sur les espaces fibrés différentiables, *C. R. Acad. Sci. Paris* **224** (1947), 1611–1612.

[Em]   Ephraim, R.: C^1-preservation of multiplicity, *Duke Math. J.* **43** (1976), 797–803.

[Es]   Esnault, H.: Fibre de Milnor d'un cone sur une courbe plane singulière, *Invent. Math.* **68** (1982) 477–496.

[FS]   Fintushel, R., Stern, R.: Instanton homology of Seifert fibered homology 3-spheres, *Proc. London Math. Soc.* **61** (1990), 109–137.

[Fr]   Fried, D.: Monodromy and dynamical systems, *Topology* **25** (1986), 443–453.

[Fj]   Fujiki, A.: Duality of mixed Hodge structures of algebraic varieties, *Publ. Res. Inst. Math. Sci. Kyoto Univ.* **16** (1980) 635–667.

[F]   Fulton, W.: On the fundamental group of the complement of a node curve, *Ann. of Math.* **111** (1980), 407–409.

[GK]   Gabrielov, A.M., Kushnirenko, A.G.: Description of deformations with constant Milnor number for homogeneous functions, *Functional Anal. Appl.* **9** (1975), 329–331.

[GW]   van Geemen, B., Werner, J.: Nodal quintics in $\mathbb{P}^4$ in: *Arithmetic of Complex Manifolds, Proceedings*, 1988, Lecture Notes in Mathematics 1399, Springer-Verlag, Berlin, 1988, pp. 48–59.

[GWPL]   Gibson, C.G., Wirthmüller, K., du Plessis, A.A., Looijenga, E.J.N.: *Topological Stability of Smooth Mappings*, Lecture Notes in Mathematics 552, Springer-Verlag, Berlin, 1976.

[GG]   Golubitsky, M., Guillemin, V.: *Stable Mappings and Their Singularities*, Graduate Text in Mathematics 14, Springer-Verlag, Berlin, 1973.

[GM]   Goresky, M., MacPherson, R.: *Stratified Morse Theory*, Ergebnisse der Math., 3. Folge, Band 14, Springer-Verlag, Berlin, 1988.

[Gn]   Greenberg, M.: *Lectures on Algebraic Topology*, W.A. Benjamin, New York, 1966.

[Gl]   Greuel, G.-M.: Der Gauss–Manin–Zusammenhang isolierter Singularitäten von vollständigen Duchschnitten, *Math. Ann.* **214** (1975), 235–266.

[G]   Griffiths, Ph.: On the periods of certain rational integrals, I, II, *Ann. of Math.* **90** (1969), 460–541.

[GH]   Griffiths, Ph., Harris, J.: *Principles of Algebraic Geometry*, Wiley, New York, 1978.

[Gr]   Grothendieck, A.: On the de Rham cohomology of algebraic varieties, *Publ. Math. IHES* **29** (1966), 351–358.

[HL]   Hà Huy Vui, Lê Dũng Tráng: Sur la topologie des polynômes complexes, *Acta Math. Vietnam* **9** (1984), 21–32.

[H1]   Hamm, H.: Lokale topologische Eigenschaften komplexer Räume, *Math. Ann.* **191** (1971), 235–252.

[H2]   Hamm, H.: Genus χ_y of quasihomogeneous complete intersections, *Functional Anal. Appl.* **11** (1978), 78–79.

[H3]   Hamm, H.: Zum Homotopietyp Steinscher Räume, *J. Reine Angew. Math.* **338** (1983), 121–135.

[H4]   Hamm, H.: Lefschetz theorems for singular varieties, *Proc. Symp. Pure Math. 40, Part I (Arcata Singularities Conference)*, American Mathematical Society, 1983, pp. 547–557.

[HLê1]   Hamm, H., Lê Dũng Tráng: Un théorème de Zariski du type de Lefschetz, *Ann. Sci. École Norm. Sup.* **6** (1973), 317–366.

[HLê2]   Hamm, H., Lê Dũng Tráng: Rectified homotopical depth and Grothendieck conjectures, in: *The Grothendieck Festschrift II*, Progress in Mathematics 87, Birkhäuser, Boston, 1990, pp. 311–351.

[Ha]   Hardt, R.: Topological properties of subanalytic sets, *Trans. Amer. Math. Soc.* **211** (1975), 57–70.

[Hs]   Harris, J.: On the Severi problem, *Invent. Math.* **84** (1986), 445–461.

[Hn]   Hartshorne, R.: *Algebraic Geometry*, Graduate Texts in Mathematics 52, Springer-Verlag, Berlin, 1977.

[Hi1]   Hironaka, H.: Subanalytic sets, in: *Number Theory, Algebraic Geometry and Commutative Algebra*, volume in honor of A. Akizuki, Kinokunyo, Tokyo, 1973, pp. 453–493.

[Hi2]   Hironaka, H.: Triangulations of algebraic sets, *Proc. Symp. Pure Math. 29 (Arcata Algebraic Geometry Conference)*, American Mathematical Society, 1974, pp. 165–185.

[Hr]   Hirsch, M.W.: *Differential Topology*, Graduate Texts in Mathematics 33, Springer-Verlag, Berlin, 1976.

[Hz1]   Hirzebruch, F.: *Topological Methods in Algebraic Geometry*, 3rd ed., Springer-Verlag, Berlin, 1966.

[Hz2]   Hirzebruch, F.: Singularities and exotic spheres, *Sém. Bourbaki 1966/67*, *Exp. 314*.

[Hz3]   Hirzebruch, F.: Some examples of threefolds with trivial canonical bundle, in: "*Collected Papers II*", Springer-Verlag, New York, 1987, pp. 757–770.

[Hu]   Husemoller, D.: *Fibre Bundles*, McGraw-Hill, New York, 1966.

[Ii]   Iitaka, S.: *Algebraic Geometry*, Graduate Texts in Mathematics 76, Springer-Verlag, New York, 1982.

[vK]   van Kampen, E.R.: On the fundamental group of an algebraic curve, *Amer. J. Math.* **55** (1933), 255–260.

[Ks]   Karchyauskas, K.K.: A generalized Lefschetz theorem, *Functional Anal. Appl.* **11** (1977), 312–313.

[Kp]   Karpishpan, Y.: Pole order filtration on the cohomology of algebraic links, *Compositio Math.* **78** (1991), 213–226.

[Ka]   Kato, M.: Topology of *k*-regular spaces and algebraic sets, in: *Manifolds, Tokyo 1973, Proceedings*, University of Tokyo Press, Tokyo, 1975, pp. 153–159.

[KM]   Kato, M., Matsumoto, Y.: On the connectivity of the Milnor fiber of a holomorphic function at a critical point, in: *Manifolds, Tokyo 1973, Proceedings*, pp. 131–136, University of Tokyo Press, Tokyo, 1975.

[KK]   Kaup, B., Kaup, L.: *Holomorphic Functions of Several Variables*, de Gruyter Studies in Mathematics 3, 1983.

[Kws]   Kawasaki, T.: Cohomology of twisted projective spaces and lens complexes, *Math. Ann.* **206** (1973), 243–248.

[KeMi]   Kervaire, M., Milnor, J.: Groups of homotopy spheres I, *Ann. of Math.* **77** (1963), 504–537.

[Kw]   Kirwan, F.: An introduction to intersection homology theory, Pitman Research Notes in Mathematics, Longman, London, 1988.

[Kl]   Kleiman, S.L.: The enumerative theory of singularities, in: *Real and Complex Singularities (Oslo 1976)*, Sijthoff and Noordhoff, Amsterdam 1977, pp. 297–396.

[Ku]   Kushnirenko, A.G.: Polyèdres de Newton et nombres de Milnor, *Invent. Math.* **32** (1976), 1–31.

[L1]   Lamotke, K.: Die Homologie isolierter Singularitäten, *Math. Z.* **143** (1975), 27–44.

[L2]   Lamotke, K.: The topology of complex projective varieties after S. Lefschetz, *Topology* **20** (1981), 15–51.

[L3]   Lamotke, K.: *Regular Solids and Isolated Singularities*, Vieweg Advanced Lectures in Mathematics, Vieweg, Braunschweig/Wiesbaden 1986.

[La]   Lang, S.: *Algebra*, Addison-Wesley, Boston, 1965.

[Lf1]   Laufer, H.: *Normal Two-Dimensional singularities*, Annals of Mathematical Studies 71, Princeton University Press, Princeton, 1971.

[Lw]   Lawson, H.B.: Codimension-one foliations of spheres, *Ann. of Math.* **94** (1971), 494–503.

[Lê1]   Lê Dũng Tráng: Sur les noeuds algébriques, *Compositio Math.* **25** (1972), 281–321.

[Lê2]   Lê Dũng Tráng: Topologie des singularités des hypersurfaces complexes, *Astérisque* **7/8** (Singularités à Cargèse) 1973, pp. 171–182.

[Lê3]   Lê Dũng Tráng: Some remarks on relative monodromy, in: *Real and Com-*

plex Singularities (Oslo 1976), Sijthoff and Noordhoff, Amsterdam, 1977, pp. 397–403.

[LêR]   Lê Dũng Tráng, Ramanujam, C.P.: The invariance of Milnor number implies the invariance of the topological type, *Amer. J. Math.* **98** (1976), 67–78.

[LêT]   Lê Dũng Tráng, Teissier, B.: Cycles évanescents et conditions de Whitney II, *Proc. Symp. Pure Math. 40, Part 2 (Arcata Singularities Conference)*, American Mathematical Society 1983, pp. 65–103.

[Lf2]   Lefschetz, S.: *Analysis Situs et la Géometrie Algébrique*, Gauthier-Villars, Paris, 1924.

[Li1]   Libgober, A.: Some properties of the signature of complete intersections, *Proc. Amer. Math. Soc.* **79** (1980), 373–375.

[Li2]   Libgober A.: Alexander polynomial of plane algebraic curves and cyclic multiple planes, *Duke Math. J.* **49** (1982), 833–851.

[Li3]   Libgober, A.: Alexander invariants of plane algebraic curves, *Proc. Symp. Pure Math. 40, Part 2 (Arcata Singularities Conference)*. American Mathematical Society, 1983, pp. 135–144.

[Li4]   Libgober, A.: Homotopy groups of the complements to singular hypersurfaces, *Bull. Amer. Math. Soc.* **13** (1985), 49–51.

[Li5]   Libgober, A.: Invariants of plane algebraic curves via representations of the braid groups, *Invent. Math.* **95** (1989), 25–30.

[Li6]   Libgober, A.: Fundamental groups of the complements to plane singular curves, *Proc. Symp. Pure Math. 46, Part 2 (Bowdoin Algebraic Geometry Conference)*, American Mathematical Society 1987, pp. 29–45.

[LiW1]   Libgober, A., Wood, J.W.: On the topological structure of even-dimensional complete intersections, *Trans. Amer. Math. Soc.* **267** (1981), 637–660.

[LiW2]   Libgober, A., Wood, J.W.: Differentiable structures on complete intersections I, *Topology* **21** (1982), 469–482.

[LiW3]   Libgober, A., Wood, J.W.: Uniqueness of the complex structure on Kähler manifolds of certain homotopy types, *J. Differential Geom.* **32** (1990), 139–154.

[LV]   Loeser, F., Vaquié, M.: Le polynome d'Alexander d'une courbe plane projective, *Topology* **29** (1990), 163–173.

[Lo]   Lojasiewicz S.: Ensembles semianalytiques, preprint IHES, 1972.

[Lg]   Looijenga, E.J.N.: *Isolated Singular Points on Complete Intersections*, London Mathematical Society Lecture Note Series 77, Cambridge University Press, Cambridge, 1984.

[LgWa]   Looijenga, E.J.N., Wahl, J.: Quadratic functions and smoothing surface singularities, *Topology* **25** (1986), 261–291.

[My]   Massey, D.B.: The Lê varieties, I, *Invent. Math.* **99** (1990), 357–376.

[Ma]   Mather, J.: *Notes on Topological Stability*, Harvard University, Cambridge, 1970.

[Ms]   Matsumura, H.: *Commutative Algebra*, W.A. Benjamin, New York, 1970.

[Mc]   McClearly, J.: *User's Guide to Spectral Sequences*, Publish or Perish, 1985.

[M1]   Milnor, J.: Construction of universal bundles II, *Ann. of Math.* **63** (1956), 430–436.

[M2]   Milnor, J.: On simply-connected 4-manifolds, *Int. Symp. Alg. Topology*, Mexico City 1958, pp. 122–128.

[M3]   Milnor, J.: *Morse Theory*, Annals of Mathematical Studies 51, Princeton University Press, Princeton, 1963.

[M4]   Milnor, J.: *Topology from the Differential Viewpoint*, The University Press of Virginia, Charlottesville, 1965.

[M5]   Milnor, J.: *Singular Points of Complex Hypersurfaces*, Annals of Mathematical Studies 61, Princeton University Press, Princeton, 1968.

[M6]   Milnor, J.: On the 3-dimensional Brieskorn manifolds $M(p, q, r)$, in: *Knots, Groups, and 3-Manifolds*—papers dedicated to the memory of R.H. Fox, Annals of Mathematical Studies 84, Princeton University Press, Princeton, 1975, pp. 175–225.

[MH]   Milnor, J., Husemöller, D.: *Symmetric Bilinear Forms*, Springer-Verlag, Berlin, 1973.

[MO]   Milnor, J., Orlik. P.: Isolated singularities defined by weighted homogeneous polynomials, *Topology* 9 (1970), 385–393.

[Mü]   Müller, G.: Reduktive Automorphismengruppen analytischer C-Algebren, *J. Reine Angew. Math.* 364 (1986), 26–34.

[Mu1]   Mumford, D.: The topology of normal singularities of an algebraic surface and a criterion for simplicity, *Publ. Math. IHES* 9, (1961), 5–22.

[Mu2]   Mumford, D.: *Algebraic Geometry I, Complex Projective Varieties*, Grundlehren der math. Wiss. 221, Springer-Verlag, Berlin, 1976.

[Ng]   Nagata, M.: Polynomial rings and affine spaces, *Reg. Conf. Series in Mathematics* 37, American Mathematical Society, 1978.

[Na]   Navarro Aznar, V.: Sur la théorie de Hodge–Deligne, *Invent. Math.* 90 (1987), 11–76.

[N1]   Némethi, A.: Théorie de Lefschetz pour les varietés algebriques affines. *C. R. Acad. Sci. Paris*, 303, Serie I, No. 12 (1986), 567–570.

[N2]   Némethi, A.: On the fundamental group of the complement of certain singular plane curves, *Math. Proc. Cambridge Philos. Soc.* 102 (1987), 453–457.

[N3]   Némethi, A.: Lefschetz theory for complex affine varieties, *Rev. Roumaine Math. Pures Appl.* 33 (1988), 233–250.

[NZ]   Némethi, A., Zaharia, A.: On the bifurcation set of a polynomial function and Newton boundary, *Publ. Res. Inst. Math. Sci. Kyoto Univ.* 26 (1990), 681–689.

[NW]   Neumann, W., Wahl, J.: Casson invariant of links of singularities, *Comment. Math. Helv.* 65 (1990), 58–78.

[Nk]   Nikulin, V.V.: Integral symmetric bilinear forms and some of their applications, *Izv. Akad. Nauk SSR* 43 (1979), 111–177.

[O1]   Oka, M.: On the homotopy types of hypersurfaces defined by weighted homogeneous polynomials, *Topology* 12 (1973), 19–32.

[O2]   Oka, M.: On the cohomology structure of projective varieties, in: *Manifolds, Tokyo 1973, Proceedings*, University of Tokyo Press, Tokyo, 1975, pp. 137–143.

[O3]   Oka, M.: The monodromy of a curve with ordinary double points, *Invent. Math.* 27 (1974), 157–164.

[O4]   Oka, M.: Some plane curves whose complements have non-abelian fundamental groups, *Math. Ann.* 218 (1975), 55–65.

[O5]   Oka, M.: On the fundamental group of the complement of certain plane curves, *J. Math. Soc. Japan* 30 (1978), 579–597.

[O6]   Oka, M.: Symmetric plane curves with nodes and cusps, preprint 1991.

[Om]   Olum, P.: Nonabelian cohomology and van Kampen's theorem, *Ann. of Math.* **68** (1958), 658–668.

[OR]   Orlik, P., Randell, R.: The monodromy of weighted homogeneous singularities, *Invent. Math.* **39** (1977), 199–211.

[OS]   Orlik, P., Solomon, L.: Singularities I, Hypersurfaces with an isolated singularity, *Adv. in Math.* **27** (1978), 256–272.

[OW]   Orlik, P., Wagreich, Ph.: Equivariant resolution of singularities with C*-action, in: *Proceedings of the Second Conference on Compact Transformation Groups II*, Lecture Notes in Mathematics 299, Springer-Verlag, Berlin, 1972, pp. 270–290.

[Pa]   Parusinski, A.: A generalization of the Milnor number, *Math. Ann.* **281** (1988), 247–254.

[Ph]   Pham, F.: Formules de Picard–Lefschetz généralisées et ramification des intégrales, *Bull. Soc. Math. France* **93** (1965), 333–367.

[P]   Prill, D.: Local classification of quotients of complex manifolds by discontinuous groups, *Duke Math. J.* **31** (1964), 375–386.

[Rm]   Ramanujam, C.P.: A topological characterization of the affine plane as an algebraic variety, *Ann. of Math.* **94** (1971), 69–88.

[Ra]   Randell, R.: On the fundamental group of the complement of a singular plane curve, *Quart. J. Math. Oxford* **31** (1980), 71–79.

[Rd]   Reid, M.: Young person's guide to canonical singularities, Proc. Amer. Math. Soc. Summer Institute Bowdoin 1985, *Proc. Sympos. Pure Math.* **46** (1987), Part I, pp. 345–414.

[Ro]   Roan, S.-S.: On Calabi–Yau orbifolds in weighted projective spaces, *Internat. J. Math.* **1** (1990), 211–232.

[R]   Rolfsen, D.: *Knots and Links*, Mathematical Lecture Series 7, Publish or Perish, 1976.

[Ru]   Russell, P.: Simple birational extensions of two-dimensional affine rational domains, *Compositio Math.* **33** (1976), 197–208.

[Sa]   Sakamoto, K.: Milnor fiberings and their characteristic maps, in: *Manifolds, Tokyo 1973, Proceedings*, University of Tokyo Press, Tokyo, 1975, pp. 145–150.

[Sy]   Sathaye, A.: On linear planes, *Proc. Amer. Math. Soc.* **56** (1976), 1–7.

[Sch]   Schoen, C.: Algebraic cycles on certain desingularized nodal hypersurfaces, *Math. Ann.* **270** (1985), 17–27.

[ST]   Sebastiani, M., Thom, R.: Un résultat sur la monodromie, *Invent. Math.* **13** (1971), 90–96.

[Se]   Serre, J.-P.: Arbres, amalgames, SL_2, *Astérisque* **46**, Soc. Math. France, 1977.

[Si1]   Siersma, D.: Classification and deformation of singularities, Ph.D. Thesis, Amsterdam, 1974.

[Si2]   Siersma, D.: Isolated line singularities, *Proc. Symp. Pure Math.* 40, Part 2 (*Arcata Singularities Conference*), American Mathematical Society, 1983, pp. 485–496.

[Si3]   Siersma, D.: Quasihomogeneous singularities with transversal type A_1, Contemporary Mathematics 90, American Mathematical Society, 1989, pp. 261–294.

[Si4]   Siersma, D.: Variation mappings on singularities with a 1-dimensional critical locus, *Topology* **30** (1991), 445–469.

[Sm]   Smale, S.: On the structure of manifolds, *Amer. J. Math.* **84** (1962), 387–399.

[Sp]   Spanier, E.H.: *Algebraic Topology*, McGraw-Hill, New York, 1966.

[S1]   Steenbrink, J.H.M.: Mixed Hodge structures on the vanishing cohomology, in: *Real and Complex Singularities (Oslo 1976)*, Sijthoff and Noordhoff, Amsterdam, 1977, pp. 525–563.

[S2]   Steenbrink, J.H.M.: Intersection form for quasi-homogeneous singularities, *Compositio Math.* **34** (1977), 211–223.

[S3]   Steenbrink, J.H.M.: Mixed Hodge structures associated with isolated singularities. *Proc. Symp. Pure Math. 40, Part II (Arcata Singularities Conference)*, American Mathematical Society, 1983, pp. 513–536.

[S4]   Steenbrink, J.H.M.: Semicontinuity of the singularity spectrum, *Invent. Math.* **79** (1985), 557–565.

[S5]   Steenbrink, J.H.M.: talk at ICM'90 (Kyoto).

[S6]   Steenbrink, J.H.M.: *Mixed Hodge Structures and Singularities* (in preparation).

[Sn]   Stein, K.: Analytische Zerlegungen Komplexer Räume, *Math. Ann.* **132** (1956), 63–93.

[Sv]   Stevens, J.: Periodicity of branched cyclic covers of manifolds with open book decomposition, *Math. Ann.* **273** (1986), 227–239.

[Sr]   van Straten, D.: On the Betti numbers of the Milnor fiber of a certain class of hypersurface singularities, Lecture Notes in Mathematics 1273, Springer-Verlag, Berlin, 1987, pp. 203–220.

[Sz]   Szczepanski, S.: Criteria for topological equivalence and the Lê–Ramanujan theorem for three complex variables, *Duke Math. J.* **58** (1989), 513–530.

[T1]   Teissier, B.: Cycles évanescents, sections planes et conditions de Whitney, *Astérisque* **718** (Singularités à Cargèse) 1973, pp. 285–362.

[T2]   Teissier, B.: Introduction to equisingularity problems, *Proc. Symp. Pure Math. 29 (Arcata Algebraic Geometry Conference)*, American Mathematical Society, 1974, pp. 593–632.

[Td]   Teodosiu, G.: A class of analytic coverings ramified over $u^3 = v^2$, *J. London Math. Soc.* (2) **38** (1988), 231–242.

[Tr]   Trotman, D.: Comparing regularity conditions on stratifications, *Proc. Symp. Pure Math. 40, Part 2 (Arcata Singularities Conference)*, American Mathematical Society, 1983, pp. 575–586.

[V]   Verdier, J.-L.: Stratifications de Whitney et théorème de Bertini–Sard, *Invent. Math.* **36** (1976), 295–312.

[Wa]   Wagreich, Ph.: The structure of quasihomogeneous singularities, *Proc. Symp. Pure Math. 40, Part 2 (Arcata Singularities Conference)*, American Mathematical Society, 1983, pp. 593–611.

[W1]   Wall, C.T.C.: Regular stratifications, in: *Dynamical Systems—Warwick 1974*, Lecture Notes in Mathematics, 468, Springer-Verlag, Berlin, 1974, pp. 332–344.

[W2]   Wall, C.T.C.: Classification of unimodal isolated singularities of complete intersections, *Proc. Symp. Pure Math. 40 Part 2 (Arcata Singularities Conference)*, American Mathematical Society, 1983, pp. 625–640.

[We]   Wells, R.O.: *Differential Analysis on Complex Manifolds*, Graduate Texts in Mathematics 65, Springer-Verlag, Berlin, 1980.

[Wn]   Werner, J.: Kleine Auflösungen spezieller dreidimensionaler Varietäten, *Bonner Math. Schriften* **186** (1987).

[Wh1]   Whitney, H.: Local properties of analytic varieties, in: *Differential and Combinatoric Topology*, Princeton University Press, Princeton, 1965, pp. 205–244.

[Wh2]   Whitney, H.: Tangents to an analytic variety, *Ann. of Math.* **81** (1965), 496–549.

[Wd1]   Wood, J.: Some criteria for finite and infinite monodromy of plane algebraic curves, *Invent. Math.* **26** (1974), 179–185.

[Wd2]   Wood, J.: A connected sum decomposition for complete intersections, *Proc. Symp. Pure Math.* **32**, Part 2, American Mathematical Society, 1978, pp. 191–193.

[XY]   Xu, Y.-J., Yau, S.S.-T.: Durfee conjecture and coordinate free characterization of homogeneous singularities, preprint 1990, University of Illinois at Chicago.

[Ym]   Yamamoto, M.: Classification of isolated algebraic singularities by their Alexander polynomials, *Topology* **23** (1984), 277–287.

[Za]   Zaidenberg, M.G.: An analytic cancellation theorem and exotic algebraic structures on $\mathbb{C}^n$, $n \geq 3$, Max-Planck-Institut, Preprint, 1991.

[Z1]   Zariski, O.: On the problem of existence of algebraic functions of two variables possessing a given branch curve, *Amer. J. Math.* **51** (1929).

[Z2]   Zariski, O.: *Algebraic Surfaces*, Springer-Verlag, Berlin, 1935, 2nd supplemented Edn. 1971.

[Z3]   Zariski, O.: On the Poincaré group of rational plane curves, *Amer. J. Math.* **58** (1936), 1–14.

[Z4]   Zariski, O.: A theorem on the Poincaré group of an algebraic hypersurface, *Ann. of Math.* **38** (1937), 131–141.

[Z5]   Zariski, O.: Some open questions in the theory of singularities, *Bull. Amer. Math. Soc.* **77** (1971), 481–491

Index

Universitext   *(continued)*

Nikulin/Shafarevich: Geometries and Groups
Øksendal: Stochastic Differential Equations
Rees: Notes on Geometry
Reisel: Elementary Theory of Metric Spaces
Rey: Introduction to Robust and Quasi-Robust Statistical Methods
Rickart: Natural Function Algebras
Rotman: Galois Theory
Rybakowski: The Homotopy Index and Partial Differential Equations
Samelson: Notes on Lie Algebras
Smith: Power Series From a Computational Point of View
Smoryński: Logical Number Theory I: An Introduction
Smoryński: Self-Reference and Modal Logic
Stillwell: Geometry of Surfaces
Stroock: An Introduction to the Theory of Large Deviations
Sunder: An Invitation to von Neumann Algebras
Tondeur: Foliations on Riemannian Manifolds
Verhulst: Nonlinear Differential Equations and Dynamical Systems
Zaanen: Continuity, Integration and Fourier Theory